Data-Driven Optimization of Manufacturing Processes

Kanak Kalita
Vel Tech Rangarajan Dr. Sagunthala R&D Institute of Science and Technology, India

Ranjan Kumar Ghadai
Sikkim Manipal Institute of Technology, Sikkim Manipal University, Majhitar, India

Xiao-Zhi Gao
University of Eastern Finland, Finland

IGI Global
PUBLISHER of TIMELY KNOWLEDGE

A volume in the Advances in Civil and Industrial Engineering (ACIE) Book Series

Published in the United States of America by
IGI Global
Engineering Science Reference (an imprint of IGI Global)
701 E. Chocolate Avenue
Hershey PA, USA 17033
Tel: 717-533-8845
Fax: 717-533-8661
E-mail: cust@igi-global.com
Web site: http://www.igi-global.com

Library of Congress Cataloging-in-Publication Data

Names: Kalita, Kanak, 1988- editor. | Ghadai, Ranjan, 1990- editor. | Gao, Xiao-Zhi, 1972- editor.
Title: Data-driven optimization of manufacturing processes / Kanak Kalita, Ranjan Ghadai, Xiao-Zhi Gao, editors.
Description: Hershey, PA : Information Science Reference, an imprint of IGI Global, [2021] | Includes bibliographical references and index. | Summary: "This book is a compilation of chapters on the application of state-of-the-art computational intelligence techniques from both predictive modeling and optimization, offering both soft computing approaches and machining processes"-- Provided by publisher.
Identifiers: LCCN 2020040225 (print) | LCCN 2020040226 (ebook) | ISBN 9781799872061 (hardcover) | ISBN 9781799872078 (paperback) | ISBN 9781799872085 (ebook)
Subjects: LCSH: Manufacturing processes--Data processing. | Engineering economy--Data processing.
Classification: LCC TS155.6 .D36 2021 (print) | LCC TS155.6 (ebook) | DDC 670.285--dc23
LC record available at https://lccn.loc.gov/2020040225
LC ebook record available at https://lccn.loc.gov/2020040226

This book is published in the IGI Global book series Advances in Civil and Industrial Engineering (ACIE) (ISSN: 2326-6139; eISSN: 2326-6155)

British Cataloguing in Publication Data
A Cataloguing in Publication record for this book is available from the British Library.

Advances in Civil and Industrial Engineering (ACIE) Book Series

Ioan Constantin Dima
University Valahia of Târgovişte, Romania

ISSN:2326-6139
EISSN:2326-6155

Mission

Private and public sector infrastructures begin to age, or require change in the face of developing technologies, the fields of civil and industrial engineering have become increasingly important as a method to mitigate and manage these changes. As governments and the public at large begin to grapple with climate change and growing populations, civil engineering has become more interdisciplinary and the need for publications that discuss the rapid changes and advancements in the field have become more in-demand. Additionally, private corporations and companies are facing similar changes and challenges, with the pressure for new and innovative methods being placed on those involved in industrial engineering.

The **Advances in Civil and Industrial Engineering (ACIE) Book Series** aims to present research and methodology that will provide solutions and discussions to meet such needs. The latest methodologies, applications, tools, and analysis will be published through the books included in **ACIE** in order to keep the available research in civil and industrial engineering as current and timely as possible.

Coverage

- Quality Engineering
- Earthquake engineering
- Coastal Engineering
- Optimization Techniques
- Productivity
- Production Planning and Control
- Urban Engineering
- Engineering Economics
- Transportation Engineering
- Construction Engineering

IGI Global is currently accepting manuscripts for publication within this series. To submit a proposal for a volume in this series, please contact our Acquisition Editors at Acquisitions@igi-global.com or visit: http://www.igi-global.com/publish/.

The Advances in Civil and Industrial Engineering (ACIE) Book Series (ISSN 2326-6139) is published by IGI Global, 701 E. Chocolate Avenue, Hershey, PA 17033-1240, USA, www.igi-global.com. This series is composed of titles available for purchase individually; each title is edited to be contextually exclusive from any other title within the series. For pricing and ordering information please visit http://www.igi-global.com/book-series/advances-civil-industrial-engineering/73673. Postmaster: Send all address changes to above address.

Titles in this Series

For a list of additional titles in this series, please visit: http://www.igi-global.com/book-series/advances-civil-industrial-engineering/73673

Driving Transformational Change in the Digital Built Environment
Jason Underwood (University of Salford, UK) and Mark Shelbourn (University of Salford, UK)
Engineering Science Reference • © 2021 • 340pp • H/C (ISBN: 9781799866008) • US $225.00

Applications and Challenges of Maintenance and Safety Engineering in Industry 4.0
Alberto Martinetti (University of Twente, The Netherlands) Micaela Demichela (Politecnico di Torino, Italy) and Sarbjeet Singh (Luleå University of Technology, Sweden)
Engineering Science Reference • © 2020 • 321pp • H/C (ISBN: 9781799839040) • US $265.00

Reconstructing Urban Ambiance in Smart Public Places
Hisham Abusaada (Housing and Building National Research Center (HBRC), Egypt) Ashraf M. Salama (University of Strathclyde, Glasgow, UK) and Abeer Elshater (Ain Shams University, Egypt)
Engineering Science Reference • © 2020 • 280pp • H/C (ISBN: 9781799838562) • US $225.00

Additive Manufacturing Applications for Metals and Composites
K.R. Balasubramanian (National Institute of Technology, Tiruchirappalli, India) and V. Senthilkumar (National Institute of Technology, Trichy, India)
Engineering Science Reference • © 2020 • 348pp • H/C (ISBN: 9781799840541) • US $205.00

Multi-Objective Optimization of Industrial Power Generation Systems Emerging Research and Opportunities
Timothy Ganesan (Royal Bank of Canada, Canada)
Engineering Science Reference • © 2020 • 233pp • H/C (ISBN: 9781799817109) • US $195.00

Impact of Industry 4.0 on Architecture and Cultural Heritage
Cecilia Maria Bolognesi (Politecnico of Milano, Italy) and Cettina Santagati (Università di Catania, Italy)
Engineering Science Reference • © 2020 • 422pp • H/C (ISBN: 9781799812340) • US $225.00

Handbook of Research on Urban-Rural Synergy Development Through Housing, Landscape, and Tourism
Aleksandra Krstić-Furundžić (University of Belgrade, Serbia) and Aleksandra Djukić (University of Belgrade, Serbia)
Engineering Science Reference • © 2020 • 437pp • H/C (ISBN: 9781522599326) • US $275.00

IGI Global
PUBLISHER of TIMELY KNOWLEDGE

701 East Chocolate Avenue, Hershey, PA 17033, USA
Tel: 717-533-8845 x100 • Fax: 717-533-8661
E-Mail: cust@igi-global.com • www.igi-global.com

Table of Contents

Detailed Table of Contents

Obtaining quality castings is of significant interest to researchers and practitioner due to the widespread use of casting components in everyday and industrial use. In this work, Taguchi method is used to optimize the process parameters involved in casting. Three casting parameters, namely pouring temperature (°C), sand particle size (AFS), and mould hardness (No), are considered. The percentage defect in the casting components is considered as the response. A L9 orthogonal array is used for design of experiments for carrying out the casting experiments. Based on the S/N ratio analysis and ANOVA, it is seen that the sand particle size has the most contribution to the linear model. Further, it is found that the level-wise optimal combination of the casting process parameters for minimizing the defects is Level 3 for pouring temperature, Level 2 for sand particle size, and Level 3 for mould hardness.

Defects associated with casting of pipes are often a main concern for the industry. In this chapter, a Taguchi analysis is carried out to understand the effect of three process parameter pouring temperature (°C), die spinning speed (rpm), and coolant flow time (mins) on the casting defect of pipes. The defect is defined in this work as the difference between the desired thickness of the pipe and the minimum actual (experimentally) achieved. A L9 orthogonal array is designed to carry out the experiments. Based on the S/N ratio analysis and ANOVA, it is seen that the die spinning speed plays the most critical role in defect of the pipes. As per the conducted experiments and Taguchi analysis, pouring temperature is seen to have the lest influence on the defects.

Incremental sheet forming (ISF) is an emerging technique of sheet metal working that comes into the picture in the last two decades. The ISF involved the forming of shapes without using the dedicated dies. ISF is suitable for customized products, rapid prototyping, and low batch production. The study aims to investigate the effect of process parameters on the surface roughness. The experiments are conducted on aluminum AA3003-O grade with six parameters, and the trials are performed according to the design of experiment (DOE). The atomic force microscopy (AFM) technique is used for measuring the surface roughness. Analysis of variance (ANOVA) is used for analyzing the effect of process parameters in ISF. The result shows that the step-down size, feed rate of the tool, and wall angle are significant process parameter and their contributions for ISF are 85.86%, 1.12%, and 12.29%, respectively.

This chapter aims to optimize centreless grinding conditions using the Taguchi method for minimizing surface roughness. The grinding operation has been performed according to the L9 orthogonal array in a centreless grinding process. The centreless grinding experiments are carried out on the crane-hook pin of C40 steel. The analysis of variance (ANOVA) and computation of signal to noise (S/N) ratio are adopted to determine the influence of grinding parameters (depth of cut [μm], regulating wheel speed

[rpm], and coolant valve opening) on surface roughness. The depth of cut (μm) is found to be the most significant among the grinding parameters on the surface roughness. The signal to noise (S/N) ratio was calculated based on smaller the best criteria. The lower level of depth of cut, medium level of regulating wheel speed, and higher-level coolant valve opening is found to be optimal grinding condition according to the mean response and signal to noise (S/N) ratio.

Bone grafting or bone implant is a typical procedure in surgery in which a missing or broken bone is replaced in order to treat bone fractures that pose a significant health risk to the patients. Several research works have been carried out in the past few years regarding various composite materials used in bone implants, their fabrication methods, and evaluation of their physical, mechanical, chemical, and thermal properties. The use of ceramic powders and ceramic-based composites in biomedical applications are steadily increasing over years mainly due to their advantages like high compressive strength, excellent hardness, etc. In this research work, organic ceramic matrix composites with varying proportions of conch shell and sea sponge are fabricated using powder metallurgy technique and their physicomechanical properties such as density, porosity, water absorption, and micro-hardness are evaluated. Finally, optimization of process parameters is done using multi-objective optimization based on ratio analysis (MOORA) to select the best possible specimen of CMCs.

In this modern world, composites are used in almost all fields due to their attractive mechanical and technological properties. They are believed to be fast replacing metal alloys thanks to their adaptability, flexibility, formability, and machinability. The present study deals with the comparative evaluation of mechanical properties between various aluminium alloy composites reinforced with boron carbide and rice husk ash at different proportions and their optimization through ranking of alternatives using TOPSIS. The composite specimens are fabricated by a liquid metallurgy technique called stir casting. The sample specimens are prepared by varying the percentage of reinforcements as per volume-based ratio with respect to the aluminium alloy Al 6061. The evaluation of mechanical properties indicates the improvement in tensile strength, hardness, impact energy, and corrosion resistance for different composite combinations compared to that of individual alloy. Finally, the best possible combination is identified among the given set of various proportions by optimization using TOPSIS.

The machining aspects of EDM sustains the discharge energy and discharge duration. This generated energy causes debris removal and evaporation of material. Therefore, in this chapter, an investigation is carried out to perform optimization of input process variables through the EDM process of ASTM A681 steel metal wielding the modified weighted TOPSIS MCDM technique. The study consists of process variables like discharge current, discharge duration, and discharge energy for the output response measurement of material removal rate (MRR) and average surface roughness. The multi-criteria decision approach is based on entropy, SDV, and mean-weighted TOPSIS and finds an output response related to the closeness coefficient between the feasible solution and the ideal solution. From the result, it is observed that the rank produced by mean weighted TOPSIS and SDV-TOPSIS are almost the same. But the rank produced by entropy-TOPSIS is slightly different as compared to the other method. However, the first, second, and third ranks are the same for all three techniques.

Steel has played critical role in the expansionist ambition of successful aristocrats and laying the foundation of enviable empires. Modern economies are built on strong infrastructure, communication, and transport networks. Industries, like automobiles, consumer durables, real estate, cannot be imagined without steel. Its synonymy with growth of an economy can be gauged by the fact that matrices, like per capital steel consumption and its contribution to the GDP, are considered as parameters of economic development. It is one of the most tracked industries by the investors, analysts, and financial institutions. In this chapter, a novel multi-criteria decision-making tool while integrating measurement alternatives and ranking according to compromise solution (MARCOS) and criteria importance through intercriteria correlation (CRITIC) methods is developed for evaluation of steel organisations which are constituents of BSE 200 index. The results derived from the implementation of integrated MARCOS-CRITIC model aid diverse stakeholders in making informed investment decisions.

In this research, a data-driven approach to metamodeling of manufacturing/machining processes is developed. Instead of the conventionally used second-order polynomial regression metamodels, a non-predefined form-free approach is discussed. The highly adaptive metamodeling strategy, called symbolic regression, is carried out by using genetic programming. A central composite design based experimental dataset on electric discharge machining is used as the training and the testing data. Four different process parameters namely (voltage, pulse on time, pulse off time, and current) are used as the independent

It is very important to select the optimal parametric values for various non-traditional machining processes (NTM) for improving their performance. The performance measures of NTM processes include material removal rate (MRR), radial overcut (ROC), heat affected zone (HAZ), etc. In this chapter, particle swarm optimization has been used to find out the optimal parametric settings for electrochemical discharge machining (ECDM) to improve its performance measure. Both single-objective as well as multi-objective optimization has been performed and the results have been compared with those obtained by other researchers.

Composite materials are preferred mostly in recent times due to their durability and ample space of applicability. Drilling is also an essential process in manufacturing, and it is frequently done to assemble products. Delamination due to drilling of the CFRP composite is considered as the primary concern in the manufacturing and assembly process. In this chapter, the empirical model for thrust force, torque, entry-delamination factor, exit-delamination factor, and eccentricity of drilling of CFRP composite is developed based on the extensive experiment. Response surface methodology is accounted to formulate a mathematical modal considering thrust force, torque, entry and exit-delamination factor, and eccentricity as response parameter and spindle speed, feed rate, and point angle as a process parameter. ANOVA is performed to check the statistical significance of the mathematical model. GA is employed to trace the optimum values of three process parameters to minimize the five response parameters. The Pareto front curve for various combinations of process parameters is also examined.

An optimized facility layout design helps to ensure a high level of machine usability along with minimum cost and good performance. However, facility layout is a multi-objective design optimization problem, and as such, it is difficult to solve. This original research work aims to design an optimal linear machine sequence using particle swarm optimization algorithm that minimizes the following: the total investment cost of machines, the total number of machines in the final sequence, the total flow time of the products, and the total flow distance of the products. The effectiveness of the proposed algorithm is demonstrated with a reasonable number of problems. Maximum of 19.1% reduction in total flow distance of products, 12.8% reduction in total investment cost of machines, 28.4% reduction in total flow time of products,

Preface

Manufacturing processes are the steps by which the raw materials are transformed into a final product. The broad classifications of manufacturing process include primary manufacturing process (which gives the primary shape to any product, for example— casting, forming, rolling etc.) and secondary manufacturing process (which gives proper finishing to the product from the primary manufacturing process, for example— drilling, milling, turning etc.). The process parameters involved in the manufacturing process significantly affects the properties of the final product from the process. So, it is of the utmost importance to find a suitable combination to all the process parameters by which the desired output response is optimized. Probably the most perpetual intellectual challenge in science and engineering is how to make the optimal decision in a given situation. The development of scientific disciplines such as operations research, management science, computer science, and statistics, in combination with the use of modern computers, is nothing but aids in assisting people in making the best decision for a given situation. Theories such as linear programming, dynamic programming, hypothesis testing, inventory control, optimization of queuing systems, and multi-criteria decision making have as a common element the search for an optimal decision (solution). In recent years various optimization and MCDM techniques have been developed to get quick and efficient results. The powerful machine-learning methods like gene expression programming (GEP), artificial neural network (ANN), support vector regression (SVM), and more can be used at an early phase of the design and optimization process to act as predictive models for the actual experiments, other metaheuristics-based methods like cuckoo search, ant colony optimization, particle swarm optimization, and others can be used to optimize these predictive models to find the optimal process parameter combination. Different multi-criteria decision making (MCDM) techniques like MOORA, TOPSIS, COPRAS, GRA, EDAS etc. can also be efficiently used to identify the optimal combination of process parameters in the various manufacturing process.

The purpose of this book is to comprehensively illustrate the endless possibilities and advantages of using data-driven approaches in manufacturing engineering. This book presents current research on various approaches ranging from Taguchi to MCDM to machine learning to metaheuristic optimization. The book is addressed towards seasoned scientists/researchers as well as practising engineers and early career researchers, who wish to explore the full potential of data-driven techniques towards improving the performance of machining and manufacturing processes. Moreover, the concepts presented in this book can be easily applied to engineering problems in other fields like mechanical, civil and electrical.

Acknowledgment

We would like to acknowledge the help of all the people involved in this project and, more specifically, the authors and reviewers that took part in the review process. Without their support, this book would not have become a reality.

We would like to thank each one of the authors for their contributions to this book, their time and expertise. We also acknowledge the valuable contributions of the reviewers regarding the improvement of quality, coherence, and content presentation of chapters.

We are also thankful to our respective institutes namely Vel Tech Rangarajan Dr. Sagunthala R&D Institute of Science and Technology, India; Sikkim Manipal Institute of Technology, India and University of Eastern Finland, Finland for their constant support and motivation.

Finally, we are very thankful to the team of IGI Global for accepting this book proposal and giving us the opportunity to work on this book project. Particularly, we are thankful to Eric Whalen (Assistant Development Editor), Josie Dadeboe (Assistant Development Editor), Crystal Moyer (Assistant Development Editor), Lindsay Wertman (Managing Director) and Jan Travers (Director of Intellectual Property and Contracts).

Kanak Kalita
Vel Tech Rangarajan Dr. Sagunthala R&D Institute of Science and Technology, India

Ranjan Kumar Ghadai
Sikkim Manipal Institute of Technology, Sikkim Manipal University, Majhitar, India

Xiao-Zhi Gao
University of Eastern Finland, Finland

Chapter 1
Minimization of Casting Defects Using Taguchi Method

Ranjan Kumar Ghadai
Sikkim Manipal Institute of Technology, Sikkim Manipal University, Majhitar, India

Ashnut Dutt
Sikkim Manipal Institute of Technology, Sikkim Manipal University, Majhitar, India

Kanak Kalita
https://orcid.org/0000-0001-9289-9495
Vel Tech Rangarajan Dr. Sagunthala R&D Institute of Science and Technology, Avadi, India

Dinesh S. Shinde
SVKM's NMIMS, Mukesh Patel School of Technology Management and Engineering, Shirpur, India

Chayut Bunterngchit
4Division of Industrial and Logistics Engineering Technology, King Mongkut's University of Technology, Thailand

ABSTRACT

Obtaining quality castings is of significant interest to researchers and practitioner due to the widespread use of casting components in everyday and industrial use. In this work, Taguchi method is used to optimize the process parameters involved in casting. Three casting parameters, namely pouring temperature (°C), sand particle size (AFS), and mould hardness (No), are considered. The percentage defect in the casting components is considered as the response. A L9 orthogonal array is used for design of experiments for carrying out the casting experiments. Based on the S/N ratio analysis and ANOVA, it is seen that the sand particle size has the most contribution to the linear model. Further, it is found that the level-wise optimal combination of the casting process parameters for minimizing the defects is Level 3 for pouring temperature, Level 2 for sand particle size, and Level 3 for mould hardness.

DOI: 10.4018/978-1-7998-7206-1.ch001

INTRODUCTION

Foundry suffers from weak quality and productivity due to a huge number of process parameters, lower automation and shortage of skilled workers. Defect-free products are demanded in the market but foundry often finds it difficult to meet the requirements. Foundry defects are analyzed by non-destructive methods and remedies must be carefully applied; otherwise, often new defects occur. This is not an easy task. For example, when a gas porosity defect occurs at high pouring temperature and the pouring temperature is decreased, it may lead to cold shut defects. Thus, casting despite its traditional usage is a very complex process from optimal parameter selection viewpoint. The success of a casting process depends greatly on the properties of the molding sand. These include strength, permeability, deformation, flowability, and refractoriness. The investigation of casting defect is the common interest of researchers, since failure in critical machines like aluminium alloy aircraft frames initiated from casting defects such as porosity (Li, Shen, & Hu, 2011). Bacaicoa et al (Bacaicoa, et al., 2017) characterized the casting defect of iron-rich aluminium-silicon-copper alloys using high-resolution scanning microscope and finite element analysis, in FEA the stress concentration was targeted and reported that the stress concentration observed to be larger on edges and inclusion of beta-Al-Si-Cu phases causes porosity. The casting products are sometimes subjected to fatigue loading for which the analysis was done by Wang et al (Wang, Apelian, & Lados, 2001) for A356-T6 casting alloy to study of the effect of casting defects on the fatigue behaviour, it was seen that the fatigue crack initiated from the casting defect i.e. porosity & oxide layers, out of which porosity is most harmful. Kunz et al (Kunz, Lukáš, Konečná, & Fintová, 2012) investigated high-temperature fatigue life of nickel super-alloy castings for the effect of casting defects and concluded that the casting defects becomes overwhelming at high-temperature applications, they should be reduced. Hamilton et al. (Hamilton, See, Butler, & Lee, 2003) used the macro scale and microscale modelling of defects in casting to predict them in aluminium alloy castings and the results were compared with experimental values and reported that the present tool can be used to reduce the casting defects. Whereas Kim et al. (Kim, Lee, & Nahm, 2006) analyzed the casting defects statistically in case of stainless-steel cast for correlating the casting defects to the fatigue life of the part. On the other hand, Borowiecki et al. (Borowiecki, Borowiecka, & Szkodzińska, 2011) used Pareto method to predict the casting defects like porosity, sand holes, slag inclusions, etc. Roy (Roy, 2013) analyzed the casting defects and presented a new approach i.e. simulation (CAE) for a possible solution in its reduction. The materials property variation also influences the casting defects, like increase in ductility of cast iron increases elongation at fracture, hence reduces casting defects (Nilsson & Vokál, 2009) and increase in tensile strength of the cast material reduces area fraction of defect in aluminium alloy castings (Caceres & Selling, 1996).

Taguchi experimental design is an optimization tool widely used in casting defects investigation. Dabade et al. (Dabade & Bhedasgaonkar, 2013) applied the Taguchi method in the analysis of casting defects with casting simulation considering casting process parameters such as moisture percentage, green compression strength, and permeability of molding sand and hardness of mold material for shrinkage porosity & yield of the cast and reported that the proper design of gating and feeding system reduces shrinkage porosity and enhanced the yield. Taguchi method was used to establish a robust casting process parameter (mold temperature, flow velocity, cooling time, etc.) model of thin-walled magnesium alloy cast for reduction of casting defects (Chang & others, 2004). Kumar et al. (Kumar, Satsangi, & Prajapati, 2013) optimize process parameters of the casting of differential housing cover for different materials using Taguchi optimization and reported that the empirical model of parameters gives a reduction in casting defects. A hybrid casting process i.e squeeze casting is a combination of casting and forging process for improved strength of the product, Vijian et al. (Vijian & Arunachalam, 2006) optimized the process parameters of squeeze casting of aluminium alloy using Taguchi method to obtain maximum mechanical characteristics of the casted part, such as tensile strength & hardness, whereas in their other reported work they optimize the casting process parameters using a similar methodology for surface finish of cast object of LM6 aluminium alloy (Vijian & Arunachalam, 2007). Mahesh et al. (Adke & Shrikant, 2014) used the technique to optimize the die-casting process parameters for casting cycle time. Ye et al. (Ye, et al., 2014) aimed to the reduction of low-pressure die-casting shrinkage porosity of ZL205A alloys using Taguchi method and reported that the pouring temperature of molten metal is the most controlling process parameter in this case; also casting defects can be reduced through control of the process parameters. Acharya et al. (Acharya, Vadher, Sheladiya, & Madhnani, 2016) improved the casting quality using Taguchi optimization technique integrated with casting simulation and concluded that the casting process can be analyzed through simulation due to high cost in its experimentation and casting process parameters control & proper design of gating system improved the characteristics of resulting cast. A casting simulation software package combined with Taguchi design; has been used by Zhizhong et al. (Sun, Hu, & Chen, 2008) for optimization of gating system parameters of magnesium alloy castings for flow velocity, shrinkage porosity and material yield with remarks that the multiple attributes can be achieved using experimental design of gating system parameters. Haq et al. (Haq, Guharaja, & Karuppannan, 2009) optimized the CO_2 casting process parameters such as the mass of gas used, the hardness of mold, the average size of sand particles, sodium silicate percentage in mold sand, mixing time of sand, molten metal pouring time, the height of pouring molten metal, molten metal temperature, and cooling time of casting for reduction of casting porosity; using Taguchi technique and reported that the selection of the proper combination of casting process parameters reduces the defects. Vijian et al. (Vijian, Arunachalam, & Charles, 2007) select best casting process parameters aluminium alloys for minimum surface roughness of resulting cast using Taguchi method. The special-purpose casting process such casting for gear blank design optimization was done by Muzammil et al. (Muzammil, Singh, & Talib, 2003) applying Taguchi robust design considering casting process parameters such as the content of clay in mold sand, moisture percentage, ramming percentage, size of sand particles, the flow of molten metal, and design of gating system for minimization of casting defects; it was observed that the selection of the combination of casting process parameters reduces defects.

Improvement of mechanical properties such as hardness, tensile strength, etc. of material after casting reduces the defects. Gangwar et al. (Gangwar, Kumar, Singh, & Singh, 2020) optimize the casting process parameters of AA-6063 alloys for the hardness of resulting cast object under microwave conditions using Taguchi optimization method and reported that mold heating can improve the output. Tiwari et al. (Tiwari, Singh, & Srivastava, 2016) observed that casting defects can be reduced by improving the material characteristics of the resulting cast; they used Taguchi method to optimize green sand casting process parameters of mild steel casting. Patil et al. (Patil, 2014) minimize the casting defects such as sand inclusion in casting, porosity, material leakage and mold conditions through optimizing the casting process parameters like the hardness of mold, percentage of moisture of mold sand, permeability percentage and green strength of mold sand for iron flange casting process.

This paper aims to identify the defects in impellers and analyze its causes. Further, the reduction of causes of these defects like pouring temperature, permeability, mold hardness and sand particles through Taguchi's method is another important goal that is achieved in this work.

Figure 1. Methodology followed in this work

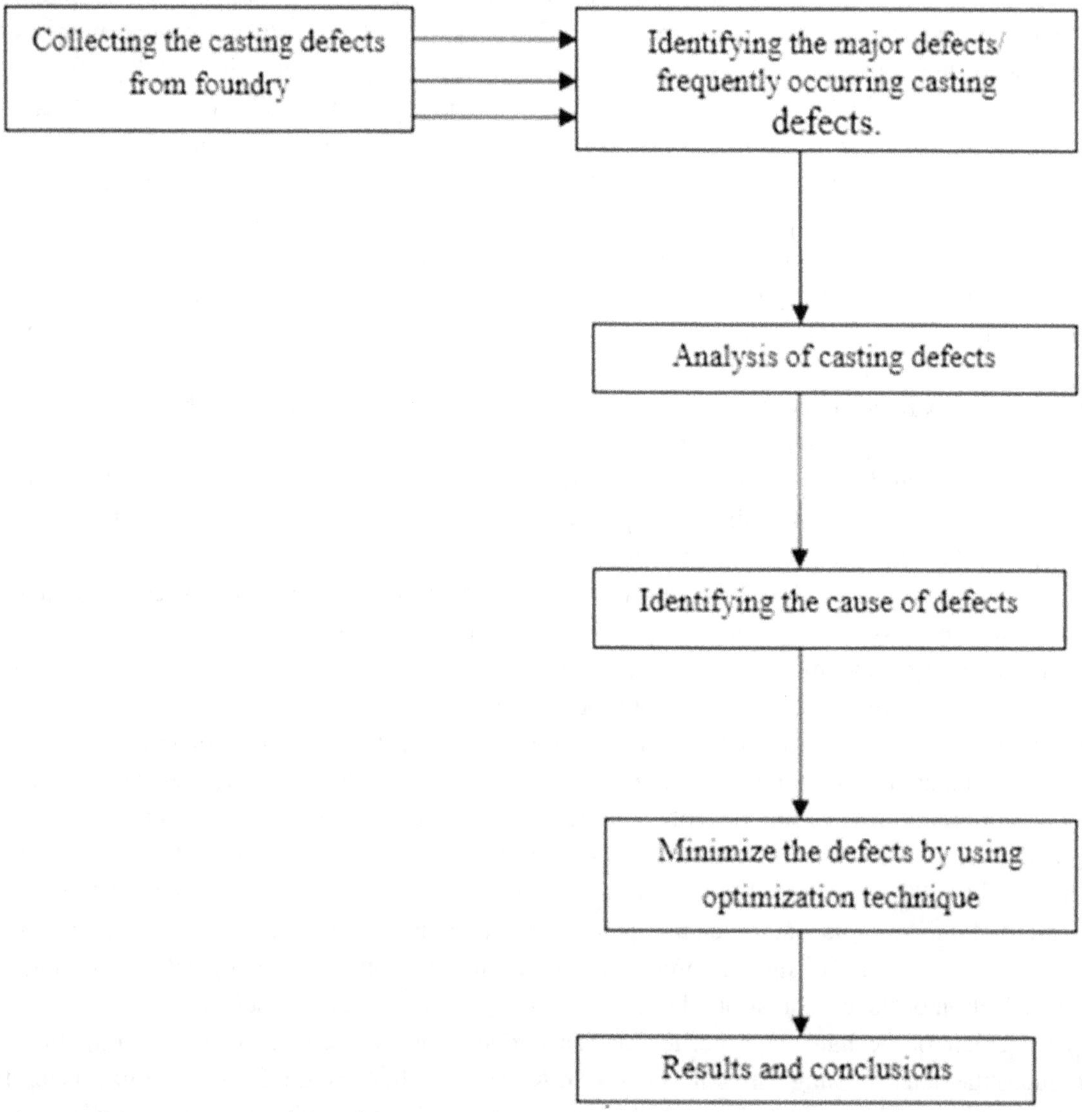

METHODOLOGY

The outline of the methodology followed in this work is shown in Figure 1. Experiments were performed in foundry producing cast iron components. The study begins with the manufacturing component impellers in the foundry where cold shut, scab, shrinkage are most important defects observed and due to these casting defects in casting the component having highest rejection rate is identified using quality tools such as total rejection sheet, defective factor level sheet, table of defects and causes with diagram. A model is also created for optimization of castings process through simulation and the major causes are predicted and its solutions are given by defect diagnostic approach. Process parameters of casting that influences the identified defects of impellers with their levels are shown in Table 1.

Table 1. Levels of process parameters

Parameter	Range	A	B	C
Pouring temperature (°C)	1570-1610	1570	1590	1610
Sand particle size (AFS)	48-55	48	51	55
Mould hardness no.	70-80	70	75	80

Casting defects were selected as a quality characteristic to be measured. The most common defects occurring in the foundry were monitored and recorded. The smaller the better number of casting defect implies better process performance. Here, the objective function to be maximized is:

$$S / \text{Nratio} = -10.\log \frac{1}{n}\left(\sum_{i=1}^{n} y_i^{\,2}\right) \quad (1)$$

Maximizing η leads to minimization of quality loss due to defects. Further, S/N ratio is used for measuring sensitivity to noise factors, n is the number of experiments orthogonal array and Yi the ith value measured.

Here three parameters with at three different levels therefore L9 orthogonal array is selected for the experimental work. As per L9 orthogonal array, nine experiments were performed randomly as shown in Table 2 and the responses are reported in Table 3.

Table 2. L9 orthogonal array used in this work

Trail No.	Pouring Temperature (°C)	Sand Particle size (AFS)	Mould Hardness no.
1	1570	48	70
2	1570	51	75
3	1570	55	80
4	1590	51	80
5	1590	55	70
6	1590	48	75
7	1610	55	75
8	1610	48	80
9	1610	51	70

The main aim of the study was to reduce the casting defects for which the ideal value is zero. The analysis was carried out by using MINITAB v16 statistical software. The S/N ratio was computed by using smaller-the-better quality characteristics. In the Taguchi method, the signal to noise ratio (S/N) is used as the data transformation method that consolidates the data for each control array row over the various noise levels into one value which computes both the mean and the variation present in the data. The equations for calculating the signals to noise ratios were based on the characteristics of the response variables being evaluated; nominal the best, smaller the better and larger the better. In the present work, smaller the best characteristic is used as the main aim is to reduce rejection in frames. The percentage rejection of check valves -PN 10 for each trial was evaluated and the report generated was obtained from MINITAB v16 statistical software.

Table 3. Percentage defects in casting

Trail No.	1	2	3	Total	Average	S/N ratio
1	8.50	7.30	8.90	24.7	8.23	-18.3080
2	7.50	5.80	7.30	20.6	6.86	-16.7265
3	9.34	7.60	9.56	26.5	8.83	-18.9192
4	5.50	6.20	6.01	17.71	5.9	-15.4170
5	8.32	6.90	9.14	24.36	8.12	-18.1911
6	9.43	8.67	8.00	26.1	8.7	-18.7904
7	9.12	7.00	7.75	23.87	7.95	-18.0073
8	5.33	9.32	6.50	21.15	7.05	-16.9638
9	5.50	7.60	3.53	16.63	5.54	-14.8702

Table 4. Response table for signal to noise ratios (SNR)

Level	Pouring Temperature (°C)	Sand particle size (AFS)	Mould Hardness no.
1	-17.98	-18.02	-17.12
2	-17.47	-15.67	-17.84
3	-16.61	-18.37	-17.10
Delta	1.37	2.70	0.74
Rank	2	1	3

RESULTS AND DISCUSSION

The Taguchi Method is used to optimize the results obtained from each trial. In the present work, an L9 orthogonal array is used for the trial purpose. The response of the S/N Ratio, contribution of different process parameters and the relation between the S/N ratio and the levels of different process parameters is studied and analyzed to obtain optimum process parameters. There are three categories of quality characteristics in the analysis of S/N ratio, i.e. smaller-the-better, larger-the-better, nominal-the-best. As the main aim of the study was to reduce the casting defects for which the ideal value is zero, the S/N ratio for each level of process parameter has been computed by using a quality characteristic smaller-the-better and reported in Figure 2. In the S/N plot, the highest value is the optimum level predicted for that casting parameter. The information regarding the analysis of the S/N ratio is also listed in Table 4. A rough estimation of the relative importance of the process parameters on minimizing defect can be made by analyzing the 'delta' value in Table 4. Delta is calculated as the difference between the highest and the lowest SNR value within each process parameter. Higher the delta value of a process parameter as

Figure 2. Main effects for SN ratio

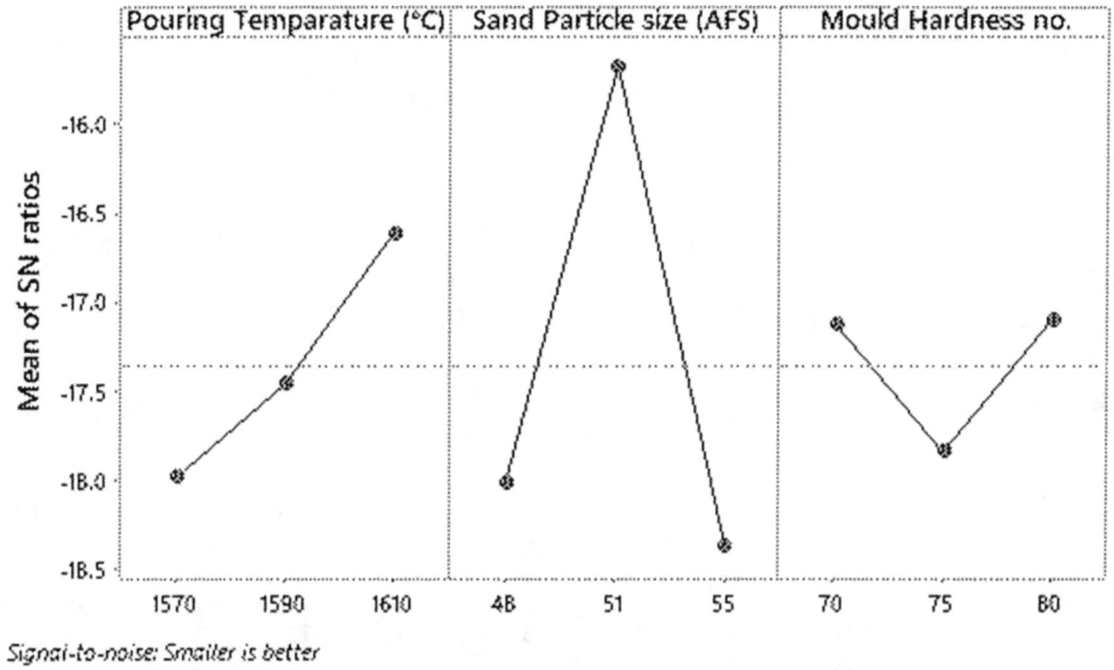

compared to the other process parameter; more is its significance in the output response. This is because even a small perturbation in the process parameter with the highest delta will have more relative effect on the output response as compared to the same disturbance in other process parameters.

Table 5. Response table for means of response

Level	Pouring Temperature (°C)	Sand particle size (AFS)	Mould Hardness no.
1	7.973	7.993	7.297
2	7.573	6.100	7.837
3	6.847	8.300	7.260
Delta	1.127	2.200	0.577
Rank	2	1	3

From the main effect plot shown in Figure 3 and Table 5, it can be concluded that the optimum value of pouring temperature is 1570°C, sand particle size is 55 AFS and mould hardness no. 80. This combination of the casting parameters gives the optimum result.

Figure 3. Main effects for means

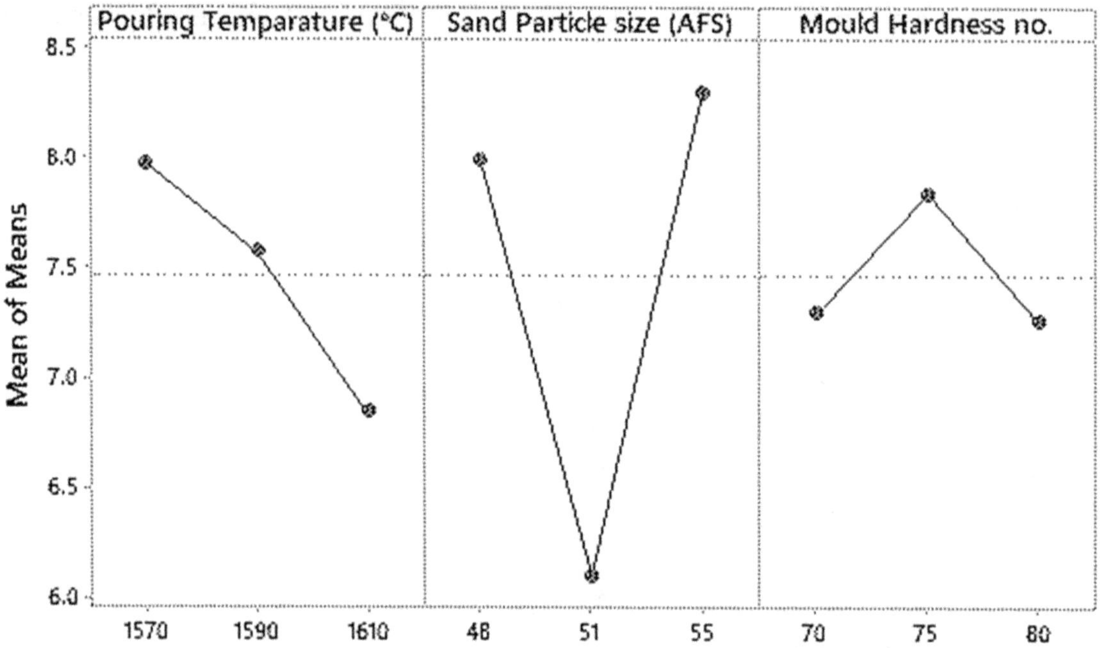

Table 6. ANOVA results for percentage defect

Source	DF	Seq SS	Contribution	Adj. SS	Adj MS	F-Value	P-Value
Pouring Temparature (°C)	2	1.9574	17.27%	1.9574	0.9787	8.53	0.105
Sand Particle size (AFS)	2	8.5188	75.18%	8.5188	4.2594	37.14	0.026
Mould Hardness no.	2	0.6255	5.52%	0.6255	0.3127	2.73	0.268
Error	2	0.2294	2.02%	0.2294	0.1147		
Total	8	11.3310	100.00%				

Table 7. Model summary of the linear model for percentage defect

S	R^2	$R^2_{adjusted}$	PRESS	$R^2_{Predicted}$
0.338641	97.98%	91.90%	4.64445	59.01%

Table 8. Coefficients of the linear model for percentage defect

Term	Coef	SE Coef	95% CI	T-Value	P-Value	VIF
Constant	7.464	0.113	(6.979, 7.950)	66.13	0.000	
Pouring Temparature (°C)						
1570	0.509	0.160	(-0.178, 1.196)	3.19	0.086	1.33
1590	0.109	0.160	(-0.578, 0.796)	0.68	0.566	1.33
Sand Particle size (AFS)						
48	0.529	0.160	(-0.158, 1.216)	3.31	0.080	1.33
51	-1.364	0.160	(-2.051, -0.678)	-8.55	0.013	1.33
Mould Hardness no.						
70	-0.168	0.160	(-0.855, 0.519)	-1.05	0.404	1.33
75	0.372	0.160	(-0.315, 1.059)	2.33	0.145	1.33

Figure 4. Percentage contribution of parameters to the response

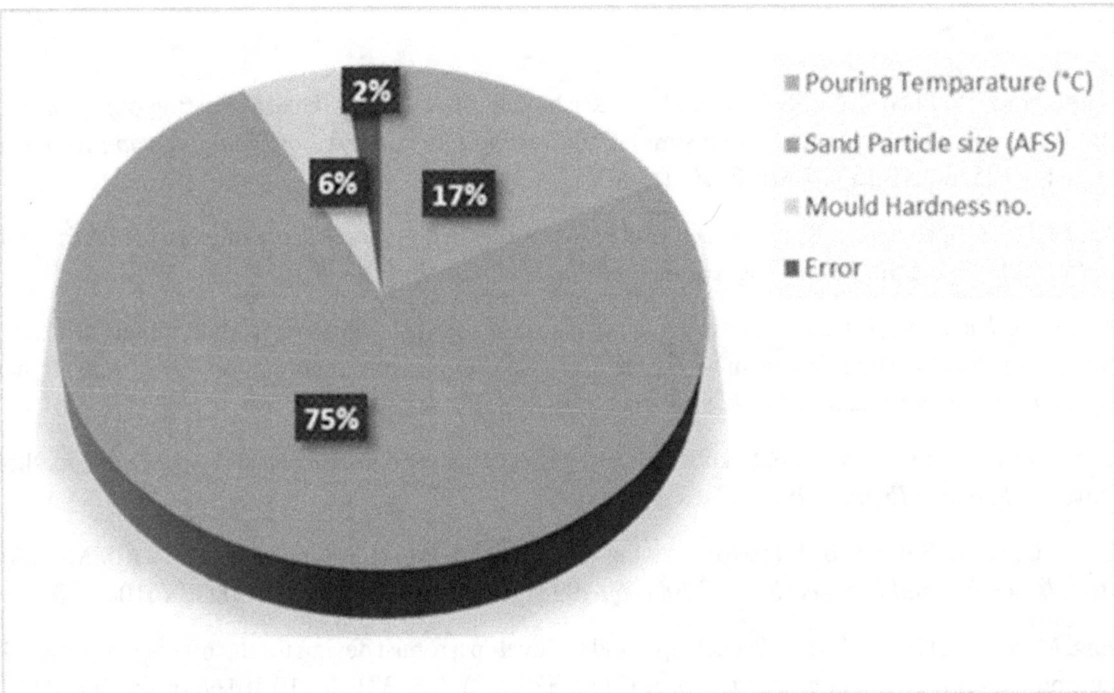

ANOVA or analysis of variance was carried out for a linear model and the findings are reported in Table 6. The ANOVA analysis was performed with a 95% (0.95) confidence level and 5% (0.05) significance level. The model summary of the linear model is reported in Table 7. It is seen that the model has a R^2 of about 98% indicating that roughly 98% of the variance in the data is explained by the linear model. Also, the $R^2_{adjusted}$ which is a measurement of the variance in the data explained by the linear model and accounting for the number of the terms is around 92%. Based on the ANOVA, it is found that sand particle size has the most contribution to the linear model. From Figure 4, it is seen that pouring temperature has around 17% contribution whereas the mould hardness has only 6% contribution towards the linear model. The coefficients of the linear model for percentage defect are reported in Table 8.

CONCLUSION

In this research work, the effect of three casting process parameters namely, pouring temperature (°C), sand particle size (AFS) and mould hardness (No) on casting defects are studied. An L9 orthogonal array is used for the design of experiments. It was seen that the percentage of casting defects is most influenced by the sand particle size and least affected by mould hardness. The contribution of the pouring temperature (°C), sand particle size (AFS) and mould hardness (No) on the linear model of casting defects is found to be 75%, 17% and 6% respectively. Based on the Taguchi analysis, the level-wise optimal combination of the casting process parameters for minimizing the defects is level 3 for pouring temperature, level 2 for sand particle size and level 3 for mould hardness.

REFERENCES

Acharya, S. G., Vadher, J. A., Sheladiya, M. V., & Madhnani, M. (2016). Quality casting of motor body using design of experiment and casting simulation. *International Journal of Manufacturing Research, 11*(2), 111–125. doi:10.1504/IJMR.2016.078243

Adke, M. N., & Shrikant, V. K. (2014). Optimization of die-casting process parameters to identify optimized level for cycle time using Taguchi method. *Int. J. Innov. Eng. Technol, 4*, 365–375.

Bacaicoa, I., Wicke, M., Luetje, M., Zeismann, F., Brueckner-Foit, A., Geisert, A., & Fehlbier, M. (2017). Characterization of casting defects in a Fe-rich Al-Si-Cu alloy by microtomography and finite element analysis. *Engineering Fracture Mechanics, 183*, 159–169. doi:10.1016/j.engfracmech.2017.03.015

Borowiecki, B., Borowiecka, O., & Szkodzińska, E. (2011). Casting defects analysis by the Pareto method. *Archives of Foundry Engineering, 11*.

Caceres, C. H., & Selling, B. I. (1996). Casting defects and the tensile properties of an AlSiMg alloy. *Materials Science and Engineering A, 220*(1-2), 109–116. doi:10.1016/S0921-5093(96)10433-0

Chang, M. S., & ... (2004). Use of Taguchi method to develop a robust design for the magnesium alloy die casting process. *Materials Science and Engineering A, 379*(1-2), 366–371. doi:10.1016/j.msea.2004.03.006

Dabade, U. A., & Bhedasgaonkar, R. C. (2013). Casting defect analysis using design of experiments (DoE) and computer aided casting simulation technique. *Procedia Cirp, 7*, 616–621. doi:10.1016/j.procir.2013.06.042

Gangwar, V., Kumar, S., Singh, V., & Singh, H. (2020). Effect of Process Parameters on Hardness of AA-6063 In-Situ Microwave Casting by Using Taguchi Method. *IOP Conference Series. Materials Science and Engineering, 804*, 012019. doi:10.1088/1757-899X/804/1/012019

Hamilton, R. W., See, D., Butler, S., & Lee, P. D. (2003). Multiscale modeling for the prediction of casting defects in investment cast aluminum alloys. *Materials Science and Engineering A*, *343*(1-2), 290–300. doi:10.1016/S0921-5093(02)00376-3

Haq, A. N., Guharaja, S., & Karuppannan, K. M. (2009). Parameter optimization of CO 2 casting process by using Taguchi method. *International Journal on Interactive Design and Manufacturing*, *3*(1), 41–50. doi:10.100712008-008-0054-4

Kim, J. Y., Lee, J. H., & Nahm, S. H. (2006). Statistical Analysis of Casting Defects in Microstructure for Understanding the Effect on Fatigue Property of 17-4PH Stainless Steel. *Key Engineering Materials*, *321*, 1503–1506. doi:10.4028/www.scientific.net/KEM.321-323.1503

Kumar, S., Satsangi, P. S., & Prajapati, D. R. (2013). Optimization of process factors for controlling defects due to melt shop using Taguchi method. *International Journal of Quality & Reliability Management*, *30*(1), 4–22. doi:10.1108/02656711311288397

Kunz, L., Lukáš, P., Konečná, R., & Fintová, S. (2012). Casting defects and high temperature fatigue life of IN 713LC superalloy. *International Journal of Fatigue*, *41*, 47–51. doi:10.1016/j.ijfatigue.2011.12.002

Li, B., Shen, Y., & Hu, W. (2011). Casting defects induced fatigue damage in aircraft frames of ZL205A aluminum alloy–A failure analysis. *Materials & Design*, *32*(5), 2570–2582. doi:10.1016/j.matdes.2011.01.039

Muzammil, M., Singh, P. P., & Talib, F. (2003). Optimization of gear blank casting process by using Taguchi's robust design technique. *Quality Engineering*, *15*(3), 351–359. doi:10.1081/QEN-120018033

Nilsson, K.-F., & Vokál, V. (2009). Analysis of ductile cast iron tensile tests to relate ductility variation to casting defects and material microstructure. *Materials Science and Engineering A*, *502*(1-2), 54–63. doi:10.1016/j.msea.2008.09.082

Patil, G. G. (2014). *Optimization of casting process parameters using Taguchi Method*. Academic Press.

Roy, T. (2013). Analysis of casting defects in foundry by computerised simulations (CAE)-A new approach along with some industrial case studies. *Transactions of 61stIndian Foundry Congress*, 1–9.

Sun, Z., Hu, H., & Chen, X. (2008). Numerical optimization of gating system parameters for a magnesium alloy casting with multiple performance characteristics. *Journal of Materials Processing Technology*, *199*(1-3), 256–264. doi:10.1016/j.jmatprotec.2007.08.036

Tiwari, S. K., Singh, R. K., & Srivastava, S. C. (2016). Optimisation of green sand casting process parameters for enhancing quality of mild steel castings. *International Journal of Productivity and Quality Management*, *17*(2), 127–141. doi:10.1504/IJPQM.2016.074446

Vijian, P., & Arunachalam, V. P. (2006). Optimization of squeeze cast parameters of LM6 aluminium alloy for surface roughness using Taguchi method. *Journal of Materials Processing Technology*, *180*(1-3), 161–166. doi:10.1016/j.jmatprotec.2006.05.016

Vijian, P., & Arunachalam, V. P. (2007). Optimization of squeeze casting process parameters using Taguchi analysis. *International Journal of Advanced Manufacturing Technology*, *33*(11-12), 1122–1127. doi:10.100700170-006-0550-2

Vijian, P., Arunachalam, V. P., & Charles, S. (2007). *Study of surface roughness in squeeze casting LM6 aluminium alloy using Taguchi method.* Academic Press.

Wang, Q. G., Apelian, D., & Lados, D. A. (2001). Fatigue behavior of A356-T6 aluminum cast alloys. Part I. Effect of casting defects. *Journal of Light Metals*, *1*(1), 73–84. doi:10.1016/S1471-5317(00)00008-0

Ye, W., Shiping, W., Lianjie, N., Xiang, X., Jianbing, Z., & Wenfeng, X. (2014). Optimization of low-pressure die casting process parameters for reduction of shrinkage porosity in ZL205A alloy casting using Taguchi method. *Proceedings of the Institution of Mechanical Engineers. Part B, Journal of Engineering Manufacture*, *228*(11), 1508–1514. doi:10.1177/0954405414521065

ADDITIONAL READING

Das, P. P., Gupta, P., Ghadai, R. K., Ramachandran, M., & Kalita, K. (2017). Optimization of turning process parameters by Taguchi-based Six Sigma. *Mechanics and Mechanical Engineering*, *21*(3), 649–656.

Guharaja, S., Haq, A. N., & Karuppannan, K. M. (2006). Optimization of green sand casting process parameters by using Taguchi's method. *International Journal of Advanced Manufacturing Technology*, *30*(11-12), 1040–1048. doi:10.100700170-005-0146-2

Kumar, S., Satsangi, P. S., & Prajapati, D. R. (2011). Optimization of green sand casting process parameters of a foundry by using Taguchi's method. *International Journal of Advanced Manufacturing Technology*, *55*(1-4), 23–34. doi:10.100700170-010-3029-0

Kumaravadivel, A., Natarajan, U., & Ilamparithi, C. (2012). Determining the optimum green sand casting process parameters using Taguchi's method. *Journal of the Chinese Institute of Industrial Engineers*, *29*(2), 148–162. doi:10.1080/10170669.2012.664789

KEY TERMS AND DEFINITIONS

Analysis of Variance: Analysis of variance is a statistical approach to analyze the differences among group means in a sample.

Casting: Casting is a commonly used manufacturing process in which the part is fabricated by pouring molten metal into a mold and then allowed to solidify.

Optimal Process Parameters: The most suitable combination of parameters involved in a process that can help attain the desired responses (outputs).

Chapter 2
Evaluation of Optimum Parameters for Casting of Birla Lance Pipes

Dinesh Shinde
SVKM's NMIMS, Mukesh Patel School of Technology Management and Engineering, Shirpur, India

Ashnut Dutt
Sikkim Manipal Institute of Technology, Sikkim Manipal University, Majhitar, India

Ranjan Kumar Ghadai
Sikkim Manipal Institute of Technology, Sikkim Manipal University, Majhitar, India

Kanak Kalita
https://orcid.org/0000-0001-9289-9495
Vel Tech Rangarajan Dr. Sagunthala R&D Institute of Science and Technology, India

Amer Nasr A. Elghaffar
Minia University, Egypt

ABSTRACT

Defects associated with casting of pipes are often a main concern for the industry. In this chapter, a Taguchi analysis is carried out to understand the effect of three process parameter pouring temperature (°C), die spinning speed (rpm), and coolant flow time (mins) on the casting defect of pipes. The defect is defined in this work as the difference between the desired thickness of the pipe and the minimum actual (experimentally) achieved. A L9 orthogonal array is designed to carry out the experiments. Based on the S/N ratio analysis and ANOVA, it is seen that the die spinning speed plays the most critical role in defect of the pipes. As per the conducted experiments and Taguchi analysis, pouring temperature is seen to have the lest influence on the defects.

DOI: 10.4018/978-1-7998-7206-1.ch002

INTRODUCTION

Casting is a basic manufacturing process in which molten metal is poured into a mold cavity and allowed to cool for obtaining the desired shape. It is the oldest production process, which nowadays is also used to get parts of intricate shape and size. Any part having several tiny cavities and irregular shapes can also be produced using casting. Due to the handling of hot molten metal, the casting process should be performed with great precision and accuracy. The casting process is widely used to manufacture automotive engine parts, irregular shape machine components, pipes, etc. Many types of casting processes are there, which are based on different principle required to prepare part using a hot molten form. Sand casting, die casting, investment casting, centrifugal casting, gravity die casting, etc. are the different casting process used in industries. Sand casting is the oldest type of casting which is preferred for a solid part of complex geometry without any external force, whereas in die casting the molten metal is forced to get the shape of the part, investment casting is also called as lost-wax casting which is used to obtain smaller size parts with a high surface finish. To manufacture symmetric parts the centrifugal casting is used, in which the molten metal is rotated and the action of centrifugal force is utilized to obtain enhanced materials properties. Each type of casting has its application field and is considered for different part to be produced. Common parameters in all types of casting are molten metal temperature also known as pouring temperature, type of coolant, flow rate of coolant, time for coolant flow etc. The major issue with the casting processes is casting defects which occurs due to uneven cooling of molten metal and shrinkage of the material. The gating system is properly designed, risers are provided and various casting process settings such as preheating of mold are used to reduce the casting defects.

The centrifugal casting process is used to manufacture pipes with good mechanical and structural properties. Special purpose pipes such as lance pipes are also manufactured using centrifugal casting. The casting process is used for mass production of the parts. Important parameters of centrifugal casting are molten metal temperature, type, time and flow rate of coolant, the rotational speed of mold, etc. and output characteristics desired are material properties, surface finish, geometrical accuracy of surfaces, etc. Stirring of molten metal during centrifugal casting improves the materials and structural properties of the cast (Martinez, Garnier, & Durand, 1987). Researchers are interested in the study of the effect of the process parameters on the resulting production outcomes such as microstructure and mechanical characteristics of the cast (Wassilkowska, 2017). Yu at al. (Pivinskii, Litovskaya, Samarina, Volchek, & Kaplan, 1991) investigated the centrifugal casting of ceramics considering the velocity of casting, the temperature during casting, centrifugal pressure and reports that the control over the casting process parameters results in better surface, material and microstructural properties. Arbabi et al. (Arbabi & Ebrahimzadeh, 2010) analyzed the effect of the wall thickness of pipes manufactured by horizontal centrifugal casting process on resulting microstructural and mechanical properties with remarks that the wall thickness pipe significantly affects the output characteristics. Mohammad et al. (Rahimipour & Sobhani, 2013) evaluated the centrifugal casting process parameters of titanium composites using microstructural and phase characteristics of the cast, whereas Jien-Wei et al. (Yeh & Jong, 1994) water cooling integrated centrifugal casting of 7075 alloys considering rotation speed, pouring temperature, water flow, and grain refiner these parameters and materials microstructure and geometry as output responses. Abdul et al. (Ganai & Singh, 2020) studied microstructural, mechanical and geometrical parameter of the resulting cast of aluminium-6061 pipes manufactured by centrifugal casting experimentally and concluded that the mold rotational speed is the most significant parameter of the process & improper flow of molten metal varied the hardness of cast.

Due to the requirement of heavy instrumentation and cost associated with the experimental study of the casting process, researchers preferred numerical simulation. Wang et al. (Wang, Wu, Xue, & Guo, 2013) studied the effect of input parameters of tilt-casting such as pouring temperature, tilt time, preheating temperature of mold on the shrinkage of the resulting part of aluminium alloy made for automotive drain pipes. Kaschnitz et al. (Kaschnitz, 2012) used flow-3D simulation package to demonstrate the molten metal flow pattern, frictional effect between it & mold cavity, turbulence in flow, velocity variation due to cooling. The simulation study carried out to optimize the process parameter of casting process parameters for copper pipes considering casting process parameters i.e. cooling water flow rate, withdrawal speed, and casting temperature and reported that structural and temperature field of metal flow can be understood to improve the defect tendency (Han, Zhang, Yu, Sun, & Gao, 2017). Guo et al. (Guo, Zheng, & Jing, 2012) simulated the centrifugal process of thick-wall stainless steel pipes in a horizontal centrifugal casting process using finite element method considering preheat of mold, thermal properties of coating for turbulence in flow, cooling of molten metal. The metal matrix composite pipes are manufacture using stir casting process for better tribological and mechanical properties. The materials compositions of these MMCs can be varied to obtain better materials properties (Christy, I. Mourad, & Arunachalam, 2019).

Since the variation in casting process parameters leads to change in resulting properties of the cast, optimization of these parameters for better output can be done. Various optimization tools have been used in literature. Saad et al. (Mahmood Ali, 2019) optimized the centrifugal casting process parameters of aluminium alloys for better tensile properties of the resulting part using response surface methodology. Jonas et al. (Nieschlag, Ruhland, Daubner, Koch, & Fleischer, 2018) employed finite element simulation for optimization of the centrifugal casting process for FRP metal pipes for reduction of its weight. Raffaella et al. (Aversa, Petrescu, Petrescu, & Apicella, 2016) optimize the centrifugal casting process parameters of composite pipes through considering thermal effects of casting process for better mechanical properties response.

Taguchi analysis is an engineering tool to obtain better outcomes by selecting the best set of input parameters. Influence of parameters on the response can also be determined using Taguchi analysis. It gives a critical evaluation of each parameter to determine its effect on the betterment of output. Guharaja et al. (Guharaja, Haq, & Karuppannan, 2006) utilize it for optimization of process parameters green strength, moisture, permeability percentage and hardness of mold of green sand casting for spheroidal graphite cast iron couplings, whereas Sushil et al. (Kumar, Satsangi, & Prajapati, 2011) used it for optimizing green strength, moisture, temperature of molten metal, and mold hardness for better quality characteristics of produced differential housing cover cast. The process parameters of die casting of aluminium were optimized by Anastasiou (Anastasiou, 2002) using the Taguchi method to reduce the porosity. G.P. Syrcos (Syrcos, 2003) optimized the die casting process parameters of aluminium alloys considering molten metal temperature, die pressure, piston velocity for maximum cast density and minimum dimensional variation. The Taguchi method was used by Shailesh et al. (Shailesh, Kumar, Sundarrajan, & Komariahia, 2012) for investigation of process parameters in the centrifugal casting of aluminium 5500 alloys and reported that increase in temperature of molten metal reduces and increase in rotational speed increases the mechanical characteristics of the material, whereas by Cavus et al. (Falamaki & Veysizadeh, 2008) for optimizing the process effect in the production of alumina membrane support using centrifugal casting with remarks that the Taguchi technique gives process parameters model which can be used to predict the outcome. The optimization of centrifugal casting of aluminium-silicon alloy using Taguchi was done by Shailesh et al. (Shailesh, Sundarrajan, & Komaraiah, 2014) for mechanical and microstructural integrity of the cast. The tensile strength of the cast is considered as output response and optimization of molten metal temperature and preheating temperatures of mold, and rotational velocity of the mold in manufacturing aluminium based-metal matrix composites using a centrifugal casting process and concluded that quality characteristics of the cast can be improved by controlling and optimizing process parameters of casting (Vajd & Samadi, 2019). The surface roughness of the cast part is one of the important concerns to the engineers, the cast surfaces should have lower surface roughness since further machining over it will increase the production cost. Munish et al. (Chhabra & Singh, 2012) used the Taguchi methods to obtain the best process parameters combination (Pouring temperature, mold wall thickness & volume of the cast) of casting for minimum surface roughness with remarks that the pouring temperature of molten metal was the most significant parameter out of them.

In literature, it was found that the casting process is a critical process in which little variation in each process parameters significantly affects the resulting properties of the cast product, which led to the need of their optimization using advanced optimization tool like Taguchi analysis. In this work, the optimal parameters necessary for obtaining defect-free or low defect casting is investigated experimentally.

MATERIALS AND METHOD

Taguchi is a statistical method developed by Genichi Taguchi to investigate the effect of different parameters on the variance of performance characteristic that determines the suitable operating conditions of the process (Kumar, Sureshkumar, & Velraj, 2015). Taguchi method making statistical design by using the orthogonal array also offers an opportunity to reduce the number of experiments (Bose, Deb, Banerjee, & Majumder, 2013) (Sarıkaya & Güllü, 2015). Taguchi technique computes a signal-to-noise (S/N) ratio based on experimental data. This ratio defines an experiment level which gives the best performance in the test variables (Bose, Deb, Banerjee, & Majumder, 2013). Also, the effect of test parameters on the experiment outputs is statistically demonstrated by ANOVA analysis. Therefore, with the help of both S/N ratio and ANOVA, optimal process parameters can be determined to provide the best performance.

Table 1. Process parameters and their levels

Process Parameters	Levels		
	1	2	3
Pouring temperature	1550	1575	1600
Die spinning speed	1650	1700	1750
Coolant flow time	1	1.25	1.5

In this work, experiments are conducted on manufacturing of Birla lance pipes through centrifugal casting. This research aims to understand the effect of process parameters which controls the thickness of the pipes and find their optimum combination so that defect-free pipes can be cast. In this regard, pouring temperature (°C), die spinning speed (rpm) and coolant flow time (mins) are considered as the input process parameters. A three-level L9 orthogonal array is selected as the design of experimentation technique. Table 1 shows the three levels of the process parameters. Casting experiments are conducted as per the L9 orthogonal array listed in Table 2. The minimum thickness of the cast pipes is recorded for each experiment. Since the desired thickness for the pipes is 8mm, the defect in casting is the difference between desired thickness and achieved minimum thickness.

Table 2. Experimental L9 orthogonal array

Exp. no.	Pouring Temperature (°C)	Die spinning speed (rpm)	Coolant flow time (minutes)	Minimum thickness (mm)	Defect (mm)
1	1550	1650	1	7.8	0.2
2	1550	1700	1.25	7.6	0.4
3	1550	1750	1.5	8	0
4	1575	1650	1.25	8	0
5	1575	1700	1.5	7.7	0.3
6	1575	1750	1	8.4	0.4
7	1600	1650	1.5	8.2	0.2
8	1600	1700	1	7.5	0.5
9	1600	1750	1.25	8.1	0.1

RESULTS AND DISCUSSION

Signal to Noise (S/N) Analysis

S/N ratios (dB) obtained from Taguchi method are used to achieve the optimization of operating parameters. S/N ratios are logarithmic functions of the expected experimental outcome which is to provide a healthy optimization (Kumar, Sureshkumar, & Velraj, 2015). There are three types of S/N ratios used in Taguchi method and these are nominal-is-best, smaller-the-better and larger-the-better (Kumar, Sureshkumar, & Velraj, 2015) (Sarıkaya & Güllü, 2014). In this study, defect is considered as a smaller-the-better type response. Thus, the following relation is used for calculating the S/N ratio.

$$S/Nratio = -10.\log\frac{1}{n}\left(\sum_{i=1}^{n} y_i^2\right) \tag{1}$$

where S/Nratio is signal to noise ratio (dB), y_i symbolizes the quality characteristic's value obtained from the tests, and n is the test number.

Figure 1. Main effects plot for S/N ratios of defect

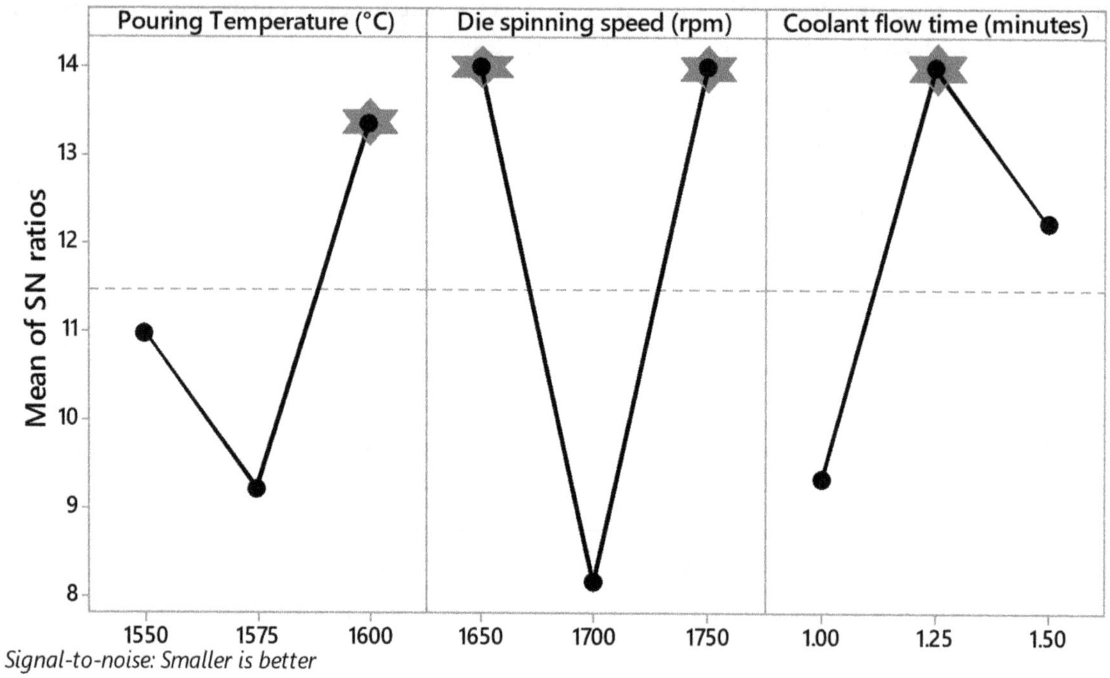

Figure 1 shows the S/N ratios for the response i.e. defect calculated using equation (1). In the S/N plot, the highest value is the optimum level predicted for that process parameter (marked in 'red' colour in Figure 1). The information regarding the analysis of the S/N ratio is also listed in Table 3. From Figure 1 and Table 3, it is seen that the optimum process parameters for obtaining low defect are— pouring temperature at 1600°C (i.e. A3), die spinning speed of 1650 rpm (B1) or 1750 rpm (B3) and coolant flow time of 1.25 minutes (i.e. C2). Thus, the optimal combination for low defect can be summarized as A3-B1-C2 or A3-B3-C2.

Table 3. Response table for Signal to Noise Ratios

Level	Pouring Temperature (°C)	Die spinning speed (rpm)	Coolant flow time (minutes)
1	10.969	13.979	9.32
2	9.208	8.146	13.979
3	13.333	13.979	12.218
Delta	4.125	5.834	4.66
Rank	3	1	2

Figure 2. Main effects plot for the mean of response (defect)

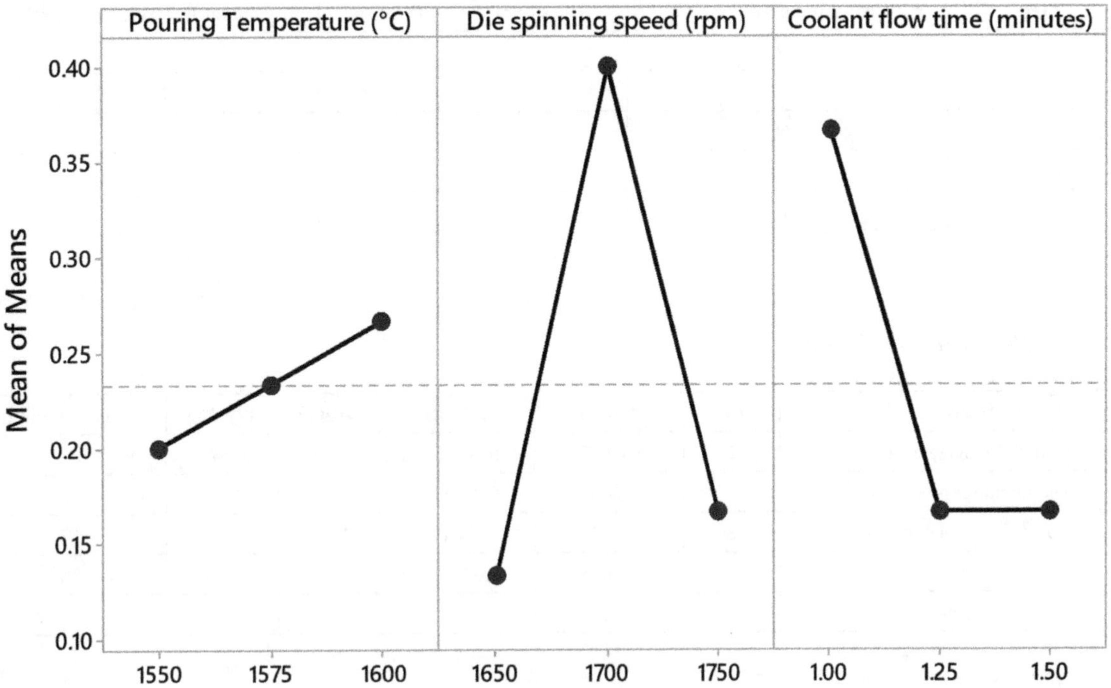

A rough estimation of the relative importance of the process parameters on minimizing defect can be made by analyzing the 'delta' value in Table 3. Delta is calculated as the difference between the highest and the lowest S/R value within each process parameter. Higher the delta value of a process parameter as compared to the other process parameters; more is its significance in the output response. This is because even a small perturbation in the process parameter with the highest delta will have more relative effect on the output response as compared to the same disturbance in other process parameters. It is seen from Table 3 that die spinning speed has the highest delta followed by coolant flow time. The analysis of the main effect of means of the process parameters is shown in Figure 2 and Table 4.

Table 4. Response table for means

Level	Pouring Temperature (°C)	Die spinning speed (rpm)	Coolant flow time (minutes)
1	0.2	0.1333	0.3667
2	0.2333	0.4	0.1667
3	0.2667	0.1667	0.1667
Delta	0.0667	0.2667	0.2
Rank	3	1	2

Table 5. ANOVA results for defect

Source	DF	Adj SS	Adj MS	F-Value	P-Value	% contribution
Pouring Temperature (°C)	2	0.006667	0.003333	0.14	0.875	3%
Die spinning speed (rpm)	2	0.126667	0.063333	2.71	0.269	49%
Coolant flow time (minutes)	2	0.08	0.04	1.71	0.368	31%
Error	2	0.046667	0.023333			18%
Total	8	0.26				

Analysis of Variance (ANOVA)

ANOVA is a statistical method which is used to define the individual interactions into the test results of all the process parameters. In this study, ANOVA analysis was performed with a 95% (0.95) confidence level and 5% (0.05) significance level. The contribution percentage of each process parameter indicates their degree of impact on defect. This means, the higher the percentage of contribution, the higher the influence of the process parameter on the defect. The results of ANOVA for defect model is listed in Table 5 along with the percentage contribution to a linear model. It is seen that die spinning speed has the greatest effect of about 49% followed by coolant flow time which contributes about 31% to the linear model. These findings of ANOVA further confirm the analysis outcome of the Taguchi S/N ratio analysis.

CONCLUSION

In this research work, the effect of three process parameters namely, pouring temperature (°C), die spinning speed (rpm) and coolant flow time (mins) on casting defect of pipes is studied. An L9 orthogonal array is used for the design of experiments. It was seen that the casting defect of pipes is unaffected by the pouring temperature, whereas die spinning speed played the most critical role. The contribution of the pouring temperature (°C), die spinning speed (rpm) and coolant flow time (mins) on casting defect of pipes was seen to 3%, 49% and 31% respectively. Based on the S/N ratio analysis, the A3-B1-C2 or A3-B3-C2 were seen to be the optimal process parameter combination to minimize defect.

REFERENCES

Anastasiou, K. S. (2002). Optimization of the aluminium die casting process based on the Taguchi method. *Proceedings of the Institution of Mechanical Engineers. Part B, Journal of Engineering Manufacture*, *216*(7), 969–977. doi:10.1243/09544050260174175

Arbabi, V., & Ebrahimzadeh, I. (2010). Effects of wall thickness on microstructures and properties of α/β brasses pipes produced by horizontal continuous casting. *International Journal of Cast Metals Research*, *23*(3), 150–157. doi:10.1179/136404609X12490478029353

Aversa, R., Petrescu, R. V., Petrescu, F. I., & Apicella, A. (2016). Smart-factory: Optimization and process control of composite centrifuged pipes. *American Journal of Applied Sciences*, *13*(11), 1330–1341. doi:10.3844/ajassp.2016.1330.1341

Bose, P. K., Deb, M., Banerjee, R., & Majumder, A. (2013). Multi objective optimization of performance parameters of a single cylinder diesel engine running with hydrogen using a Taguchi-fuzzy based approach. *Energy*, *63*, 375–386. doi:10.1016/j.energy.2013.10.045

Chhabra, M., & Singh, R. (2012). Obtaining desired surface roughness of castings produced using ZCast direct metal casting process through Taguchi's experimental approach. *Rapid Prototyping Journal*, *18*(6), 458–471. doi:10.1108/13552541211272009

Christy, J. V., Mourad, I. A.-H., & Arunachalam, R. (2019). Mechanical and Tribological Evaluation of Aluminum Metal Matrix Composite Pipes Fabricated by Gravity and Squeeze Stir Casting. *Pressure Vessels and Piping Conference*, *58974*, V06AT06A018. 10.1115/PVP2019-93857

Falamaki, C., & Veysizadeh, J. (2008). Taguchi design of experiments approach to the manufacture of one-step alumina microfilter/membrane supports by the centrifugal casting technique. *Ceramics International*, *34*(7), 1653–1659. doi:10.1016/j.ceramint.2007.07.020

Ganai, A. R., & Singh, B. (2020). Study of Microstructure, Hardness and Dimensional Accuracy in Al-6061 Centrifugally Cast Pipe. In Manufacturing Engineering (pp. 51–60). Springer.

Guharaja, S., Haq, A. N., & Karuppannan, K. M. (2006). Optimization of green sand casting process parameters by using Taguchi's method. *International Journal of Advanced Manufacturing Technology*, *30*(11-12), 1040–1048. doi:10.100700170-005-0146-2

Guo, E. Y., Zheng, Q., & Jing, T. (2012). Numerical Simulation of Solidification of Thick-Wall Stainless Steel Pipe in Horizontal Centrifugal Casting Process. *Materials Science Forum*, *706*, 1427–1432. doi:10.4028/www.scientific.net/MSF.706-709.1427

Han, Y., Zhang, X.-B., Yu, E., Sun, L., & Gao, Y. (2017). Numerical analysis of temperature field and structure field in horizontal continuous casting process for copper pipes. *International Journal of Heat and Mass Transfer*, *115*, 294–306. doi:10.1016/j.ijheatmasstransfer.2017.08.037

Kaschnitz, E. (2012). Numerical simulation of cenrifugal casting of pipes. *IOP Conf. Series: Materials Science and Engineering*, 33.

Kumar, R. S., Sureshkumar, K., & Velraj, R. (2015). Optimization of biodiesel production from Manilkara zapota (L.) seed oil using Taguchi method. *Fuel*, *140*, 90–96. doi:10.1016/j.fuel.2014.09.103

Kumar, S., Satsangi, P. S., & Prajapati, D. R. (2011). Optimization of green sand casting process parameters of a foundry by using Taguchi's method. *International Journal of Advanced Manufacturing Technology*, *55*(1-4), 23–34. doi:10.100700170-010-3029-0

Mahmood Ali, S. (2019). Optimization of Centrifugal Casting Parameters of AlSi Alloy by using the Response Surface Methodology. *International Journal of Engineering*, *32*, 1516–1526.

Martinez, G., Garnier, M., & Durand, F. (1987). Stirring phenomena in centrifugal casting of pipes. In *Modelling the Flow and Solidification of Metals* (pp. 225–239). Springer. doi:10.1007/978-94-009-3617-1_14

Nieschlag, J., Ruhland, P., Daubner, S., Koch, S.-F., & Fleischer, J. (2018). Finite element optimisation for rotational moulding with a core to manufacture intrinsic hybrid frp metal pipes. *Production Engineering*, *12*(2), 239–247. doi:10.100711740-017-0788-6

Pivinskii, Y. E., Litovskaya, T. I., Samarina, O. N., Volchek, I. B., & Kaplan, F. S. (1991). Centrifugal casting of ceramics. Main parameters and the regularities of the process. *Refractories*, *32*(11-12), 551–558. doi:10.1007/BF01280846

Rahimipour, M. R., & Sobhani, M. (2013). Evaluation of centrifugal casting process parameters for in situ fabricated functionally gradient Fe-TiC composite. *Metallurgical and Materials Transactions. B, Process Metallurgy and Materials Processing Science*, *44*(5), 1120–1123. doi:10.100711663-013-9903-z

Sarıkaya, M., & Güllü, A. (2014). Taguchi design and response surface methodology based analysis of machining parameters in CNC turning under MQL. *Journal of Cleaner Production*, *65*, 604–616. doi:10.1016/j.jclepro.2013.08.040

Sarıkaya, M., & Güllü, A. (2015). Multi-response optimization of MQL parameters using Taguchi-based GRA in turning of difficult-to-cut alloy Haynes 25. *Journal of Cleaner Production*, *91*, 347–357. doi:10.1016/j.jclepro.2014.12.020

Shailesh, P., Kumar, B. P., Sundarrajan, S., & Komariahia, M. (2012). Experimental investigation on centrifugal casting of 5500 alloy: A Taguchi approach. *Scientific Research and Essays*, *7*, 3797–3808.

Shailesh, P., Sundarrajan, S., & Komaraiah, M. (2014). Optimization of process parameters of Al-Si alloy by centrifugal casting technique using Taguchi design of experiments. *Procedia Materials Science*, *6*, 812–820.

Syrcos, G. P. (2003). Die casting process optimization using Taguchi methods. *Journal of Materials Processing Technology*, *135*(1), 68–74. doi:10.1016/S0924-0136(02)01036-1

Vajd, A., & Samadi, A. (2019). Optimization of Centrifugal Casting Parameters to Produce the Functionally Graded Al–15wt% Mg 2 Si Composites with Higher Tensile Properties. *International Journal of Metalcasting*, 1–12.

Wang, Y., Wu, S. P., Xue, X., & Guo, J. J. (2013). Optimization of Tilt-Casting of Aluminum Alloy Automobile Drain Pipes through Numerical Simulation. *Materials Science Forum*, *762*, 633–638. doi:10.4028/www.scientific.net/MSF.762.633

Wassilkowska, A. (2017). Microstructure inhomogeneity of centrifugally cast ductile iron pipes and its effect on mechanical properties. *Kovove Mater*, *55*(05), 311–316. doi:10.4149/km_2017_5_311

Yeh, J.-W., & Jong, S.-H. (1994). The cast structure of a 7075 alloy produced by a water-cooling centrifugal casting method. *Metallurgical and Materials Transactions. A, Physical Metallurgy and Materials Science*, *25*(3), 643–650. doi:10.1007/BF02651606

ADDITIONAL READING

Dabade, U. A., & Bhedasgaonkar, R. C. (2013). Casting defect analysis using design of experiments (DoE) and computer aided casting simulation technique. *Procedia Cirp*, *7*, 616–621. doi:10.1016/j.procir.2013.06.042

Das, P. P., Gupta, P., Ghadai, R. K., Ramachandran, M., & Kalita, K. (2017). Optimization of turning process parameters by Taguchi-based Six Sigma. *Mechanics and Mechanical Engineering*, *21*(3), 649–656.

Haq, A. N., Guharaja, S., & Karuppannan, K. M. (2009). Parameter optimization of CO 2 casting process by using Taguchi method. *International Journal on Interactive Design and Manufacturing*, *3*(1), 41–50. doi:10.100712008-008-0054-4

Kumaravadivel, A., Natarajan, U., & Ilamparithi, C. (2012). Determining the optimum green sand casting process parameters using Taguchi's method. *Journal of the Chinese Institute of Industrial Engineers*, *29*(2), 148–162. doi:10.1080/10170669.2012.664789

KEY TERMS AND DEFINITIONS

Analysis of Variance: Analysis of variance is a statistical approach to analyze the differences among group means in a sample.

Casting: Casting is a commonly used manufacturing process in which the part is fabricated by pouring molten metal into a mold and then allowed to solidify.

Optimal Process Parameters: The most suitable combination of parameters involved in a process that can help attain the desired responses (outputs).

Chapter 3
Investigating the Effect of Process Parameters in Incremental Sheet Forming Process

Manish Oraon
Birla Institute of Technology, Mesra, India

Manish Kumar Roy
https://orcid.org/0000-0002-3323-8619
Sikkim Manipal Institute of Technology, Sikkim Manipal University, Majhitar, India

Vinay Sharma
Birla Institute of Technology, Mesra, India

ABSTRACT

Incremental sheet forming (ISF) is an emerging technique of sheet metal working that comes into the picture in the last two decades. The ISF involved the forming of shapes without using the dedicated dies. ISF is suitable for customized products, rapid prototyping, and low batch production. The study aims to investigate the effect of process parameters on the surface roughness. The experiments are conducted on aluminum AA3003-O grade with six parameters, and the trials are performed according to the design of experiment (DOE). The atomic force microscopy (AFM) technique is used for measuring the surface roughness. Analysis of variance (ANOVA) is used for analyzing the effect of process parameters in ISF. The result shows that the step-down size, feed rate of the tool, and wall angle are significant process parameter and their contributions for ISF are 85.86%, 1.12%, and 12.29%, respectively.

DOI: 10.4018/978-1-7998-7206-1.ch003

INTRODUCTION

Incremental sheet forming (ISF) is a recently developed sheet forming technique in which the metal sheets are shaped without using any dedicated dies. The ISF branch included single point incremental forming (SPIF) in which no support is needed and two-point incremental forming (TPIF) is which metal sheet is supported behind with a partial/full die. The ISF is performed at normal temperature and the geometrical accuracy of the product depends upon the efficiency of the computer numerical control (CNC) machining centre. Extensive research is going on in ISF but still, adequate solutions are required to estimate the formability of sheet metal, tool path optimization, force, surface roughness, etc. the surface roughness of end product is one the deciding factor for the acceptability of the ISF. Several researchers reported that the controlled environment i.e. controlling the process parameter during the ISF enhanced the formability of thin metal sheet and also improved the surface roughness. From a customer point of view, the product should have sufficient smoothness for its acceptance. Therefore, the process capability of ISF included the forming of the thin metal sheet without its failure and achieving the desired surface roughness. In the extensive review of ISF, it is observed that the rate of deformation is a major process parameter in the ISF process (Park and Kim, 2003), (Amborgio et al., 2007), (Bahoul, Arfa, & Belhadj, 2014), (Oraon & Sharma, 2010) (Pohlak, Majak, & Kuttner, 2007). The preliminary ISF experiment with varying wall angle on SPCC steel reported springback error and dimensional inaccuracy at the edges of formed parts (Luo, He, & Du, 2010). In the succession, the authors formed the pyramid frustum on AA7075 grade aluminum through ISF and concluded the effects of process parameters amplitudes Durante, Formisano, and Langella, 2010). In SPIF, the external surface roughness is measured (orange peel effect) for the estimation of the proposed model (Hamilton, 2010). ISF is also introduced in medical implants manufacturing. The partial resurfacing of the femoral condylar surface (used at the knee) from titanium is formed through ISF successfully and its surface roughness is estimated by developed mathematical models that define different roughness parameters. Finally, the authors highlighted the significant process parameters which affected the surface roughness of titanium alloy in ISF (Oleksik et al., 2010). The dimensional accuracy, surface roughness, and microstructure of ISF parts made from carbon steel DC01, stainless steel 304, and aluminum A1050 are analyzed statistically and discussed (Radu & Cristea, 2013). The artificial intelligence e.g. three-layer back propagation neural network and genetic algorithm are proposed for the optimization of surface roughness and wall angle in the ISF of aluminum Al5052 (Varthini et al., 2014). Further, the Grey Relational Analysis (GRA) is applied by the authors for the prediction of surface roughness in ISF of aluminum alloy of Indian standard (Patel et al., 2015). In the ISF of stainless steel SS304, it is reported that the step depth with 63.56% contribution for the surface roughness (Shah & Chaudhary, 2016) whereas in ISF of aluminum alloy, the contribution of process parameters on the surface roughness are as wall angle 69%, step depth 27%, Spindle speed 3% and tool diameter 1% (Luo, He, & Du, 2010). The vertical force component (Oraon & Sharma, 2018) and the average surface roughness in ISF of aluminum AA3003-O by using artificial neural network (AAN) (Oraon & Sharma, 2018). The authors reported that mean absolute error of 0.215 and 1.068 for vertical force and average surface roughness respectively. The body parts of car such as mudguard and bumper was prepared successfully through ISF of aluminum alloy sheet (Soren et al., 2019).

EXPERIMENTAL INVESTIGATION

The ISF is conducted on thin sheets of aluminum AA3003-O with varying the process parameters. The six process parameters are considered during the experiment which is as step down size (Δz), feed rate (f), spindle speed (R), sheet thickness (T), wall angle (Θ), and lubricant (L). The two levels of process parameters are set after an extensive review of literatures work and preliminary ISF experiments done in the department of Production Engineering, Birla Institute of Technology, Mesra Ranchi. The numeric code of the process parameters is shown in Table 1.

Table 1. Extreme values of the experimental plane

Process Parameter	Δz (mm)	f (mm)	R	T (mm)	Θ (^{o})	L
level 1	0.1	20	500	0.2	15	MoS_2 Grease
level 2	0.7	100	2000	0.4	45	Graphite Powder

The machine, forming tool and aluminum sheet is used in ISF are detailed below

Machine

The ISF is conducted on a multifunctional CNC machining center DT-110 (Mikrotool Pvt. Ltd., Singapore) (Fig. 1) placed in the laboratory of the production engineering department, BIT Mesra. The specifications of the DT-110 are as Table size: 810×400 mm, Travel: x-axis:200 mm, y-axis:100 mm and z-axis:100 mm, Spindle speed: 0 -3000, and Position accuracy: +/- 1 micron/100mm

Figure 1. Incremental sheet forming set-up on DT-110

Forming Tool

The customized ISF forming tool is prepared to mild steel rod of diameter 07mm and length 50mm. A face of the rod is grooved and a 06mm diameter stainless steel bearing ball inserted to form a hemispherical end. The bearing ball is high wear-resistant and thermal shock properties. The advantage of this tool is to replace only the ball after it's worn out despite of replacing or re-engineering the entire tool. The customized tool is shown in Figure 2.

Figure 2. The customized ISF tool

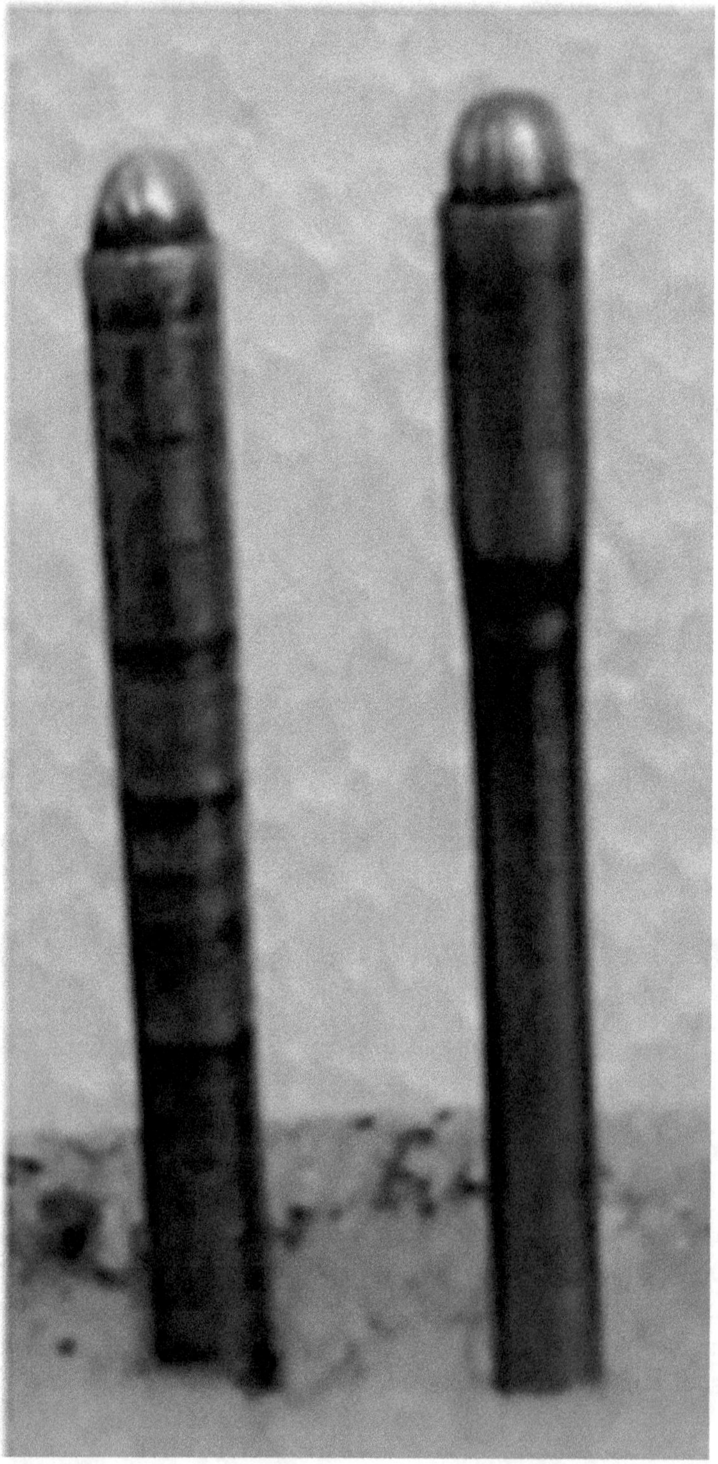

Working Sheet

The ISF is more suitable for the sheets of low thickness since only a point load is applied during processing. Also, it is difficult to the ISF of thick metal sheet without special arrangement. Therefore domestic purpose aluminum AA 3003-O is taken as the worksheet. The purity of the working sheet is measured by using scan electron microscopy (SEM). The test results include the chemical composition of AA3003-O and are summarized in Table 2.

Table 2. Compositions of other elements in AA 3003-O

	Composition (%)						
AA3003-O	Al	Si	Fe	Cu	Co	Mg	Zr
	98.12	0.628	0.705	0.05	0.08	1.2	0.05

The tensile test is carried out in the 'INSTRON' Series IX Automated Materials Testing System in the department of polymer engineering, BIT Mesra, Ranchi. The samples are prepared as per American society for testing and materials (ASTM) standard E8 and the tensile loading is applied at the crosshead speed 5mm/min. The mechanical properties are given in Table 3.

Table 3. Mechanical properties of Aluminium AA3003-O

Material	3003-O
Ultimate tensile strength	166.8MPa
Yield strength	125.2MPa
Density	2.73 g/cm^3
% Elongation	18
Poisson ratio	0.33
Melting point	655^0 C

Since the output response i.e. absolute surface roughness (*Ra*) is unknown in the ISF, therefore DOE is adopted for experiments in which orthogonal array L8 is chosen. It is the combination of process parameters to form a minimum number of experiments set with six process parameters (Table 4). The process parameters are chosen after the review of ISF journals and the preliminary test conducted in the laboratory of BIT Mesra, Ranchi. Two levels of the process parameters (low and high) and the enough differences of inputs are taken during the experiment to analyzed the effect of process parameters.

Table 4. Orthogonal array L8 design

Exp. No.	Δz (mm)	f (mm)	R	T (mm)	θ (°)	L
1	1	1	1	1	1	1
2	1	1	1	2	2	2
3	1	2	2	1	1	2
4	1	2	2	2	2	1
5	2	1	2	2	1	1
6	2	1	2	1	2	2
7	2	2	1	2	1	2
8	2	2	1	1	2	1

The square pyramid frustum of side 35mm*35mm (Fig. 3) is developed whereas the height of frustum is dependent upon the combination of process parameters. The test samples are prepared by cutting 5mm*5mm section carefully from each part without damaging the formed texture i.e. impression of tool mark. The *Ra* is measured in the precise machine NT-MDT solver-4 with the magnification of 5X. The atomic force microscopy (AFM) technique is used for measuring the *Ra*-value in which an optical probe scanned the particular area. The measured roughness's i.e. pick to pick roughness (*Ry*), average ten peaks (*Rz*), root mean square roughness (*Rq*), and absolute roughness (*Ra*) along with 2D and 3D micro images are generated. The measured *Ra*-value along with the numeric value of process parameters are tabulated in Table 5.

Figure 3. The formed square pyramid and sample

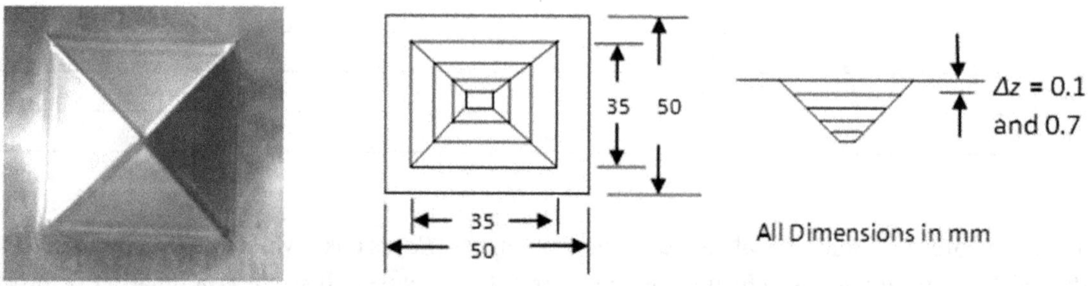

The statistical tool MINITAB 17.0.1 is used for analysis in which the data are processed with 95% confidence that means the value of P should be £ 0.05 (fishers test). The less value of P indicated the significance of the process parameter otherwise insignificant individually but the interaction with other process parameters may show their significance. In the present study, higher *Ra*-value is indicated the average rough surface of the end product therefore during statistical analysis 'smaller is best' approach is taken. The signal to noise ratio (SN) for the *Ra* is shown in equation 1.

Figure 4. 3D micro-images of each test sample

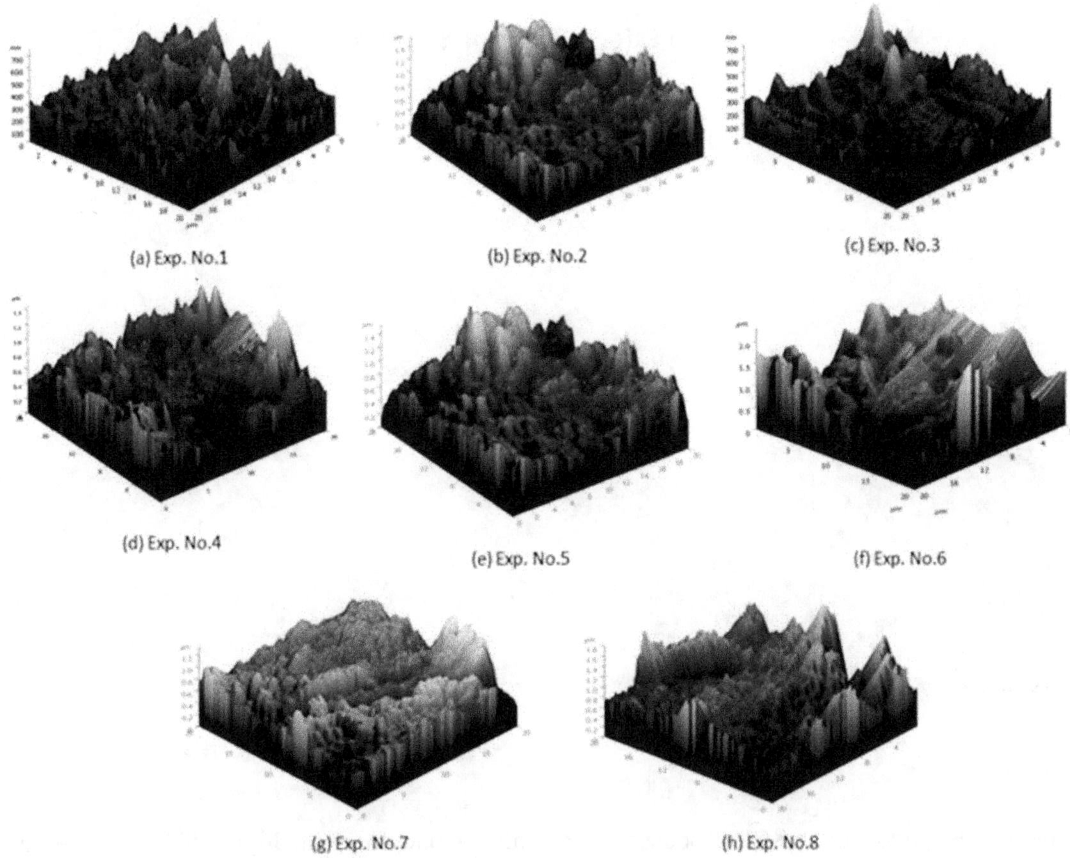

$$S/N\,of\,Ra = -10log\frac{1}{n}\sum_{i=1}^{n} y_{ij}^{2} \ldots\ldots 1$$

Where; n is number of replications, and y_{ij} is observed response value.

RESULT AND DISCUSSION

The 3D micrographs generated in the AFM test of each sample are presented in Figure 4a-h. These micrographs are associated with the surface roughness e.g. Peak-to-peak (*Ry*), Ten point height (*Rz*), Root Mean Square (*Rq*), and Average Roughness (*Ra*). Since the *Ra* is represented the average roughness of entire product, therefore *Ra* is considered for further analysis.

Figure 5 showed the effect of the individual process parameter on the *Ra*. The step down size Δz, tool feed rate *f*, and wall angle Θ is showing greater slop. It is indicated that these are the controlling factors during ISF of AA3003-O.

Figure 5. Main effect plot of Process parameter on Ra

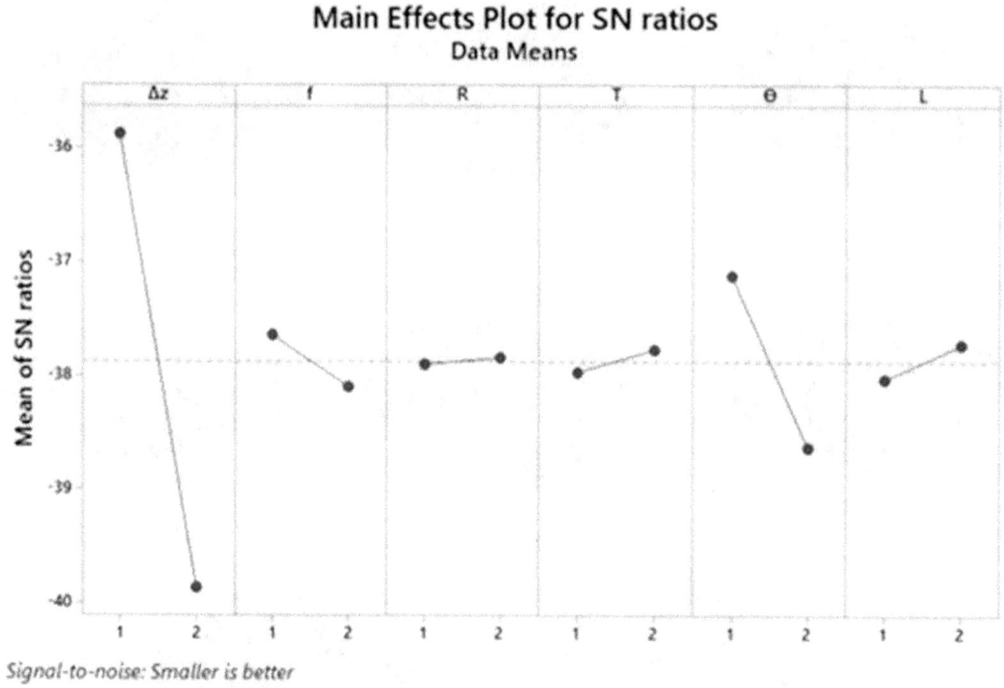

Table 6 corresponds to the S/N responses for *Ra*-value. Concerning the higher SN ratio, the rank of process parameters for lower *Ra* are as Δz=0.1 mm followed by f=20mm/min, R= 2000, T= 0.4 mm, Θ= 45^0, and L= Graphite Powder respectively. ANOVA is the result showing the conformity information about the system. It states the spread of measured value from the mean value that should be acceptable or not. The ANOVA result of the present study is tabulated in Table 7.

Table 5. Measured Ra with numeric value process parameters

Exp. No.	Δz *(mm)*	*f* *(mm)*	*R*	*T* *(mm)*	Θ *(°)*	*L*	*Ra*
1	0.1	20	500	0.2	15	MoS_2 Grease	57.232
2	0.1	20	500	0.4	45	Graphite	64.562
3	0.1	100	2000	0.2	15	Graphite	58.156
4	0.1	100	2000	0.4	45	MoS_2 Grease	69.744
5	0.7	20	2000	0.2	45	MoS_2 Grease	107.621
6	0.7	20	2000	0.4	15	Graphite	85.058
7	0.7	100	500	0.2	45	Graphite	109.831
8	0.7	100	500	0.4	15	MoS_2 Grease	93.755

Table 6. Rank of process parameters for Ra-value

Level	Δz	f	R	T	θ	L
1	-35.88	-37.65	-37.90	-37.97	-37.12	-38.03
2	-39.87	-38.10	-37.85	-37.78	-38.63	-37.73
Delta	3.99	0.46	0.65	0.20	1.51	0.30
Rank	1	3	6	5	2	4

Table 7. ANOVA of Ra-value

Source	DF	Seq SS	Adj MS	F	P	Significance
Δz	1	31.8892	31.8892	13485.45	0.005	Yes
f	1	0.4195	0.4195	177.38	0.048	Yes
R	1	0.0056	0.0056	2.39	0.366	No
T	1	0.0787	0.0787	33.26	0.109	No
θ	1	4.5651	4.5651	1930.51	0.014	Yes
L	1	0.1802	0.1802	76.18	0.073	No
Residual Error	1	0.0024	0.0024			
Total	7	37.1406				

Table 8. Percentage contribution of significant parameters

Input Variable	P	Significance	Contribution (%)
Δz	0.005	Yes	85.86
f	0.048	Yes	1.12
θ	0.014	Yes	12.29

Figure 6. 3D surface plot of significant process parameters

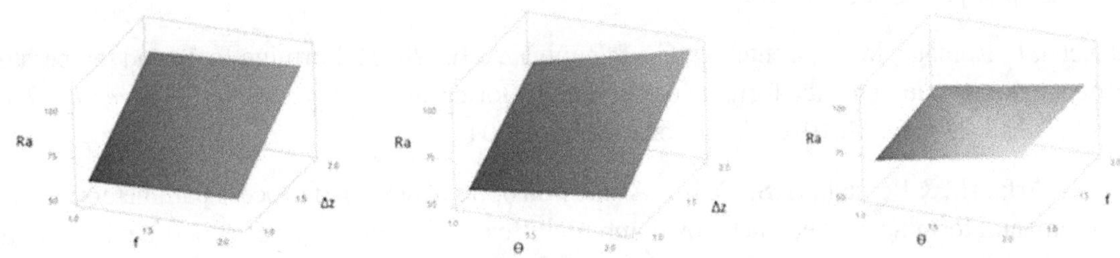

(a) Surface Plot of Ra vs Δz, f (b) Surface Plot of Ra vs Δz, θ (c) Surface Plot of Ra vs f, θ

Table 7 confirmed that the process parameter which is significant in the ISF of aluminum AA3003-O. The significant process parameters for controlling the Ra-value are step down size Δz (P=0.005), tool feed f (P=0.048), and wall angle θ (P=0.014). The percentage contribution of the significant input variables is shown in Table 8.

The response surface plot is showing the effect of variation of the significant process parameters (Fig. 6a-c). Figure 6a showing that the Ra is improved by processing the AA3003-O with lower Δz value at a lower feed rate of the tool f whereas the wall angle Θ should be low when processing is done at low Δz (Fig. 6b). The interactions of feed rate of tool f and wall angle Θ are not too affect the *Ra*-value wherever Δz is involved (Fig. 6c)

CONCLUSION

The part is formed successfully in ISF of AA3003-O even with the high step down size and low feed rate. The higher order of spindle speed retarded the blank strength which results in the metal sheet that deformed with lower force. The surface roughness of the parts is improved by keeping the wall angle as low as possible. The low wall inclination reduced the tool feed impression and improved the forming limit of thin metal sheets. With this analysis, the following conclusions are made:

- The statistical analysis confirmed that step down size Δz, feed rate f, and wall angle Θ are significant process parameters in ISF of AA3003-O.
- The contribution of Δz, f, and Θ are 85.86%, 1.12%, and 12.29% respectively.
- The *Ra* is improved by conducting ISF with Δz=0.1mm, f=20 mm/min, and $\Theta=15^0$ respectively.

This paper reveals only the significance of the individual process parameter on the *Ra*-value. The interaction of process parameters will show the significance of individually insignificant process parameters.

REFERENCES

Ambrogio, G., Cozza, V., Filice, L., & Micari, F. (2007). An analytical model for improving precision in single point incremental forming. *Journal of Materials Processing Technology*, *191*(1-3), 92–95. doi:10.1016/j.jmatprotec.2007.03.079

Bagudancha, I., Romeua, M. L., Centenob, G., & Zuniga, A. E. (2015). Forming force and temperature effects on single point incremental forming of polyvinylchloride. *Journal of Materials Processing Technology*, *219*, 221–229. doi:10.1016/j.jmatprotec.2014.12.004

Bahoul, R., Arfa, H., & Belhadj, H. S. (2014). A study on optimal design of process parameters in single point incremental forming of sheet metal by combining Box–Behnken design of experiments, response surface methods and genetic algorithms. *Int J Adv Manuf, 74,* 163-185.

Durante, M., Formisano, A., & Langella, A. (2010). Comparison between analytical and experimental roughness values of components created by incremental forming. *Journal of Materials Processing Technology*, *210*(14), 1934–1941. doi:10.1016/j.jmatprotec.2010.07.006

Hamilton, K. A. (2010). *Friction and External Surface Roughness in Single Point Incremental forming: A study of Surface friction, Contact Area and The 'Orange Peel' Effect.* Queen's University.

Khatal, G. D., Borkar, B. R., & M, G. M. (2016). Analytical Study of SPIF Process. *IJARIIE, 2*(4), 359-363.

Luo, Y., He, K., & Du, R. (2010). A new sheet metal forming system based on incremental punching, part 2: Machine building and experiment results. *International Journal of Advanced Manufacturing Technology, 54*, 481–491. doi:10.100700170-010-2634-2

Oleksik, V., Pascu, A., Deac, C., Fleaca, R., Bologa, O., & Racz, G. (2010). Experimental Study on The Surface Quality Of The Medical Implants Obtained By Single Point Incremental Forming. *International Journal of Material Forming, 3*(1), 935–938. doi:10.100712289-010-0922-x

Oraon, M., & Sharma, V. (2010). Sheet Metal Micro Forming: Future ResearchPotentials. *Int. J. on Production and Industrial Engineering, 1*, 31–35.

Oraon, M., & Sharma, V. (2018). Predicting force in single point incremental forming by using artificial neural network. *International Journal of Engineering, 3*(1), 88-95.

Oraon, M., & Sharma, V. (2018). Prediction of surface roughness in single point incremental forming of AA3003-O using artificial neural network. *International Journal of Materials Engineering. Innovation, 9*, 1–19.

Park, J. J., & Kim, Y. H. (2003). Fundamental studies on the incremental sheet metal forming technique. *Journal of Materials Processing Technology, 140*(1-3), 447–453. doi:10.1016/S0924-0136(03)00768-4

Patel, J. R., Samvatsar, K. S., Prajapati, H. P., & Rangrej, S. S. (2015). Optimization of Process Parameters for Reducing Surface Roughness Produced During Single Point Incremental Forming Process. *International Journal on Recent Technologies in Mechanical and Electrical Engineering*, 2(9), 19–23.

Pohlak, M., Majak, J., & Kuttner, R. (2007). Manufacturability and limitations in incremental sheet forming. *Proc. Estonian Acad. Sci. Eng, 13*(2), 129–139.

Radu, M. C., & Cristea, I. (2013). Processing Metal Sheets by SPIF and Analysis of Parts Quality. *Materials and Manufacturing Processes, 28*(3), 287–293. doi:10.1080/10426914.2012.746702

Shah, H., & Chaudhary, S. (2016). Optimization of Process Parameters for Incremental Sheet Metal Forming Process. *International Journal For Technological Research In Engineering, 3*(7), 1432–1435.

Soren, S., Alexander, P., Peter, S., Sebastian, M., Dieter, W., & Zdenek, R. (2019). Incremental sheet metal forming on the example of car exterior skin part. *Proceedia Manufacturing, 29*, 105–111. doi:10.1016/j.promfg.2019.02.112

Varthini, R., & Gandhinathan, R., Pandivelan, C., & Jeevanantham, A. K. (2014). Modelling and optimization of process parameters of The single point incremental forming of aluminium 5052 alloy sheet using genetic algorithm-back Propagation neural network. *International Journal of Mechanical And Production Engineering*, 2(5), 55–62.

ADDITIONAL READING

Gheysarian, A., & Honarpisheh, M. (2019). Process parameters optimization of the explosive-welded Al/Cu bimetal in the incremental sheet metal forming process. *Iranian Journal of Science and Technology. Transaction of Mechanical Engineering*, *43*(1), 945–956. doi:10.100740997-018-0205-6

Kim, Y. H., & Park, J. J. (2002). Effect of process parameters on formability in incremental forming of sheet metal. *Journal of Materials Processing Technology*, *130*, 42–46. doi:10.1016/S0924-0136(02)00788-4

Li, J., Geng, P., & Shen, J. (2013). Numerical simulation and experimental investigation of multistage incremental sheet forming. *International Journal of Advanced Manufacturing Technology*, *68*(9-12), 2637–2644. doi:10.100700170-013-4870-8

Li, Y., Chen, X., Zhai, W., Wang, L., Li, J., & Guoqun, Z. (2018). Effects of process parameters on thickness thinning and mechanical properties of the formed parts in incremental sheet forming. *International Journal of Advanced Manufacturing Technology*, *98*(9-12), 3071–3080. doi:10.100700170-018-2469-9

Liu, Z. B., Li, Y. L., Daniel, W. J. T., & Meehan, P. (2014). Taguchi optimization of process parameters for forming time in incremental sheet forming process. *Materials Science Forum*, *773*, 137–143.

Vahdati, M., Mahdavinejad, R., & Amini, S. (2017). Investigation of the ultrasonic vibration effect in incremental sheet metal forming process. *Proceedings of the Institution of Mechanical Engineers. Part B, Journal of Engineering Manufacture*, *231*(6), 971–982. doi:10.1177/0954405415578579

KEY TERMS AND DEFINITIONS

ANOVA: Analysis of Variance, or ANOVA for short, is a statistical test that looks for significant differences between means on a particular measure.

DOE: Design of Experiments (DOE) is an approach for planning, conducting, analyzing, and interpreting controlled tests to evaluate the factors that control the value of a parameter or group of parameters.

Optimization: A mathematical technique for finding a maximum or minimum value of a function of several variables subject to a set of constraints.

Chapter 4
Optimization of Surface Roughness in Centreless Grinding Process Based on Taguchi Method

Prosun Mandal
https://orcid.org/0000-0003-3530-5616
Indian Institute of Engineering Science and Technology, Shibpur, India

ABSTRACT

This chapter aims to optimize centreless grinding conditions using the Taguchi method for minimizing surface roughness. The grinding operation has been performed according to the L9 orthogonal array in a centreless grinding process. The centreless grinding experiments are carried out on the crane-hook pin of C40 steel. The analysis of variance (ANOVA) and computation of signal to noise (S/N) ratio are adopted to determine the influence of grinding parameters (depth of cut [µm], regulating wheel speed [rpm], and coolant valve opening) on surface roughness. The depth of cut (µm) is found to be the most significant among the grinding parameters on the surface roughness. The signal to noise (S/N) ratio was calculated based on smaller the best criteria. The lower level of depth of cut, medium level of regulating wheel speed, and higher-level coolant valve opening is found to be optimal grinding condition according to the mean response and signal to noise (S/N) ratio.

INTRODUCTION

In machining industries, high-quality surface finishes very important in order to improve component life. Centreless grinding is one of the crucial surface finishing processes. The centreless grinding is a complex, nonlinear, and sensible machining process, a large number of input factors, e.g., depth of cut, regulating wheel speed, and coolant supply are influenced system stability and output performance (Mandal and Mondal, 2017). Khan et al. (2012) considered three process parameters, namely grinding speed, grinding feed, and dressing feed as independent variables and diametral deviation, out-of-round-

DOI: 10.4018/978-1-7998-7206-1.ch004

ness, and surface roughness of the work-piece is considered as dependent performance characteristics in the centreless grinding process. Modelling and optimization are useful techniques for obtaining the best machining parameters (Mandal & Mondal, 2014; Mandal & Mondal, 2015; Mondal, Mandal, & Ghosh, 2017; Ghosh, Mandal, & Mondal, 2019; Ghosh, Mandal, & Mondal, 2019; Mandal & Mondal, 2020). Bhattacharya et al. (2009) were employed the Taguchi technique to investigate the influence of machining conditions on surface finish and power consumption. The cutting speed has a significant effect on surface roughness and power consumption. The optimal cutting condition was obtained. Zhang and Chen (2009) utilized the Taguchi method to optimize the surface quality for drilling. The statistical analysis of the response variable and signal-to-noise ratios were helped to identify the optimal machining condition. The predicted result was verified by the confirmation test. Günay and Yücel (2013) applied the Taguchi method to obtain the optimal cutting parameters for the reduction of surface roughness. The calculation of signal-to-noise (S/N) ratio for surface roughness based on smaller-the better approach led to determine optimal cutting parameters. The analysis of variance (ANOVA) was used to evaluate the effects of the cutting parameters on the surface roughness. Aslan et al. (2011) performed experiments according to Taguchi's L27 orthogonal array and analyzed the Ra and cutting tool wear for machining of AISI 4140 (63 HRC) steel. Davim and Figueira (2007) also performed experiments according to Taguchi's L27 orthogonal array for investigating the machinability turning operation of a cold work tool steel. Gopalsamy et al. (2009) utilized analysis of variance (ANOVA) for investigating the effects of machining parameters on tool life and surface roughness (Ra). The cutting speed is found to be the most influential process parameter. The experiment was performed according to the L18 orthogonal array. Sharma et al. (2005) obtained optimal process parameters for strontium ferrite sintered magnets using Taguchi L9 design.

Various work materials and tool materials have a very crucial influence on machining performance (Mandal & Mondal, 2019a; 2019b; 2019c). The crane-hook pin is normally made of C40 steel, used in the complex centreless grinding process. A very high surface finish is very important for the product like crane-hook-pin (Mandal & Mondal, 2017). The statistical analysis and optimization of process parameters is the important requirement for investigating very complex centreless grinding operation. This present study aims to obtain optimal grinding conditions (depth of cut, regulating wheel speed, and coolant valve opening) for minimizing surface roughness while grinding operation performed on C40 steel pin. Taguchi L9 design of experiment (DOE) was employed in order to reduce the number of experiments (Bendell, 1989; Lochner, & Matar, 1990). The analysis of variance (ANOVA) was used to investigate the statistical significance of the grinding parameters.

TAGUCHI METHOD

Taguchi developed a powerful statistical method for designing high-quality systems (Taguchi, 1990). With this method, a system is easily optimized for performance, quality, and cost in a simple, efficient, and systematic manner (Yang & Tarng, 1998). Quality engineering aims to develop robust products against all noise factors. The selection of control factors is the most important stage for the design of an experiment. The signal-to-noise (S/N) ratio is termed as a quality characteristic in the Taguchi technique (Asiltürk, & Akkuş, 2011). The loss function is the calculation of the deviation between the experimental value and the desired value. This loss function further transformed into a signal-to-noise (S/N) ratio (Park, 1996; Nalbant, Gökkaya & Sur, 2007). Three different performance characteristics in the analysis of signal-to-noise (S/N) ratio, namely, the lower-the-better, the higher-the-better, and the nominal- the-better (Shetty, Pai, Rao, & Nayak, 2009; Gupta, Singh, & Aggarwal, 2011) are calculated by Equation (1) to (3).

Nominal the best:

$$[S/N]_{NB} = 10\log\left(\frac{\bar{y}}{S_y^2}\right) \tag{1}$$

Larger is the better (Maximize):

$$\left[\frac{S}{N}\right]_{LB} = -10log\left(\frac{1}{n}\sum_{i=1}^{n}\frac{1}{y_i^2}\right) \tag{2}$$

Smaller is the better (minimize):

$$\left[\frac{S}{N}\right]_{SB} = -10log\left(\frac{1}{n}\sum_{i=1}^{n}y_i^2\right) \tag{3}$$

In the above Equation, $\bar{y}$ is denoted average of the observed data, S_y^2 is denoted the variance of y, n is the number of observations, and y is the observed data. The signal-to-noise (S/N) ratio is calculated based on the signal-to-noise (S/N) analysis for each level of process parameters. The large signal-to-noise (S/N) ratio indicates the better performance characteristic for each category of the performance characteristic (Mandal, Doloi, Mondal, & Das, 2011; Souissi, Souissi, Niniven, Amar, Bradai, & Elhalouani, 2014; Hasçalık, & Çaydaş, 2008). Therefore, the optimal level of process parameters can be obtained by the signal-to-noise (S/N) ratio. Furthermore, analysis of variance (ANOVA) needs to be performed to find the contribution of process parameters on the performance. The optimal process parameters can be predicted with the help of signal-to-noise (S/N) analysis and analysis of variance (ANOVA). In this study, the aim of the research to minimize the surface roughness for centreless grinding operation. So the smaller-the-better performance characteristic was used to compute S/N ratio in order to find optimal process parameters.

EXPERIMENTATION

A high level of micro-finishing accuracy is required for the crane-hook pin in order to improve machining life. The centreless grinding process is the best choice for achieving a high level of surface finish. The crane-hook pins of C40 steel (17 mm diameter and 100 mm length) were grinned using centreless grinding process. The grinned pins are shown in Figure 3. The experiments were performed in centreless grinding machine-tool (Machine No-616 and Model: GCL 63, MIC machine tool and industries, Ghaziabad). The range of the diameter of the work-piece that can be ground is 2–100 mm. The surface roughness tester (Mitutoyo SJ-301) is used to measure the roughness of the grinned surface.

Figure 1. Grinned crane-hook-pin

Three independent centreless grinding process parameters such as depth of cut (Dc), regulating wheel speed (N), and coolant flow valve opening (Cf) were chosen for experimentation. Three different levels (low, medium, and high) of centreless grinding parameters were selected by the influence of machined specification. Table 1 shows the centreless grinding parameters and three levels. A wide range of experiments needs to be performed by varying three process parameters when the simple factorial design or central composite design matrix was followed. For reducing the number of experiments, Taguchi L9 orthogonal array was used in the present study. The L9 standard orthogonal array experimental design matrix is shown in Table 2.

Table 1. Centreless grinding parameters and their levels

Factors	Low level	Medium level	High level
Depth of cut (Dc)	20	25	30
Wheel speed (N)	12	14	16
Coolant flow valve opening (Cf)	1/3	2/3	1

RESULTS AND DISCUSSION

The experiment runs according to the Taguchi L9 design matrix, and corresponding surface roughness is measured for each of the specimens. Table 3 shows the surface roughness for each of the experiments. The final column of Table 3 denoted the signal-to-noise (S/N) ratio for surface roughness which is computed by using Equation (3). Figure 2 and Table 4 shows the average signal-to-noise (S/N) ratio graph and signal-to-noise (S/N) ratio response Table for surface roughness. The maximum S/N ratio is

Table 2. Taguchi L9 design matrix

Experiment No	Depth of cut (Dc) (μm)	Wheel speed (N) (rpm)	Coolant flow (Cf) (fraction of valve open)
1	20	12	1/3
2	20	14	2/3
3	20	16	1
4	25	12	2/3
5	25	14	1
6	25	16	1/3
7	30	12	1
8	30	14	1/3
9	30	16	2/3

to represent the optimal machining condition for centreless grinding. The highest S/N ratios are observed for the depth of cut, regulating wheel speed and coolant flow valve opening at the level of low, medium, and high, respectively. So the values of optimal cutting parameters are found to be the depth of cut= 20 μm, regulating wheel speed= 14 rpm, and coolant flow valve opening= full opening. The main effects of the process parameters on the response are shown in Figure 3. The average value of the performance characteristics is referred to as the mean response for each factor at different levels. The mean response analysis indicates the optimum level of parameters as similar to the S/N ratio analysis.

Table 3. The surface roughness and signal-to-noise ratio

Experiment No	Surface Roughness (Ra)	Calculated S/N ratio for Ra
1	0.1	20
2	0.11	19.1721463
3	0.18	14.8945499
4	0.24	12.39577517
5	0.12	18.41637508
6	0.25	12.04119983
7	0.2	13.97940009
8	0.27	11.37272472
9	0.22	13.15154638

Table 5 presents the analysis of variance (ANOVA) for surface roughness in centreless grinding process. The F value for depth of cut, regulating wheel speed, and coolant valve opening are 1.54, 0.39, and 0.23 respectively indicated that the depth of cut (Dc) had a specific contribution. The calculated by percentage contribution also indicated that the depth of cut (Dc) is the most contributing process parameter on the surface roughness.

Table 4. Signal-to-noise response table for surface roughness (Ra)

Level	Dc	N	Cf
1	18.02	15.46	14.47
2	14.28	16.32	14.91
3	12.83	13.36	15.76
Delta	5.19	2.96	1.29
Rank	1	2	3

Figure 2. Effect of process parameters on average S/N ratio

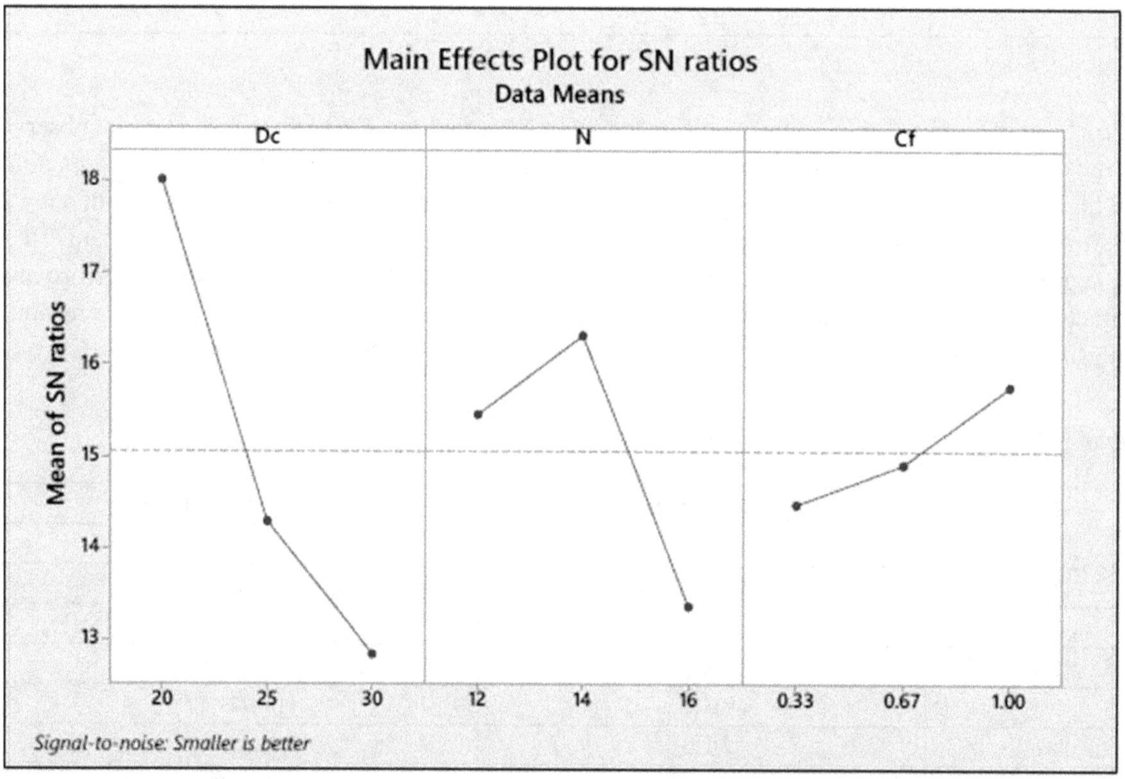

CONCLUSION

In this paper, the centreless grinding operation was performed on the crane-hook pin made of C40 steel according to the Taguchi L9 orthogonal design matrix. The Taguchi method and the ANOVA technique were employed to find the best machining condition. The experiments were performed according to Taguchi L9 orthogonal array by the varying depth of cut, regulating wheel speed, and coolant valve opening. To minimize the surface roughness, the signal-to-noise ratio was computed based on smaller the better criteria. According to the maximum signal-to-noise ratio, optimum machining parameters were found to be the low level of depth of cut (20 μm), medium level of regulating wheel speed (14 rpm), and high level of valve opening (full open). The obtained results can be used in the future for the improvement of machining performance.

Figure 3. Effect of process parameters on mean response characteristics

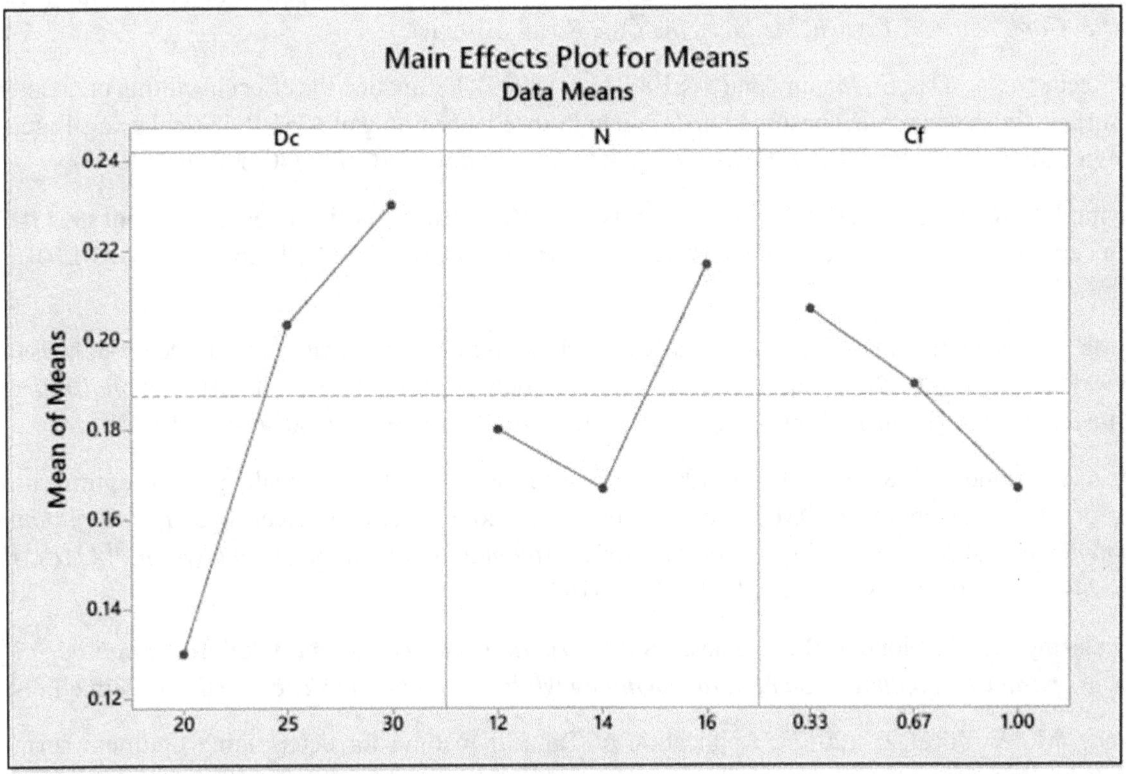

Table 5. Results of the ANOVA for surface roughness

Source	Degree of Freedom (DF)	Sum of Squares (SS)	Mean Square (MS)	F-Value	Contribution
Dc	2	0.016089	0.008044	1.54	48.82
N	2	0.004022	0.002011	0.39	12.20
Cf	2	0.002422	0.001211	0.23	7.34
Error	2	0.010422	0.005211		31.63
Total	8	0.032956			

ACKNOWLEDGMENT

The author gratefully acknowledges the kind support and cooperation provided by the technical staffs at Workshop in IIEST, Shibpur, Howrah.

REFERENCES

Asiltürk, I., & Akkuş, H. (2011). Determining the effect of cutting parameters on surface roughness in hard turning using the Taguchi method. *Measurement*, *44*(9), 1697–1704.

Bendell, T. (Ed.). (1989). *Taguchi Methods: Proceedings of the 1988 European Conference; the 1. European Conference on Taguchi Methods, the Cafe Royal*. Elsevier.

Bhattacharya, A., Das, S., Majumder, P., & Batish, A. (2009). Estimating the effect of cutting parameters on surface finish and power consumption during high speed machining of AISI 1045 steel using Taguchi design and ANOVA. *Production Engineering*, *3*(1), 31–40. doi:10.100711740-008-0132-2

Davim, J. P., & Figueira, L. (2007). Machinability evaluation in hard turning of cold work tool steel (D2) with ceramic tools using statistical techniques. *Materials & Design*, *28*(4), 1186–1191. doi:10.1016/j.matdes.2006.01.011

Ghosh, G., Mandal, P., & Mondal, S. C. (2019). Modeling and optimization of surface roughness in keyway milling using ANN, genetic algorithm, and particle swarm optimization. *International Journal of Advanced Manufacturing Technology*, *100*(5), 1223–1242. doi:10.100700170-017-1417-4

Ghosh, S., Mandal, P., & Mondal, S. C. (2017). Application of simulated annealing for the optimization of process parameters in WEDM process for machining 201LN stainless steel. In *2017 International Conference on Advances in Mechanical, Industrial, Automation and Management Systems (AMIAMS)* (pp. 24-29). IEEE. 10.1109/AMIAMS.2017.8069183

Gopalsamy, B. M., Mondal, B., & Ghosh, S. (2009). *Taguchi method and ANOVA: An approach for process parameters optimization of hard machining while machining hardened steel*. Academic Press.

Günay, M., & Yücel, E. (2013). Application of Taguchi method for determining optimum surface roughness in turning of high-alloy white cast iron. *Measurement*, *46*(2), 913–919. doi:10.1016/j.measurement.2012.10.013

Gupta, A., Singh, H., & Aggarwal, A. (2011). Taguchi-fuzzy multi output optimization (MOO) in high speed CNC turning of AISI P-20 tool steel. *Expert Systems with Applications*, *38*(6), 6822–6828. doi:10.1016/j.eswa.2010.12.057

Hasçalık, A., & Çaydaş, U. (2008). Optimization of turning parameters for surface roughness and tool life based on the Taguchi method. *International Journal of Advanced Manufacturing Technology*, *38*(9-10), 896–903. doi:10.100700170-007-1147-0

Khan, Z. A., Siddiquee, A. N., & Sheikh, M. H. (2012). Selection of optimal condition for finishing of centreless-cylindrical ground parts using grey relational and principal component analyses. *International Journal of Materials & Product Technology*, *43*(1-4), 2–21. doi:10.1504/IJMPT.2012.047636

Kıvak, T. (2014). Optimization of surface roughness and flank wear using the Taguchi method in milling of Hadfield steel with PVD and CVD coated inserts. *Measurement*, *50*, 19–28. doi:10.1016/j.measurement.2013.12.017

Köksoy, O., & Muluk, F. Z. (2004). Solution to the Taguchi's problem with correlated responses. *Gazi University Journal of Science, 17*(1), 59–70.

Lochner, R. H., & Matar, J. E. (1990). *Designing for quality: an introduction to the best of Taguchi and western methods of statistical experimental design*. Springer.

Mandal, N., Doloi, B., Mondal, B., & Das, R. (2011). Optimization of flank wear using Zirconia Toughened Alumina (ZTA) cutting tool: Taguchi method and Regression analysis. *Measurement, 44*(10), 2149–2155. doi:10.1016/j.measurement.2011.07.022

Mandal, P., & Mondal, S. C. (2017). An application of artificial neural network and particle swarm optimisation technique for modelling and optimisation of centreless grinding process. *International Journal of Productivity and Quality Management, 20*(3), 344–362. doi:10.1504/IJPQM.2017.082637

Mandal, P., & Mondal, S. C. (2019). Surface characteristics of mild steel using EDM with Cu-MWCNT composite electrode. *Materials and Manufacturing Processes, 34*(12), 1326–1332. doi:10.1080/10426914.2019.1605179

Mandal, P., & Mondal, S. C. (2019). Development and application of Cu-SWCNT nanocomposite–coated 6061Al electrode for EDM. *International Journal of Advanced Manufacturing Technology, 103*(5-8), 3067–3076. doi:10.100700170-019-03710-5

Mandal, P., & Mondal, S. C. (2019). Investigation on the Performance of Copper-Coated 6061 Aluminium Alloy Electrode in Electric Discharge Machining. In *Research into Design for a Connected World* (pp. 345–355). Springer. doi:10.1007/978-981-13-5974-3_30

Mondal, S. C., & Mandal, P. (2014). Application of artificial neural network for modeling surface roughness in centerless grinding operation. In *5th International & 26th All India Manufacturing Technology, Design and Research Conference*. IIT Guwahati.

Mondal, S. C., & Mandal, P. (2015). An Application of Particle Swarm Optimization Technique for Optimization of Surface Roughness in Centerless Grinding Operation. In *ICoRD'15–Research into Design Across Boundaries* (Vol. 2, pp. 687–697). Springer. doi:10.1007/978-81-322-2229-3_59

Mondal, S. C., Mandal, P., & Ghosh, G. (2017). Application of genetic algorithm for the optimization of process parameters in keyway milling. In *International Conference on Research into Design* (pp. 71-86). Springer. 10.1007/978-981-10-3518-0_7

Nalbant, M., Gökkaya, H., & Sur, G. (2007). Application of Taguchi method in the optimization of cutting parameters for surface roughness in turning. *Materials & Design, 28*(4), 1379–1385. doi:10.1016/j.matdes.2006.01.008

Park, S. (1996). *Robust design and analysis for quality engineering*. Boom Koninklijke Uitgevers.

Sharma, P., Verma, A., Sidhu, R. K., & Pandey, O. P. (2005). Process parameter selection for strontium ferrite sintered magnets using Taguchi L9 orthogonal design. *Journal of Materials Processing Technology, 168*(1), 147–151. doi:10.1016/j.jmatprotec.2004.12.003

Shetty, R., Pai, R. B., Rao, S. S., & Nayak, R. (2009). Taguchi's technique in machining of metal matrix composites. *Journal of the Brazilian Society of Mechanical Sciences and Engineering*, *31*(1), 12–20. doi:10.1590/S1678-58782009000100003

Souissi, N., Souissi, S., Niniven, C. L., Amar, M. B., Bradai, C., & Elhalouani, F. (2014). Optimization of squeeze casting parameters for 2017 a wrought al alloy using Taguchi method. *Metals*, *4*(2), 141–154. doi:10.3390/met4020141

Taguchi, G. (1990). *Introduction to Quality Engineering. Asian Productivity Organization*. APO.

Yang, W. P., & Tarng, Y. S. (1998). Design optimization of cutting parameters for turning operations based on the Taguchi method. *Journal of Materials Processing Technology*, *84*(1-3), 122–129. doi:10.1016/S0924-0136(98)00079-X

Zhang, J. Z., & Chen, J. C. (2009). Surface roughness optimization in a drilling operation using the Taguchi design method. *Materials and Manufacturing Processes*, *24*(4), 459–467. doi:10.1080/10426910802714399

ADDITIONAL READING

Lodhi, B. K., & Agarwal, S. (2014). Optimization of machining parameters in WEDM of AISI D3 Steel using Taguchi Technique. *Procedia CIRP*, *14*, 194–199. doi:10.1016/j.procir.2014.03.080

Mandal, V., Mohan, Y., & Hemalatha, S. (2008). Microwave assisted extraction of curcumin by sample–solvent dual heating mechanism using Taguchi L9 orthogonal design. *Journal of Pharmaceutical and Biomedical Analysis*, *46*(2), 322–327. doi:10.1016/j.jpba.2007.10.020 PMID:18309573

Rowe, W. B. (1979). Research into the mechanics of centreless grinding. *Precision Engineering*, *1*(2), 75–84. doi:10.1016/0141-6359(79)90137-5

Rowe, W. B., Bell, W. F., Brough, D., & Davies, B. J. (1986). Optimization studies in high removal rate centreless grinding. *CIRP Annals*, *35*(1), 235–238. doi:10.1016/S0007-8506(07)61878-2

Rowe, W. B., Miyashita, M., & Koenig, W. (1989). Centreless grinding research and its application in advanced manufacturing technology. *CIRP Annals*, *38*(2), 617–625. doi:10.1016/S0007-8506(07)61129-9

Udupa, N. S., Shunmugam, M. S., & Radhakrishnan, V. (1987). Influence of workpiece position on roundness error and surface finish in centreless grinding. *International Journal of Machine Tools & Manufacture*, *27*(1), 77–89. doi:10.1016/S0890-6955(87)80041-X

Vankanti, V. K., & Ganta, V. (2014). Optimization of process parameters in drilling of GFRP composite using Taguchi method. *Journal of Materials Research and Technology*, *3*(1), 35–41. doi:10.1016/j.jmrt.2013.10.007

KEY TERMS AND DEFINITIONS

ANOVA: Analysis of Variance, or ANOVA for short, is a statistical test that looks for significant differences between means on a particular measure.

Centreless Grinding: Centreless grinding is the process of removing material from the outside diameter of a work piece using an abrasive wheel. In its simplest form, a centreless grinder consists of the machine base, grinding wheel, regulating wheel and work blade.

Optimization: A mathematical technique for finding a maximum or minimum value of a function of several variables subject to a set of constraints.

Process Parameter: The different independent variables involved in a machining process or referred to as process parameters. In this study, depth of cut (μm), regulating wheel speed (rpm) and coolant valve opening are considered as process parameter.

Process Response: The measured output which is dependent on the process is known as its response. In this study, surface roughness is considered as process response.

Taguchi Method: The Taguchi method is a powerful problem-solving technique for improving process performance, yield and productivity.

Chapter 5
MOORA-Driven Decision Making to Select the Optimal Specimen of Organic CMCs

Rajesh P. V.
Saranathan College of Engineering, India

ABSTRACT

Bone grafting or bone implant is a typical procedure in surgery in which a missing or broken bone is replaced in order to treat bone fractures that pose a significant health risk to the patients. Several research works have been carried out in the past few years regarding various composite materials used in bone implants, their fabrication methods, and evaluation of their physical, mechanical, chemical, and thermal properties. The use of ceramic powders and ceramic-based composites in biomedical applications are steadily increasing over years mainly due to their advantages like high compressive strength, excellent hardness, etc. In this research work, organic ceramic matrix composites with varying proportions of conch shell and sea sponge are fabricated using powder metallurgy technique and their physicomechanical properties such as density, porosity, water absorption, and micro-hardness are evaluated. Finally, optimization of process parameters is done using multi-objective optimization based on ratio analysis (MOORA) to select the best possible specimen of CMCs.

INTRODUCTION

The term Composite is now widely used in manufacturing and processing industries. It is described as the combination of two or more materials joined together at flexible proportions. It is generally construed to be having two phases viz., matrix, the major constituent and reinforcement, the minor constituent. The matrix or outer cover may be metals, ceramics or polymers. The reinforcement may be inducted into the matrix in the form of powders, whiskers, short or continuous fibres.

Ceramic Matrix Composites (CMCs) have ceramic powders as major constituents. The CMCs are having increased applications in structural, aviation, marine and medical industries such as engine cylinder blocks, piston heads, combustion passages, etc. due to their high hardness, high compressive

DOI: 10.4018/978-1-7998-7206-1.ch005

strength, good toughness, excellent thermal stability, better electrical insulation and superior corrosion and wear resistance. More specifically, a number of ceramic bio materials and their associated composites such as Alumina, Zirconia, Silica and Tri Calcium Phosphate (TCP) enjoy immense usage in the field of medical sciences, thanks to their bio-compatibility, resistance to organic solvents, bio- inertness and good crack growth initiation. Zirconia toughened Alumina is used abundantly in anthroplasty (Roualdes et al., 2010). Applications of these bio ceramics range from artificial teeth, filling the gap between teeth, orthopedic implants, prosthesis to hip bone replacement, maxillo-facial figuration, plastic surgery and so on. Goswami et al. assessed the physico-mechanical and surface wear properties of Magnesium oxide filled ceramic composites for hip implant applications. In their review paper, they explained the advantages of using ceramic based composites over conventional metal inserts (Goswami et al., 2019). Afzal and Adeel in their survey paper, reviewed various implantable zirconia based bio ceramics for bone repair and replacement (Afzal & Adeel, 2014). To be more precise, CMCs are on the verge of replacing conventional metals, which are traditionally used to join broken bones, exponentially. Nag and Banerjee in their book chapter, explained the properties, processing techniques, development and failure modes of various medical implant materials as well as the responsiveness of host materials to those implants (Nag & Banerjee, 2012). So, the biomaterials can be natural, synthetic or organic ceramics and their composites.

Organic ceramics are a promising variety of bio materials, which are extensively used nowadays to replace broken or disfigured femoral, collar or hip bones. Bone tissues resemble Hydroxy Apatite (HA) structure. Hence, organic ceramics, which are rich in HA can be used in artificial prostheses (Rivera-Muñoz, 2011). Usage of body parts of dead marine organisms for bone tissue engineering is encouraged by Clarke et al. in their review paper (Clarke et al., 2011). They also briefed about the merits of using conch shell in broken human bone replacement. As these organic ceramics are processed from animal body parts, particularly the shells and bones of dead sea organisms like oysters, sea sponges, conches, crabs, snails and cuttlefish, they are less expensive, more environmentally friendly, have more calcium content and provide good adhesion, when more than one such ceramics are used together (Organic CMCs). The advantages of obtaining high crack growth initiation, cell attachment and development in scaffolds, when a sea sponge based artificial fiber skeleton is used are clearly elaborated by Green et al. in their study (Green et al., 2003). Due to their attractive attributes, their usage is phenomenally increasing in bone restructuring and allied activities.

When two or more conflicting or contrasting characteristics, subjected to certain constraints are optimized simultaneously, that particular optimization process is called multi-objective optimization or multi-attribute optimization. Multi Objective Optimization based on Ratio Analysis (MOORA) is one of the novel and recent analytical Multi Criteria Decision Making (MCDM) techniques used in conditions where multi objective parametric optimization is required to select the best set of process parameters and properties among various alternatives. Gadakh indicated the applicability, versatility and flexibility of MOORA, while solving various complex decision-making problems in recent manufacturing scenario (Gadakh, 2011). Karande and Chakraborty observed that, as far as MOORA method is concerned, it is very simple to understand, easy to implement and provide precise ranking to the alternatives, when investigating a problem, that deals with selection of materials (Karande & Chakraborty, 2012).

MATERIALS AND METHOD

Selection of Materials

It is very important to decide the ingredient materials to be joined together for fabricating composites as well as the processing mechanism before starting the research. The selection of materials should be done by reviewing various literatures and having prior knowledge about the pros and cons of each and every material. Here, for fabricating CMCs, two organic ceramics namely conch and sea sponge are selected with varying proportions (Bauer et al., 2013). Poly Vinyl Chloride (PVC) is used as a binder to provide adhesive action between the two ceramics.

Conch

Strombus Gigas, commonly known by its domestic name Conch is the other constituent in the composite. It can be commonly seen in shallow under water as well as sea shores. Its shell has a spiral profile and a noticeable siphonal canal. It is also rich in calcium and phosphorus, which are very much useful for bone implants (Laonapakul, 2013).

Sea Sponge

Porifera, commonly known by its domestic name Sea sponge is one of the two constituents in the composite. It is abundantly available in deep sea in and around Indian coastal areas. It is a multi-cellular organism that has a body full of pores and channels allowing water to circulate through it (Zhang & Vecchio, 2013). Its hard outer epidermis is calcium and silica rich and exhibits good fracture toughness (Cunningham, et al., 2010).

Poly Vinyl Chloride

Poly Vinyl Chloride or PVC, as it is commonly known is the most popular polymer material generally used in both solid and liquid state as pipes, fittings, taps, tubes, cables, binding solution, etc. Its biocompatibility and non-reactivity makes it as an ideal binder for providing sticking action to the mixture of ceramic materials. The purpose of binder is to reduce compaction and ejection pressure and to increase particles packaging in powder compaction. They are also used to increase the green strength as well as improve interfacial bonding of the powder compacts. They act as lubricants to reduce the friction between mixed powders and the die wall.

Fabrication Method

As far as fabrication of CMCs is considered, the common technique employed widely is Powder Metallurgy (PM) (Elsen et al., 2014). The advantages of this technique are low material wastage, excellent binding strength among the constituents and good surface finish with negligible porosity. This material forming technique uses ceramic powders as subsequent raw materials. The binded raw materials are kept inside a metallic die slot of required shape and cross section and compacted with the help of a plunger, pressed against the top open die. This green compact is heated to a temperature, less than the melting point of the powders called sintering temperature to form rigid setting.

The PM technique comprises of four steps namely,

- Pulverization
- Mixing
- Compaction
- Sintering

Initially, the raw materials sponge and conch are converted into powder form by crushing them into smaller and finer particles through ball milling machine shown in Fig. 1 for a long time. This phase is called pulverization. The raw materials which are converted into granules through ball milling process are further grinded to powder form of micron level size through stone grinder. They are then classified and grouped into uniform micron sizes by passing through a set of sieves having different mesh size shown in Fig. 2.

Figure 1. Ball milling machine

Figure 2. Set of sieves with different mesh sizes

The second stage is mixing, in which the powders are mixed together thoroughly by adding binder. They form a colloidal state, so that flowing inside the die is easy.

The third stage is compaction, in which a punch or plunger is pushed inside the die opening from top side having cylindrical cross section, thereby pressing the colloidal ceramic powders against the walls of die opening as shown in Fig. 3. The entire process is done in compression part of Universal Testing Machine (UTM) shown in Fig.4.

Figure 3. Plunger and die assembly

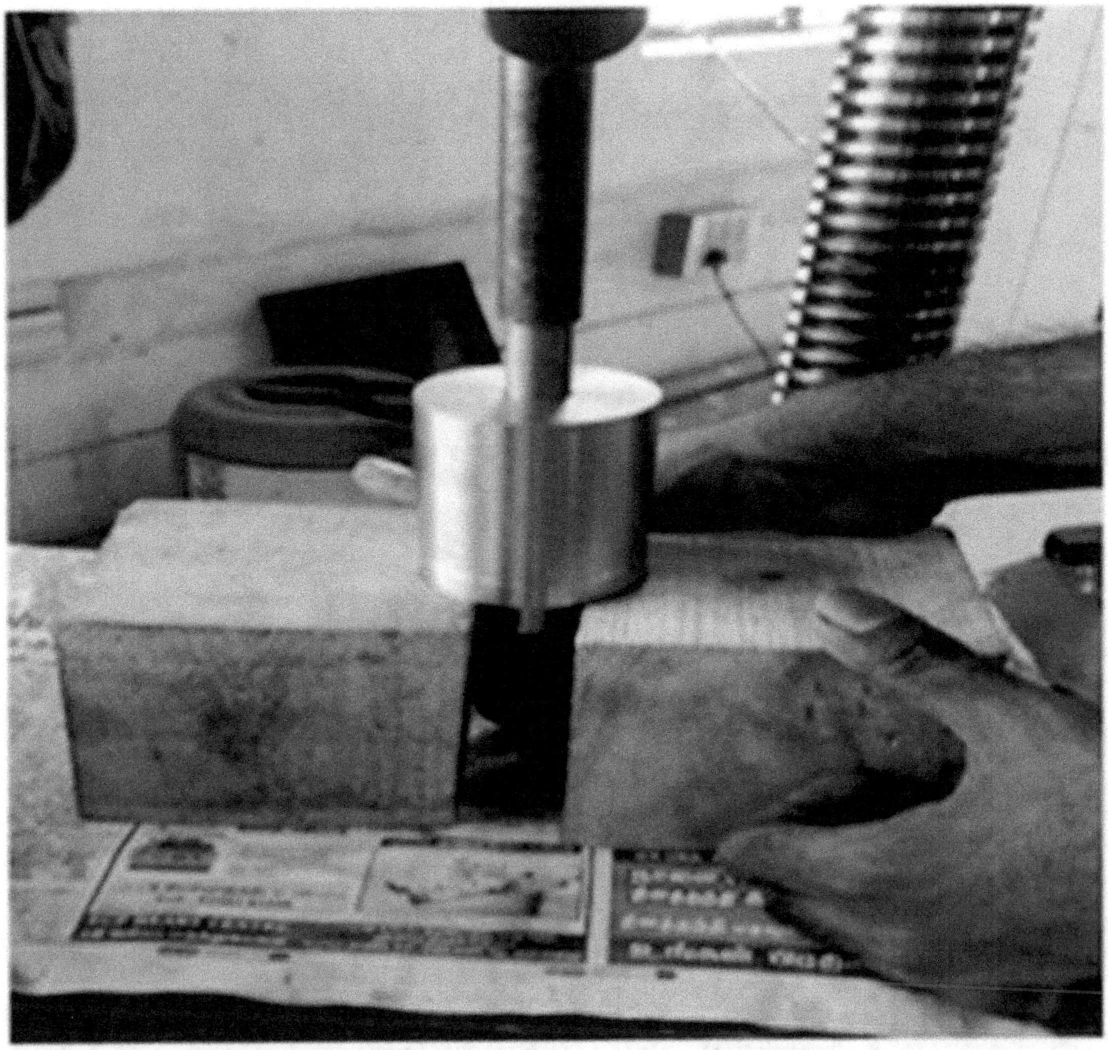

Figure 4. Universal testing machine

Finally, the green compacts of ceramics having cylindrical shape, produced as a result of punch and die pressure are taken out carefully and heated in a closed furnace shown in Fig. 5 at a temperature, lower than that of the melting point of ceramic powders. This process is called sintering.

Figure 5. Muffle furnace

Table 1. Process parameters of composite specimens and their values

Specimens	Particle size	Composition of Conch	Composition of Sea Sponge	Compaction Pressure	Sintering Temperature	Volume of Binder
	microns	wt%	wt%	KN	°C	ml
1	75	25	75	35	400	2
2		50	50	35	400	
3		75	25	35	400	
4	90	25	75	35	400	
5		50	50	35	400	
6		75	25	35	400	
7	150	25	75	35	400	
8		50	50	35	400	
9		75	25	35	400	

Thus, ceramic composites of required shape and size are obtained from powder metallurgy. The different process parameters instrumental for producing various combinations of ceramic composite specimens in powder metallurgy and their values are represented in Table 1 below:

EXPERIMENTAL WORK

This part of the research work deals with the evaluation of macro properties such as density, porosity, water absorption rate and hardness of the CMCs by conducting physical and mechanical tests (Takakuda et al., 2007).

Physical Tests

Physical tests are conducted to identify bulk properties of the composites. Normally, they include density, porosity and water absorption tests (Elsen et al., 2014).

Measurement of Density

For CMCs to be applied in bio medical field, the weight of the composite specimen should be as less as possible. In order to ensure this, measurement of apparent density as well as particle density is carried out.

Apparent density (or) Theoretical Density

It can be calculated from the ratio between the mass of the particular composite sample weighed in air and the volume of the sample as shown in Eq. (1). The mass of the sample can be measured using a digital mass balance of high precision. As the resultant specimen is cylindrical in shape:

$$\rho_T = \frac{mass\ of\ the\ sample}{(\grave{A}/4) * d^2 * l} \tag{1}$$

where,

ρT = theoretical density in g/cm3·
d = diameter of the composite sample in cm,
l = length of the composite specimen in cm.

Particle density (or) Experimental Density

It is the true density of the composite specimen, as during measurement, it removes the air bubbles present due to sub surface porosities, when immersed in water (Otten, 1997). It is denoted by ρE and can be calculated from the 'Archimedes principle' formula shown in Eq. (2) below:

$$\rho_E = \left[m_A / (m_A - m_w)\right] * \rho_{H2O} \tag{2}$$

where,

ρE = experimental density in g/cm3·

m_A = mass of the sample in air in g,

m_w = mass of the sample suspended in water in g,

ρH_{2O} = density of water in g/cm3

Calculation of Porosity

Porosity is the prime reason for the failure of the specimen during tests. A composite should be less porous in order to ensure the close packing and strong cohesive bonding of ceramic particles, thereby increasing strength. Porosity can be calculated from the below mentioned formula stated as Eq. (3):

$$\%Porosity = \frac{(\rho_T - \rho_E)*100}{\rho_T} \tag{3}$$

where,

ρT = theoretical density in g/cm3·

ρE = experimental density in g/cm3·

Determination of Water Absorption

It is the measure of water retention capacity of the composite sample, after the sample has been soaked in water for a relatively long time. This test is done to identify small microscopic cracks, pores and holes at the surface as well as sub surface of the specimen, which make the specimen weak. Hence, water absorption attribute should be as minimum as possible. The specimen is soaked in water for about 5 minutes. Then it is taken out and dried for a while. Water absorption or percolation is calculated from the Eq. (4) given below:

$$Water\ absorption = \frac{V_W}{T_S} \tag{4}$$

where,

V_W = volume of water entrapped inside the specimen in ml

T_S = specimen soaking time inside water in minutes.

Mechanical Test

Mechanical test or destructive test is conducted by deforming the composite specimens to find out the surface mechanical capabilities of the particular specimen.

Figure 6. Micro hardness tester

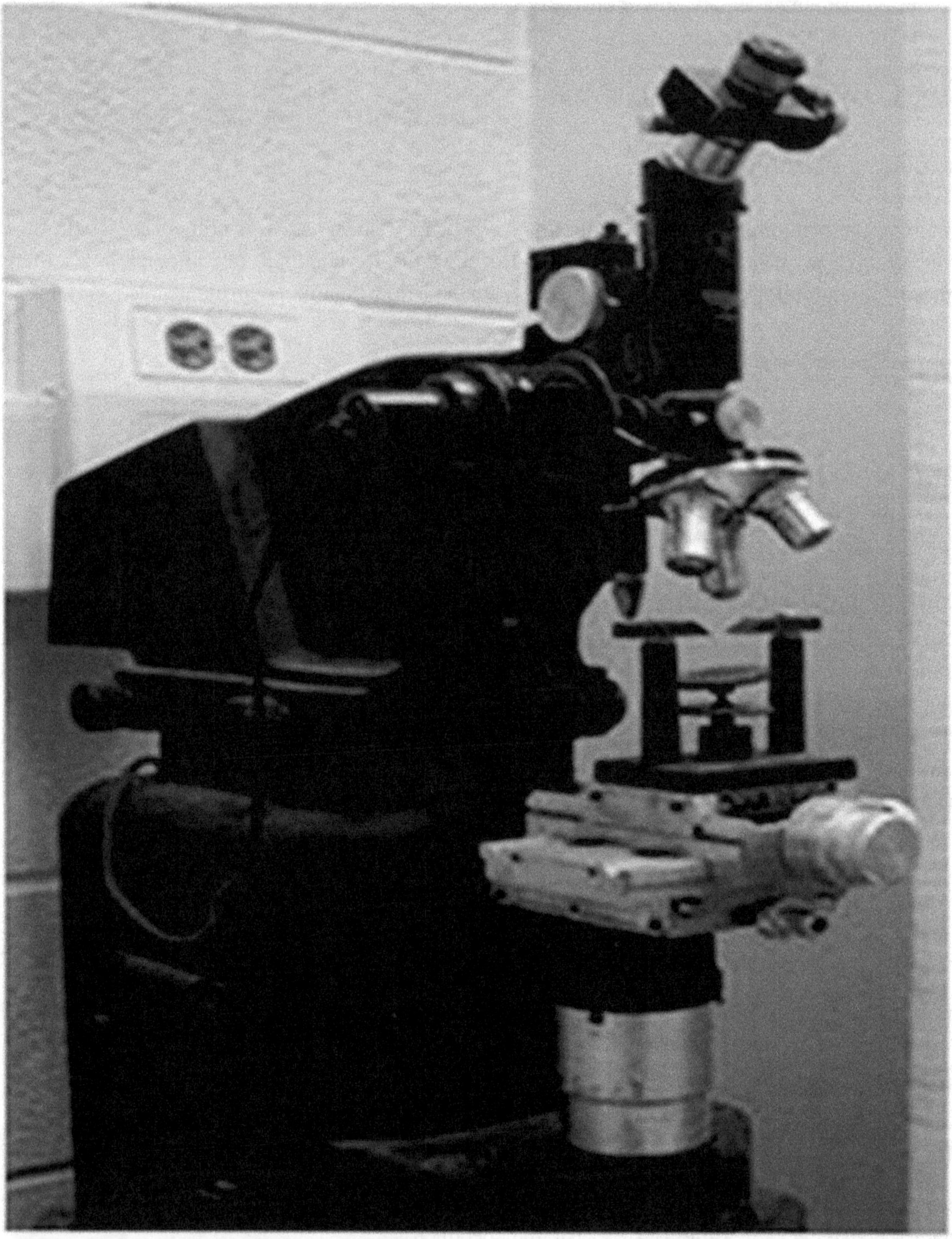

Table 2. Theoretical density values

Specimens	Theoretical density (g/cm^3)
1	1.89
2	2.03
3	2.07
4	**1.86**
5	1.95
6	1.94
7	1.87
8	1.99
9	2.05

Vickers Hardness

Vickers hardness or micro hardness test is done to identify the ability of the material to resist wear, indentation or abrasion. Hence, the hardness value of the composite should be high, so that the material can withstand heavy load and pressure without deformation. This test is done in micro hardness testing machine (shown in Fig. 6) by placing the sample vertically in the work holding device and pressing the sample using a pyramid-shaped diamond indenter. A 10gf load is applied for 10 seconds by the indenter. Then the specimen is removed from the machine, after releasing point load and the diameter of impression made by the indenter is measured using optical microscope. Normally, for powder metallurgy components, only micro hardness is measured.

Table 3. Experimental density values

Specimens	Experimental density (g/cm^3)
1	2.06
2	2.00
3	1.92
4	**1.91**
5	2.07
6	1.93
7	2.09
8	1.97
9	2.10

Figure 7. Theoretical density values

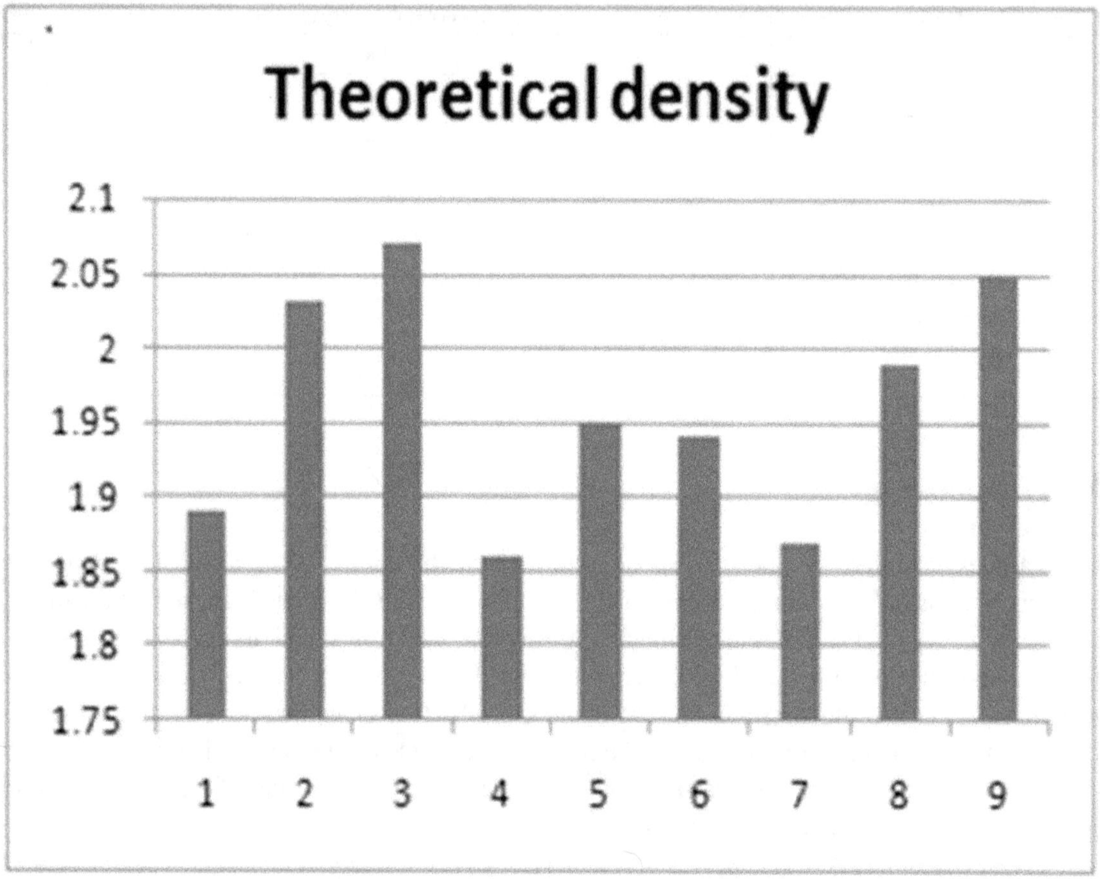

RESULTS AND DISCUSSION

This part of research work deals with the results obtained from conducting physicomechanical tests and inferences got from those test results.

Density

The theoretical density and experimental density are measured by weighing the composite specimens in both air and water. Theoretical density is calculated through dividing the mass of the specimen by volume with air as the measuring medium. Experimental density is obtained by dividing the mass of the specimen in air with the difference between the masses of specimens in air and water and finally multiplying the entire ratio by density of water. The test results for theoretical as well as experimental density are tabulated in Tables 2 and 3 and are shown as bar graphs in Fig. 7 and 8 respectively below:

From the obtained results, it is clear that particle size plays a major role in the determination of theoretical density of composite samples. The density reduces with increasing particle size initially and then increases. The intermediate particle size of 90 microns seems to be ideal, as the composite samples having that particle size are less dense. The density tends to increase with increase in the % composition of conch.

Figure 8. Experimental density values

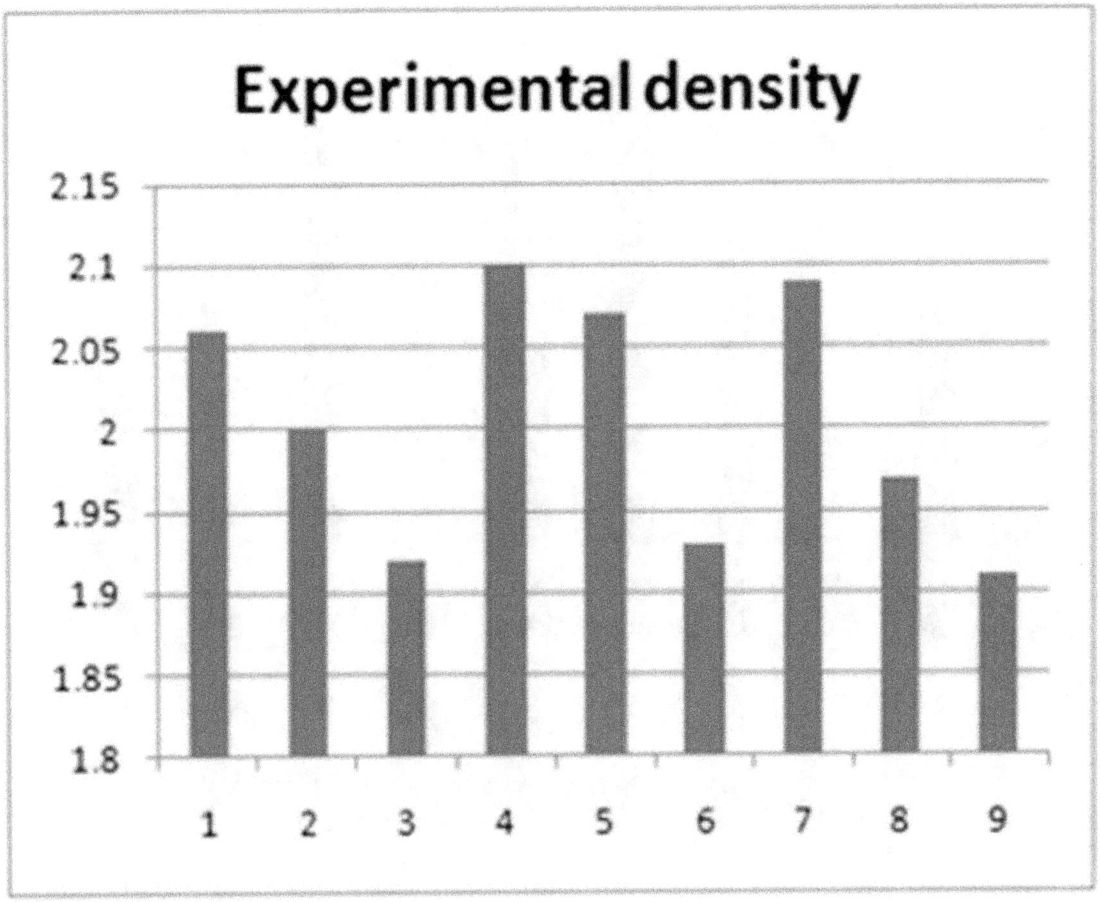

Table 4. Calculated results for porosity

Specimens	% Porosity
1	6.492
2	1.332
3	7.722
4	11.344
5	5.885
6	1.077
7	10.398
8	**1.056**
9	6.868

Figure 9. Calculated results for porosity

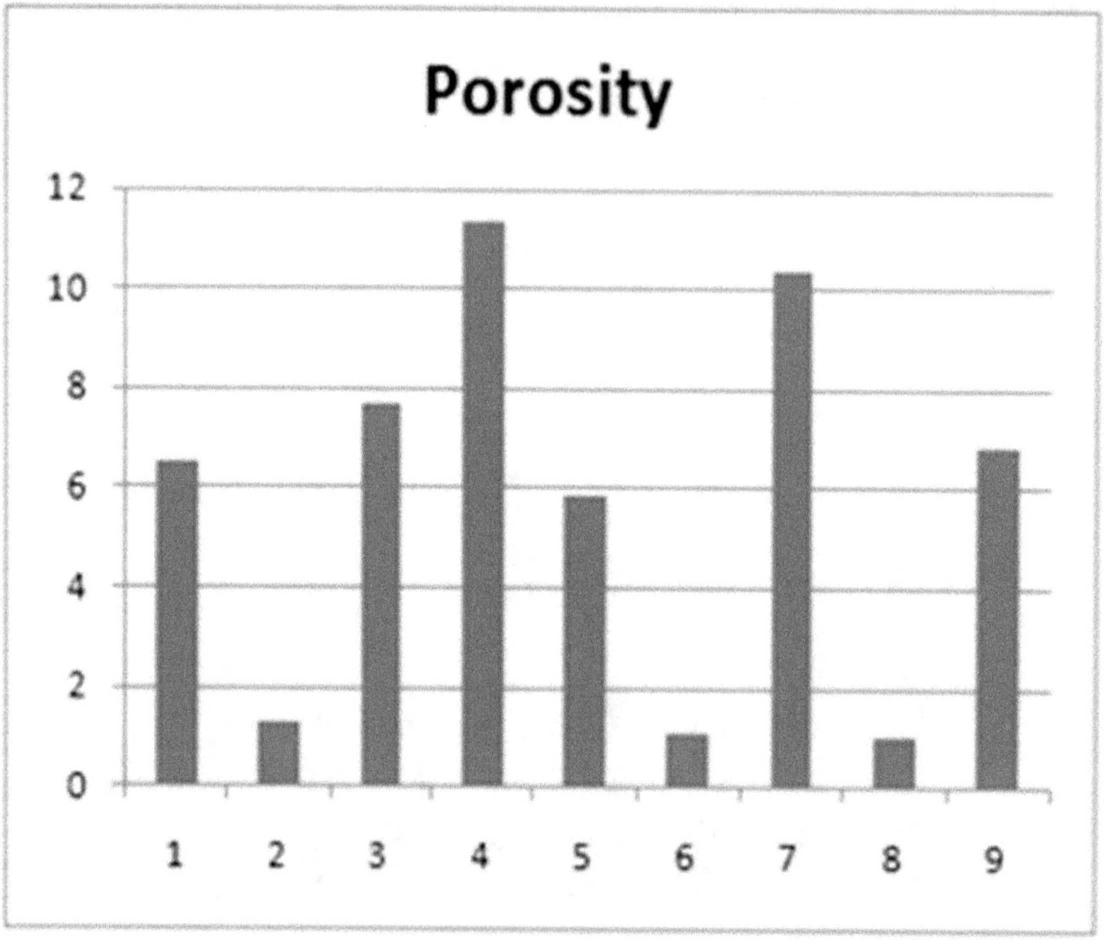

Table 5. Water absorption rates

Specimens	Initial weight	Final weight	Increased weight	Water retained	Water absorption rate
	g	g	g	ml	(ml/min)
1	69.3	69.5	0.2	0.22	0.044
2	69.1	69.4	0.3	0.33	0.066
3	68.9	69.1	0.2	0.22	0.044
4	68.7	68.9	0.2	0.22	0.044
5	68.3	68.7	0.4	0.44	0.088
6	68.1	68.4	0.3	0.33	0.066
8	**67.7**	**67.8**	**0.1**	**0.11**	**0.022**
9	67.5	67.8	0.3	0.33	0.066

Figure 10. Water absorption test results

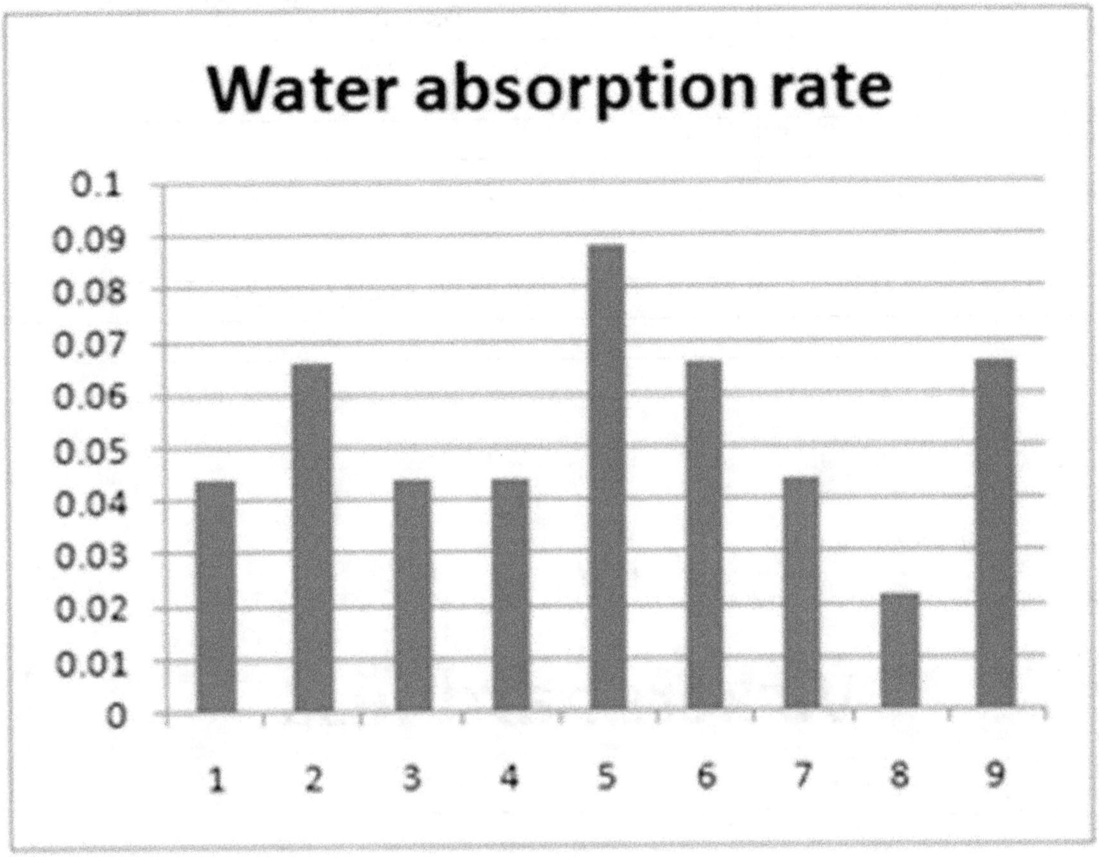

The results of experimental density show a mixed view, as it is hard to determine the influence of process parameters on experimental density. But, here also the specimen 4 is giving the lowest value, which is similar to theoretical density.

Porosity

Porosity is calculated by dividing the difference between theoretical density and experimental density or vice versa with theoretical density or experimental density respectively (whichever is greater) and multiplying the ratio with 100%. The experimental results obtained for various combinations of composite specimens are represented in Table 4 and shown in Fig. 9.

The porosity test results clearly indicate the effect of process parameters over porosity values. The porosity increases with increase in particle size initially and then decreases. At the same time, porosity reduces with increase in the proportion of conch and then increases. The specimen 8 is giving the least value.

Table 6. Hardness test results

Specimens	Vickers Hardness, H_v
1	147.00
2	141.33
3	141.67
4	133.67
5	138.33
6	126.66
7	146.33
8	135.67
9	136.67

Figure 11. Hardness test results

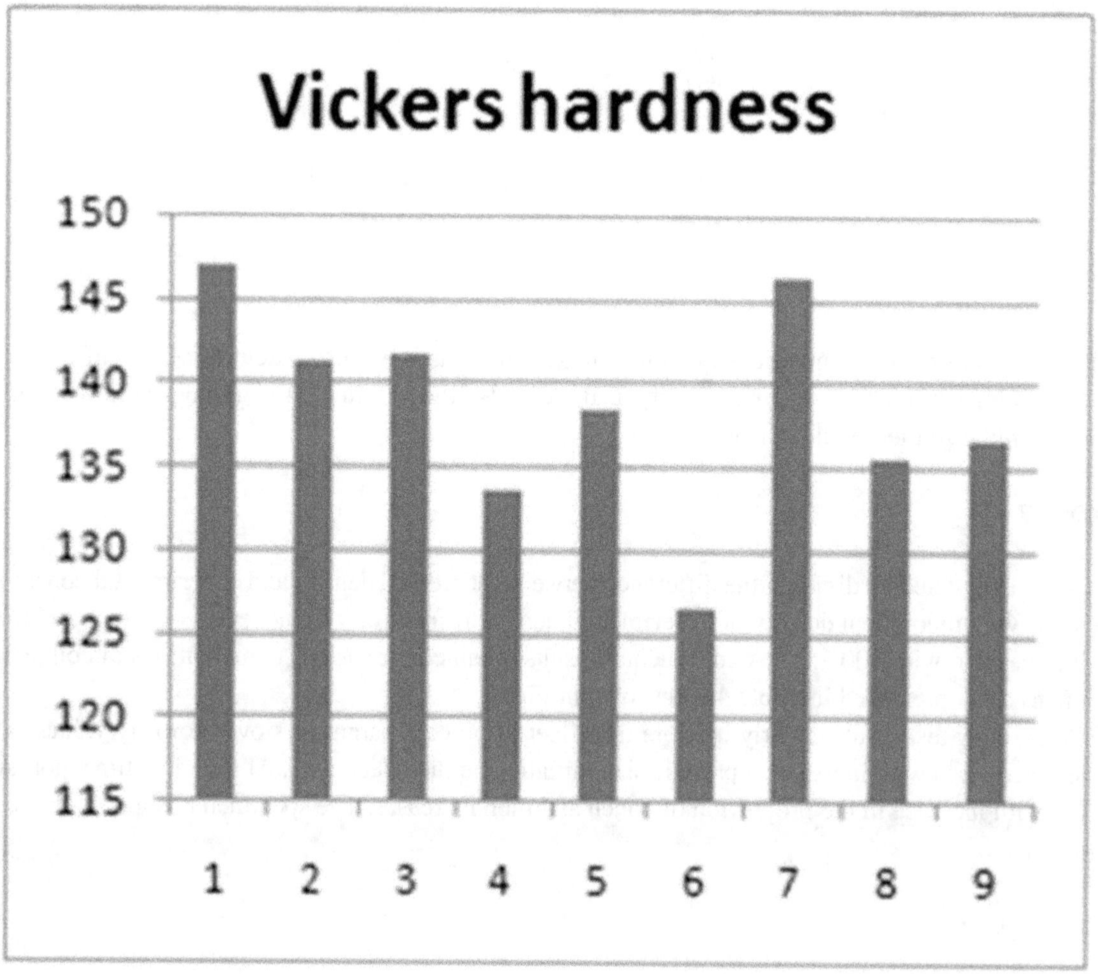

Water Absorption

Water absorption test is done by measuring the mass of the specimen in air initially. Then, the specimen is immersed in water for 5 minutes. After that, the wet specimen is taken out, dried thoroughly. Later, the mass of the specimen is measured in air again. The difference between the two masses (g) gives the volume of water retained inside the specimen (ml).

For converting mass into volume, multiply it by 1.1, according to below conversion:

1g = 1.1 ml.

Finally, this value is divided by immersion time (5 minutes) to get water absorption rate, which is represented in Table 5.

Similarly, the water absorption rate values for all the 9 specimens are shown in Fig. 10.

From the water absorption test results, it is understood that the water absorption rate is inversely proportional to particle size. The effect of volume fraction of conch or sea sponge is insignificant, as seen in the results. The specimen 8 gives the lowest water absorption rate among the 9 specimens.

Vickers Hardness

Vickers hardness test is carried out in micro hardness tester by applying 10gf of point load on the circular surface of the specimen for 10 seconds and measuring the diameter of the impression created by the point load exerted by diamond tip ball indenter. The results for all the 9 different proportions of composite specimens are shown as Table 6 and Fig. 11 below:

From the above represented hardness test values, it is clearly revealed that hardness values are less in the particle size of 90 microns, when compared to 75 and 150 microns. Further, the hardness values decrease with increase in the weight% of conch. The lowest value of hardness is registered at particle size of 90 microns and volume fraction of 75% conch. Likewise, highest value is obtained at particle size of 75 microns and volume fraction of 25% conch. First specimen gives the highest hardness value.

OPTIMIZATION

Optimization and research are inseparable nowadays. Optimization is the methodology of finding out the best possible result from a given set of alternatives, while taking the effects of all available process or testing parameters into consideration. The objective of optimization can be either maximizing or minimizing the optimal solution.

Multi Criteria Decision Making

Multi Criteria Decision Making (MCDM) is one of the sub-branches of optimization, where there are multiple criteria in determining the ideal solution from a given set of alternatives. It is finding an indispensible place in resources, logistics, inventory and production planning, control and management.

Multi Objective Optimization Based on Ratio Analysis

Multi Objective Optimization based on Ratio Analysis (MOORA) is one of the most popular and recent MCDM technique used in optimization (Chakraborty, 2011). This technique is more suitable for the problems, where both maximizing and minimizing characteristics or attributes exist. The maximizing property is called beneficiary criterion and the minimizing property is called cost criterion.

Procedural Steps in MOORA

Step 1: Creation of a Decision matrix comprising of 'm' alternatives or combinations and 'n' criteria or properties, in which the intersection of each alternative and criteria is denoted as x_{ij}, shown as Eq. (5):

$$\begin{pmatrix} x_{11} & x_{12} & \cdots & x_{1n} \\ x_{21} & x_{22} & \cdots & x_{2n} \\ \vdots & \vdots & \cdots & \vdots \\ x_{m1} & x_{m2} & \cdots & x_{mn} \end{pmatrix} \quad (5)$$

Step 2: Normalization of decision matrix should be done, as the numbers represent different values with different measuring units and different multiplication factors (Khan & Maity, 2016).

Normalized value of each variable in the criteria can be obtained from the below Eq. (6):

$$r_{ij} = \frac{x_{ij}}{\sqrt{\sum_{i=1}^{m} x_{ij}^{2}}} \quad (6)$$

where,

i = 1,2,…,m

j = 1,2,…,n.

The normalized matrix R is represented as Eq. (7):

$$\begin{pmatrix} r_{11} & r_{12} & \cdots & r_{1n} \\ r_{21} & r_{22} & \cdots & r_{2n} \\ \vdots & \vdots & \cdots & \vdots \\ r_{m1} & r_{m2} & \cdots & r_{mn} \end{pmatrix} \quad (7)$$

Step 3: Each criterion should be assigned a weight factor. The sum of all weight factors must be equal to 1.

Table 7. Criteria and their values (Decision matrix)

Specimens	Particle size	Comp. of conch	Comp. of sea sponge	Theo. density	Exp. density	Porosity	Water absorption rate	Vickers hardness
	microns	wt%	wt%	g/cm³	g/cm³	%	ml/min.	H_v
1	75	25	75	1.89	2.06	6.492	0.044	147
2	75	50	50	2.03	2	1.332	0.066	141.33
3	75	75	25	2.07	1.92	7.722	0.044	141.67
4	90	25	75	1.86	1.91	11.344	0.044	133.67
5	90	50	50	1.95	2.07	5.885	0.088	138.33
6	90	75	25	1.94	1.93	1.077	0.066	126.66
7	150	25	75	1.87	2.09	10.398	0.044	146.33
8	150	50	50	1.99	1.97	1.056	0.022	135.67
9	150	75	25	2.05	2.1	6.868	0.066	136.67

The values in each criterion should be multiplied with that weight factor or weightage, as shown in Eq. (8):

$$a_{ij} = w_j * r_{ij} \tag{8}$$

The resultant weighted normalized decision matrix is represented in Eq. (9) below:

$$\begin{pmatrix} a_{11} & a_{12} & \cdots & a_{1n} \\ a_{21} & a_{22} & \cdots & a_{2n} \\ \vdots & \vdots & \cdots & \vdots \\ a_{m1} & a_{m2} & \cdots & a_{mn} \end{pmatrix} \tag{9}$$

Step 4: Calculation of Assessment value, y_i row-wise, which is the difference between the sum of variables in maximizing criteria (of each row) from weighted normalized decision matrix and the sum of variables in minimizing criteria (of each row).

Table 8. Weightage given to criteria

Criteria	Type	Objective	Description	Weightage (%)
1	Cost	Minimize	Theoretical density	12.5
2	Cost	Minimize	Experimental density	12.5
3	Cost	Minimize	Porosity	12.5
4	Cost	Minimize	Water absorption rate	12.5
5	Beneficiary	Maximize	Vickers hardness	50

Table 9. Normalized decision matrix

Theoretical density	Experimental density	Porosity	Water absorption rate	Vickers hardness
0.3210	0.3422	0.3151	0.2582	0.3532
0.3448	0.3322	0.0647	0.3873	0.3396
0.3516	0.3189	0.3748	0.2582	0.3404
0.3159	0.3172	0.5506	0.2582	0.3212
0.3312	0.3438	0.2857	0.5164	0.3324
0.3295	0.3206	0.0523	0.3873	0.3043
0.3176	0.3471	0.5047	0.2582	0.3516
0.3380	0.3272	0.0513	0.1291	0.3260
0.3482	0.3488	0.3334	0.3873	0.3284

Step 5: Ranking of assessment values, y_i in the descending order shows the final stage of optimization. Thus, the best preferential alternative will be having the highest y_i value.

RANKING BASED OPTIMIZATION OF ABOVE PRESENTED WORK

As far as the above conducted work involving optimization of process parameters in the evaluation of physicomechanical properties is concerned, the ranking based optimization using MOORA is carried out for the identification of best proportion or composition of organic CMCs from given set of alternatives.

Let us consider, the nine different combinations of composite samples as alternatives. The different physical and mechanical properties such as theoretical density, experimental density, porosity, water absorption rate and Vickers hardness, respectively obtained as a result of conducting respective tests are considered as criteria.

Table 10. Weighted normalized decision matrix

Theoretical density	Experimental density	Porosity	Water absorption rate	Vickers hardness
0.0401	0.0428	0.0394	0.0323	0.1766
0.0431	0.0415	0.0081	0.0484	0.1698
0.0439	0.0399	0.0469	0.0323	0.1702
0.0395	0.0397	0.0688	0.0323	0.1606
0.0414	0.0430	0.0357	0.0645	0.1662
0.0412	0.0401	0.0065	0.0484	0.1522
0.0397	0.0434	0.0631	0.0323	0.1758
0.0423	0.0409	0.0064	0.0161	0.1630
0.0435	0.0436	0.0417	0.0484	0.1642

Table 11. Assessment values

Assessment values
0.0220
0.0287
0.0073
-0.0197
-0.0184
0.0160
-0.0027
0.0573
-0.0130

Here, higher value is expected for Vickers hardness (maximizing objective), while lower values are required for theoretical density, experimental density, porosity and water absorption rate (minimizing objective).

The decision matrix, which contains different specimens fabricated as a result of various process parameters (alternatives) and different criteria is represented as Table 7.

This research work has four physical properties against one mechanical property. The mechanical property (ie.) hardness is a major criterion to reckon with. All the physical properties are mutually interdependent. Hence, weight factor should be considered accordingly. One half of the weightage is given to mechanical property entirely and the other half is shared equally by all the physical properties. The sum of all weights must be equal to 1. The weightage given to the criteria is represented in Table 8.

Table 12. Rank order based on assessment values

Assessment values	**Rank**	**Theo. density**	**Exp. density**	**Porosity**	**Water absorption rate**	**Vickers hardness**
		g/cm³	**g/cm³**	**%**	**ml/min.**	**H_V**
0.0220	3	1.89	2.06	6.492	0.044	147
0.0287	2	2.03	2	1.332	0.066	141.33
0.0073	5	2.07	1.92	7.722	0.044	141.67
-0.0197	9	1.86	1.91	11.344	0.044	133.67
-0.0184	8	1.95	2.07	5.885	0.088	138.33
0.0160	4	1.94	1.93	1.077	0.066	126.66
-0.0027	6	1.87	2.09	10.398	0.044	146.33
Assessment values	Rank	Theo. density	Exp. density	Porosity	Water absorption rate	Vickers hardness
		g/cm³	g/cm³	%	ml/min.	H_V
0.0573	1	1.99	1.97	1.056	0.022	135.67
-0.0130	7	2.05	2.1	6.868	0.066	136.67

DECISION MAKING BASED ON PROCEDURAL STEPS OF MOORA

Normalized Decision Matrix

The normalized matrix is formed from the normalizing formula stated above in the procedural steps and represented as Table 9.

Weighted Normalized Decision Matrix

The element or value belonging to each criterion of the normalized decision matrix table is multiplied by the corresponding weightage, mentioned above to form the weighted normalized decision matrix table stated as Table 10.

Assessment Value (y_i)

Assessment value is the difference between the sum of beneficiary criteria values and the sum of cost criteria values in weighted normalized matrix (row-wise). The assessment values for all rows are represented as Table 11.

Ranking of Alternatives Based on Assessment Values

Ranking of the different combinations of composite specimens based on assessment values are represented in Table 12.

Top ranked proportion of composites based on physicomechanical properties gives the optimized process parameters. The respective process parameters and criteria/properties of the particular specimen, which attained Rank 1 is given in Table 13.

Table 13. Top most ranked composite specimen

Specimen	Particle size	Comp. of conch	Comp. of sea sponge	Theo. density	Exp. density	Porosity	Water absorption rate	Vickers hardness
	microns	wt%	wt%	g/cm³	g/cm³	%	ml/min.	H_V
8	150	50	50	1.99	1.97	1.056	0.022	135.67

CONCLUSION

From the above experimental work, we have the following conclusions:

- Ceramics and their composites are having profound usage in biomedical applications.
- Organic ceramics are a special category of biomaterials, which can be processed from the outer body parts of dead marine organisms.

- Organic CMCs are combinations of two or more organic ceramics joined together at varied proportions to enhance biocompatibility and properties required for bone implants.
- Conch shells and sea sponges are the most common and popular organic ceramic materials, which are easily available at sea shores, beaches and shallow regions in the sea.
- Powder metallurgy is the apt technique for fabricating CMCs.
- The composite samples fabricated using powder metallurgy have closely packed particles and rigid.
- For orthopedic and artificial prosthetic applications, the composite specimens should be less dense, less porous, tough and hard. They should neither absorb nor retain water or any other liquid.
- MOORA is a robust and most effective MCDM technique for multi objective optimization.

REFERENCES

Afzal, A. (2014). Implantable zirconia bioceramics for bone repair and replacement: A chronological review. *Materials Express*, *4*(1), 1–12. doi:10.1166/mex.2014.1148

Bauer, S., Schmuki, P., von der Mark, K., & Park, J. (2013). Engineering biocompatible implant surfaces Part I: Materials and surfaces. *Progress in Materials Science*, *58*(3), 261–326. doi:10.1016/j.pmatsci.2012.09.001

Chakraborty, S. (2011). Applications of the MOORA method for decision making in manufacturing environment. *International Journal of Advanced Manufacturing Technology*, *54*(9-12), 1155–1166. doi:10.100700170-010-2972-0

Clarke, S. A., Walsh, P., Maggs, C. A., & Buchanan, F. (2011). Designs from the deep: Marine organisms for bone tissue engineering. *Biotechnology Advances*, *29*(6), 610–617. doi:10.1016/j.biotechadv.2011.04.003 PMID:21527337

Cunningham, E., Dunne, N., Walker, G., Maggs, C., Wilcox, R., & Buchanan, F. (2010). Hydroxyapatite bone substitutes developed via replication of natural marine sponges. *Journal of Materials Science*, *21*, 2255–2261. PMID:20012771

Gadakh, V. S. (2011). Application of MOORA method for parametric optimization of milling process. *International Journal of Advanced Engineering Research*, *1*(4), 743–758.

Goswami, C., Bhat, I. K., Bathula, S., Singh, T., & Patnaik, A. (2019). Physico-mechanical and surface wear assessment of magnesium oxide filled ceramic composites for hip implant applications. *Silicon*, *11*(1), 39–49. doi:10.100712633-018-9880-6

Green, D., Howard, D., Yang, X., Kelly, M., & Oreffo, R. O. C. (2003). Natural Marine Sponge Fiber Skeleton: A Biomimetic Scaffold for Human Osteoprogenitor Cell Attachment, Growth and Differentiation. *Tissue Engineering*, *9*(6), 1159–1166. doi:10.1089/10763270360728062 PMID:14670103

Karande, P., & Chakraborty, S. (2012). Application of multi-objective optimization on the basis of ratio analysis (MOORA) method for materials selection. *Materials & Design*, *37*, 317–324. doi:10.1016/j.matdes.2012.01.013

Khan, A., & Maity, K. (2016). Parametric Optimization of Some Non-Conventional Machining Processes Using MOORA Method. *International Journal of Engineering Research in Africa*, *20*, 19–40. doi:10.4028/www.scientific.net/JERA.20.19

Laonapakul, T. (2015). Synthesis of hydroxyapatite from biogenic wastes. *KKU Engineering Journal*, *42*(3), 269–275.

Nag & Banerjee. (2012). Fundamentals of Medical Implant Materials. ASM Handbook: Materials for Medical Devices, 23, 6-17.

Otten, K. (1997). *Method for producing Ceramic Implant materials including Hydroxyl Apatite.* US Patent: 5,667,796, 1-4.

Renold Elsen, S., Ramesh, T., & Aravinth, B. (2014). Optimization of process parameters of zirconia reinforced alumina by powder forming process using Response Surface Method. *Advanced Materials Research*, *984-985*, 129–139. doi:10.4028/www.scientific.net/AMR.984-985.129

Rivera-Munoz. (2011). Hydroxyapatite-Based Materials: Synthesis and Characterization. *Biomedical Engineering - Frontiers and Challenges*, 75-98.

Roualdes, O., Duclos, M.-E., Gutknecht, D., Frappart, L., Chevalier, J., & Hartmann, D. J. (2010). In vitro and in vivo evaluation of an alumina-zirconia composite for arthoplasty applications. *Biomaterials*, *31*(8), 2043–2054. doi:10.1016/j.biomaterials.2009.11.107 PMID:20053439

Takakuda, K., Koyama, Y., Matsumoto, H. N., Shirahama, N., Akita, K., Shoji, D., Ogawa, T., Kikuchi, M., & Tanaka, J. (2007). Material Design of Bioabsorbable Inorganic / Organic Composites for Bone Regeneration. *Journal of Nanoscience and Nanotechnology*, *7*(3), 738–741. doi:10.1166/jnn.2007.502 PMID:17450826

Zhang, X., & Vecchio, K. S. (2013). Conversion of natural marine skeletons as scaffolds for bone tissue engineering. *Frontiers of Materials Science*, *7*(2), 103–117. doi:10.100711706-013-0204-x

ADDITIONAL READING

Alvarez, K., Camero, S., Alarcón, M. E., Rivas, A., & González, G. (2002). Physical and mechanical properties evaluation of *Acropora palmata* coralline species for bone substitution applications. *Journal of Materials Science. Materials in Medicine*, *13*(5), 509–515. doi:10.1023/A:1014787209506 PMID:15348605

Dee, K. C., & Bizios, R. (1996). Proactive biomaterials and bone tissue engineering. *Biotechnology and Bioengineering*, *50*(4), 438–442. doi:10.1002/(SICI)1097-0290(19960520)50:4<438::AID-BIT11>3.0.CO;2-F PMID:18626993

Figueiredo, M., Henriques, J., Martins, G., Guerra, F., Judas, F., & Figueiredo, H. (2010). Physicochemical characterization of biomaterials commonly used in dentistry as bone substitutes—Comparison with human bone. *Journal of Biomedical Materials Research. Part B, Applied Biomaterials*, *92*(2), 409–419. doi:10.1002/jbm.b.31529 PMID:19904820

Hooper, J. N. A., & Van Soest, R. W. (2002). *Systema Porifera: A Guide to the Classification of Sponges* (Vol. 1). Kluwer Academic Plenum. doi:10.1007/978-1-4615-0747-5

Kelly, E. B. (2000). New frontiers in bone grafting. *Orthop Technol Rev*, *2*(9), 28–35.

Lemons, J. E. (1996). Ceramics: Past, present, and future. *Bone*, *19*(1), 121–128. doi:10.1016/S8756-3282(96)00128-7 PMID:8831003

Moore, W. R., Graves, S. E., & Bain, G. I. (2001). Synthetic bone graft substitutes. *ANZ Journal of Surgery*, *71*(6), 354–361. doi:10.1046/j.1440-1622.2001.02128.x PMID:11409021

Nomura, T., Katz, J. L., Powers, M. P., & Saito, C. (2005). Evaluation of the micromechanical elastic properties of potential bone-grafting materials. *Journal of Biomedical Materials Research. Part B, Applied Biomaterials*, *73*(1), 29–34. doi:10.1002/jbm.b.30201 PMID:15672390

Peppas, N. A., & Langer, R. (1994). New challenges in biomaterials. *Science*, *263*(5154), 1715–1720. doi:10.1126cience.8134835 PMID:8134835

Zhang, Y., Ni, M., Zhang, M., & Ratner, B. (2003). Calcium phosphate chitosan composite scaffolds for bone tissue engineering. *Tissue Engineering*, *9*(2), 337–345. doi:10.1089/107632703764664800 PMID:12740096

KEY TERMS AND DEFINITIONS

Beneficiary Criteria: Beneficiary criteria are the properties of a material, which provide benefit when they are maximized (larger-the-better).

Bone Grafting: Bone grafting is the technique used in surgery to fix broken bones by replacing the broken original part with a synthetic bone.

Cost Criteria: Cost criteria are the properties of a material, which provide benefit when they are minimized (smaller-the-better).

Micro Hardness: Micro hardness is the ability of a soft material to withstand light to moderate load for a specific period of time.

Organic Ceramics: Organic ceramics are unique kind of biomaterials, which are processed from body parts obtained from various land and sea animals (either living or dead).

Powder Metallurgy: Powder metallurgy is one of the most popular material forming process in which two or more powdered ceramics and metals are compressed to form composites.

Water Absorption: Water absorption is the capability of water to percolate inside any material through fine pores.

Chapter 6
TOPSIS-Based Selection of Optimal Proportion Among Different Combinations of Hybrid AMCs

Rajesh P. V.
Saranathan College of Engineering, India

ABSTRACT

In this modern world, composites are used in almost all fields due to their attractive mechanical and technological properties. They are believed to be fast replacing metal alloys thanks to their adaptability, flexibility, formability, and machinability. The present study deals with the comparative evaluation of mechanical properties between various aluminium alloy composites reinforced with boron carbide and rice husk ash at different proportions and their optimization through ranking of alternatives using TOPSIS. The composite specimens are fabricated by a liquid metallurgy technique called stir casting. The sample specimens are prepared by varying the percentage of reinforcements as per volume-based ratio with respect to the aluminium alloy Al 6061. The evaluation of mechanical properties indicates the improvement in tensile strength, hardness, impact energy, and corrosion resistance for different composite combinations compared to that of individual alloy. Finally, the best possible combination is identified among the given set of various proportions by optimization using TOPSIS.

INTRODUCTION

Composites are generally a combination of materials, in which two or more metals or metals with non-metals are joined together at any proportion. Ceramics, industrial and agricultural wastes are chosen more often to be a part of these composites. Metal Matrix Composites (MMCs) form a predominant part in the fabrication of composites for various uses, as they are produced and used more excessively than its peers Ceramic Matrix and Polymer Matrix Composites. Bodunrin et al. presented a review on the overall reinforcement philosophies, mechanical, corrosion and tribological properties of Aluminium

DOI: 10.4018/978-1-7998-7206-1.ch006

Hybrid Metal Matrix Composites. Their findings are: Aluminium hybrid composites are a new generation of metal matrix composites that have the potentials of satisfying the recent demands of advanced engineering applications (Bodunrin, et al., 2015). In Metal Matrix Composites, metals and their alloys are the major constituents.

Aluminium, which is one of the most abundant materials in the Earth's crust has been used in various forms, shapes and sizes over a long period of time. It is traditionally an inevitable metal, which provides its contribution to each and every industrial manufacturing industry. Normally, it is called 'researcher's delight', as its alloys as well as composites account to more than half of the total researches done across the world. Aluminium Matrix Composites (AMCs) are extensively used in a number of fields, due to their fascinating attributes like low cost, less weight, high strength-to-weight ratio, high hardness and excellent resistance to corrosion and wear, to name a few. Mali and Sonawane, in their paper dealt with the effect of hybrid reinforcement on mechanical behaviour of Aluminium Matrix Composites. The paper deals with the fabrication of aluminum-based hybrid metal matrix composite and then characterized their mechanical properties such as hardness, toughness and tensile strength (Mali & Sonawane, 2014).

Generally, more quantum of researches is being made in paradigms like effect of process parameters in casting AMCs and their optimization, Experimental investigation of mechanical properties in AMCs, Study of corrosion resistance and tribological behavior in AMCs, etc. Bandare and Sonawane described the preparation of Metal Matrix Composites by Stir Casting method. Stir casting process is mainly used for manufacturing of particulate reinforced metal matrix composite (PMMC). Manufacturing of aluminum alloy based casting composite by stir casting is one of the most economical method of processing MMC (Bandare & Sonawane, 2013). Krishna et al. made an attempt to enhance the mechanical properties like tensile strength and hardness of AMCs by reinforcing 6061Al matrix with B_4C particles. By stir casting route, aluminium matrix was reinforced with boron carbide particulates of 37, 44, 63, 105, 250μ sizes respectively (Krishna et al., 2013). Lancaster et al. elaborated that the application of agro-industrial wastes in Aluminum Metal Matrix Composites, which is been getting more attention in modern days as they can enhance the strength properties of the composites with various case studies (Lancaster, et al., 2013). The effect of inclusion of rice husk ash as second reinforcement in the mechanical and tribological properties of Alumina reinforced AMC is discussed by Gupta and Tahi in their research paper (Gupta & Tahi, 2015). The results clearly showed that the tensile strength, micro hardness, flexural strength and wear resistance is more in Al-Al_2O_3-RHA hybrid composite than single reinforced Al- Al_2O_3 composite.

Zavadskas et al. in their paper dealing with the state of art surveys of overviews on MCDM/MADM methods, explained the types, importance of MCDM in operations research and the application of different MCDM techniques in various fields (Zavadskas, et al., 2014). Olsen in his introductory paper about comparison of weights in TOPSIS models provided an eye opener about the step-by-step procedure of ranking the alternatives by TOPSIS, when dissimilar criteria are used (Olson, 2004). Diyaley et al. compared the effectiveness of PSI and TOPSIS based selection of process parameters in WEDM. Both optimization methodologies were found to yield the same desired results (Diyaley et al., 2017). Their paper tried to help those researches that are going on in the fabrication, processing, evaluation and characterization of AMCs by providing an insight regarding the selection of a feasible and optimal solution (best combination of AMCs) from a given set of alternatives (different proportions of AMCs) through one of the Multi Criteria Decision Making (MCDM) techniques called Technique for Order Preference by Similarity to the Ideal Solution (TOPSIS).

SELECTION OF MATERIALS

Hybrid composites are distinctive variety of composite materials in which more than one reinforcement is used for superior properties (Shankar et al., 2013). As far as AMCs are concerned, selection of major and minor constituents to be joined together to form composite materials are very important as they are going to play a vital role in the enhancement of required properties.

Meticulous selection of reinforcements in the form of particulates, fibres or whiskers to embed inside metallic matrix can improve its characteristics, that suits practical applications in various streams. The materials selected for matrix and reinforcements are:

Matrix

Matrix or the outer cover is the major constituent in the composite within which the reinforcement is embedded the matrix is important to keep the reinforcements intact without disassociation. The metal chosen to be matrix here is Al 6061.

Al 6061 is a wrought aluminium alloy commonly known as Al-Mg-Si alloy, as these three metals contribute more to the performance of that alloy (Bodunrin et al., 2015). It is durable, lustrous, and malleable which has good formability, excellent weldability and resistance to corrosion and rust. It is one of the widely popular alloys and commonly used in home appliances, utensils, bicycle frames, window and door panels as well as aircraft interior structures and ship hulls (Vengadesh & Chandramohan, 2014) (Peters & Layens, 2004).

The chemical composition which is obtained through EDAS and general properties of Al 6061 are presented in Tables 1 & 2 respectively.

Table 1. Composition of Al6061

Constituents	Percentage
Manganese (Mn)	0.15%
Iron (Fe)	0.70%
Copper (Cu)	0.40%
Magnesium (Mg)	0.15%
Silicon (Si)	0.8%
Zinc (Zn)	0.25%
Chromium (Cr)	0.4%
Others (Total)	0.15%
Aluminium (Al)	95.8%

Table 2. General Properties of Al6061

Properties	Unit	Value
Density	g/cm^3	2.7
Melting point	°C	582-652
Brinell Hardness	-	45
Ultimate Tensile Strength	MPa	130
Yield Strength	MPa	276
Modulus of Elasticity	MPa	68.9
Thermal conductivity	W/m-K	167
Coefficient of Thermal Expansion	m/°C	23.6×10-6

Reinforcements

Reinforcements or strengthening materials are the minor constituents that are embedded inside the metallic matrix in the form of powders, granules, long or discontinuous fibres or flakes essentially to strengthen the matrix (Gowrishankar, et al., 2013). For hybrid composites, more than one reinforcement is used (Mali & Sonawane, 2014).

The reinforcements used here are Boron Carbide with a particle size of 105 μm (Rama Rao & Padmanabhan, 2012), a ceramic and Rice Husk Ash with a particle size of 75 μm, an agricultural waste (Lancaster, 2013). The particle size of the reinforcements is determined by sieve analysis.

Boron Carbide (B_4C)

Boron Carbide is a well-known ceramic, that exists in the form of fine powder with coarse particles. It is the third hardest abrasive next to diamond and cubic boron nitride. It has low density, high degree of chemical inertness and excellent thermoelectric properties as a non-conductor (Krishna et al., 2013).

The lower density, high elastic modulus, high refractoriness and higher hardness of B_4C than SiC and Al_2O_3 make it better reinforcement for high performance MMCs (Raviteja et al., 2014). It has also been stated that the interfacial bonding between the aluminum matrix and the B_4C reinforcement seems to be much better than that of SiC (Dhinakaran & Moorthy, 2014).

The common properties of boron carbide are represented in Table 3 below:

Table 3. Properties of Boron carbide

Properties	Units	Values
Density	g/cm^3	2.52
Melting point	°c	2450
Fracture toughness	$MPa\sqrt{m}$	3.0
Thermal conductivity	W/mK	40
Vickers hardness	-	3850
Tensile strength	MPa	350
Elastic modulus	GPa	450

Rice Husk Ash

Rice husk ash is an agricultural waste, which is believed to increase the mechanical properties of AMCs such as hardness and wear resistance, when used as a reinforcement (Usman et al., 2014) (Fatile et al., 2014). Rice husk is unusually high in ash compared to other agricultural wastes in the range 10-20%. The ash is 87-97% silica, which is highly porous and light weight, with a very high adsorption area (Subrahmanyam, 2015). Presence of high amount of silica, lignin and cellulose content makes it a valuable material for industrial applications (Kumar et al., 2012).

The raw material for RHA is rice husk, which is procured from paddy fields after harvesting (Kumar, et al., 2013). The grainless rice husk is washed thoroughly with water and made to dry in sunlight for about 48 hours. It is then burnt at a temperature of more than 300°C for 2 hours to remove the moisture and organic matters. The carbonaceous material is removed, when it is continuously heated at a temperature of 600°C, till ash is formed.

Both Boron Carbide and Rice Husk Ash are used together as hybrid reinforcements (Krishna et al., 2015), because:

- Boron Carbide is expensive
- Agricultural wastes along with ceramics give improved mechanical properties [16]
- Agricultural wastes can give only marginal increase in strength and hardness, if dispersed in Aluminium matrix individually [15,16].
- Rice Husk Ash has low elastic modulus and good absorption capacity.

FABRICATION METHOD

After selecting suitable materials for matrix and reinforcements, fabrication of composites is done. Stir casting is the most widely used cost efficient process for fabricating AMCs (Bandare & Sonawane, 2013). In this method, pre heated ceramic particulates are incorporated into the vortex of the molten matrix created by a rotating impeller driven either by hand or motor. This conventional method is also called as vortex method and liquid metallurgy route.

The stir casting process starts with the preheating of graphite crucible in a gas-fired furnace for 20 minutes. The rice husk ash and boron carbide particles were initially preheated separately at a temperature of 250°C to remove moisture and to help even distribution within Al6061 alloy. The Al6061 alloy billets were charged into the furnace, fitted with a temperature probe and heated to a temperature of 750°C (i.e) above the liquidus temperature of the alloy to ensure that the alloy melts completely. When it attains liquid stage, 5% weight of solid dry hexa-chloro-Methane tablets are put to degas the molten alloy and ensure elimination of any porosity. Slag is removed using scum powder. Now with the help of electrical stirrer, the molten alloy is stirred at a constant speed of 450 rpm to create vortex or whirlpool. The preheated Rice Husk Ash and Boron Carbide particles are then charged into the melt at constant pour rate and stirring of the slurry was performed manually for 5-10 minutes. Magnesium of about 2% of weight is added to ensure good wettability for all proportions of the reinforcements and to avoid the formation of intermetallics. The composite slurry was superheated to 800°C and a second stirring performed using a mechanical stirrer. The stirring operation was performed at a speed of 400 rpm for 10 minutes before casting into prepared sand moulds. Meanwhile the mould is preheated to avoid shrinkage of casting material. Then the melted matrix and reinforced particles are poured into the preheated mould of required dimension. The entire process is done with either nitrogen gas or inert gas surrounding it to avoid contamination from atmosphere. Now the composite can be cut into different shapes and sizes. An electrical furnace with stirrer setup has been shown in Fig. 1.

Figure 1. Electrical stirrer setup

A total of 9 specimens have been made with different compositions as represented in Table 4.

Table 4. Compositions of composite specimens

Specimens	Weight % of Al 6061	Weight % of B_4C	Weight % of RHA
1	100	0	0
2	90	10	0
3	95	5	0
4	90	7.5	2.5
5	90	5	5
6	90	2.5	7.5
7	95	3	2
8	95	2.5	2.5
9	95	2	3

TESTS CONDUCTED AND RESULTS

To compare and evaluate mechanical properties such as ultimate tensile strength, hardness, impact strength, etc. and chemical properties like corrosion resistance, it is mandatory to conduct various tests in conformance with the ASTM Standards among the different compositions of composites.

Tensile Test

Tension means pulling force. Tensile test is conducted in a specimen by elongating it as a measure of its ductility. Normally metals like aluminium exhibit excellent ductility after plastic deformation. It is an important criterion in most of the structural parts, where the materials tend to yield over the application of external force without breaking (Saravanan & Senthilkumar, 2014).

The specimen to be tested is fastened to the two end-jaws of the UTM shown in Fig. 2, in which one end is fixed and the other is movable. Now the load is applied gradually on the specimen by means of pulling the movable crosshead, till the specimen fractures. The maximum load, it withstands by elongating, before cup and cone fracture takes place is the ultimate tensile strength. The test is done as per ASTM standards for tension testing of metallic materials E8.

Figure 2. Universal testing machine

All the nine composite specimens are subjected to tensile test and the table and graph showing the results are given as Table 5 & Fig. 3. respectively.

Table 5. Tensile test results

Specimens	1	2	3	4	5	6	7	8	9
Tensile strength (MPa)	125	129	80	92	98	107	102	85	89

From the above graph, depicting the results of the tensile test, it is understood that, the tensile strength of the individual aluminium alloy (specimen 1) is higher than most of the composite combinations, barring the second specimen (90% Al 6061 + 10% B_4C), which is an exception. But there is a fluctuation in the tensile strength among other composite proportions, which is not clear. The hybrid combinations give marginally good tensile strength when compared to only Al 6061+ B_4C combinations.

Figure 3. Tensile test results

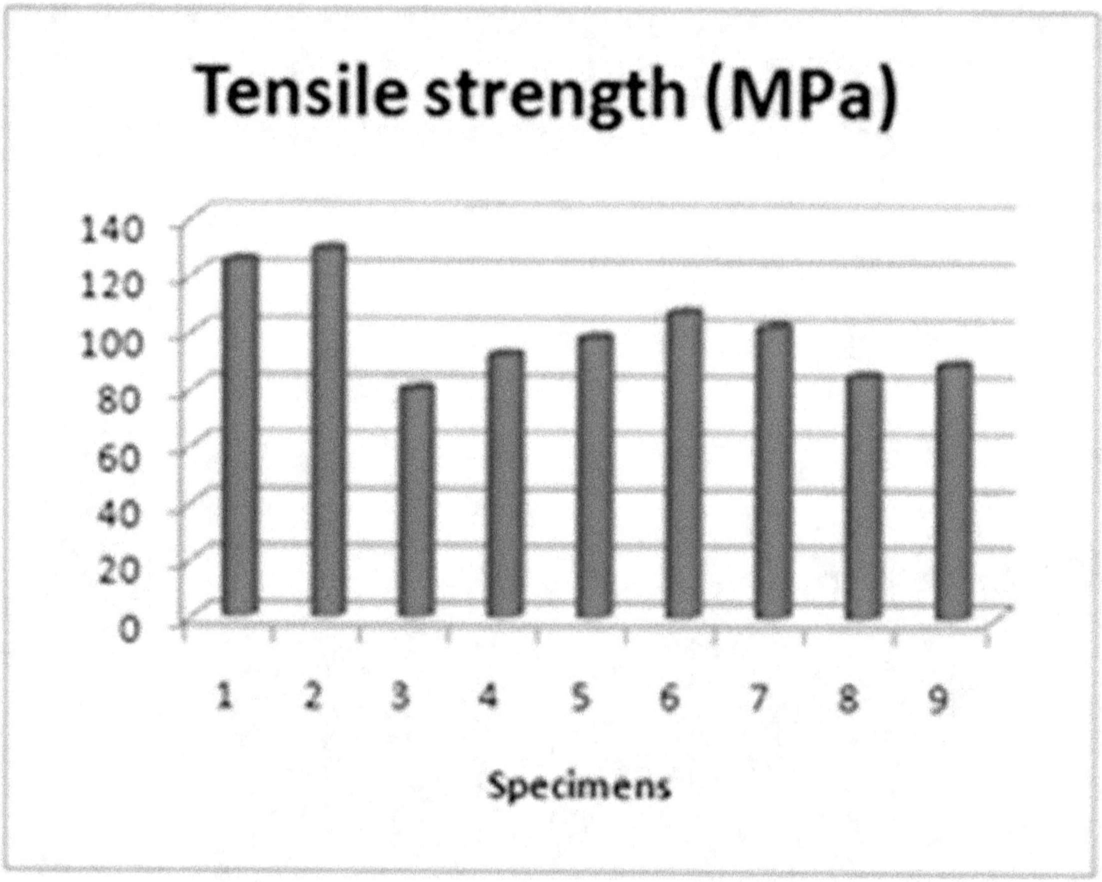

Normally, the tensile strength decreases with addition of nonmetallic reinforcements, as the ductility reduces. This is due to the existence of metal alloy and reinforcements at different distinctive phases. They are only joined together as a composite, not mixed together as in the case of alloy. Hence, their physical and chemical properties do not change. Due to the different phases in composite, porosity is created during melting of the metal alloy at the interface of matrix- reinforcement interaction. The void created inside molten metal during casting is not properly filled by reinforcement particles, due to the difference in physical properties such as density, melting point, etc.

Figure 4. Hardness testing machine

Table 6. Hardness test results

Specimens	1	2	3	4	5	6	7	8	9
Brinell Hardness (BHN)	44.02	66.81	56.20	59.86	54.75	51.25	52.79	50.28	48.41

Hardness Test

Hardness test is done on a specimen to know its ability to withstand or resist wear, abrasion and indentation (Babu et al., 2014). For metallic materials and their allied components, especially aluminium based materials, Brinell hardness test is taken. The Brinell hardness is measured using Brinell hardness tester shown in Fig. 4 from an indentation produced in the composite by applying a constant load of 250 kgf on steel ball indenter of diameter 5 mm in contact with the surface of the specimen for 10 seconds. The test is conducted as per ASTM standards for Brinell hardness testing E10. The indentation in the specimen surface is calculated from the Eq. (1):

Figure 5. Hardness test results

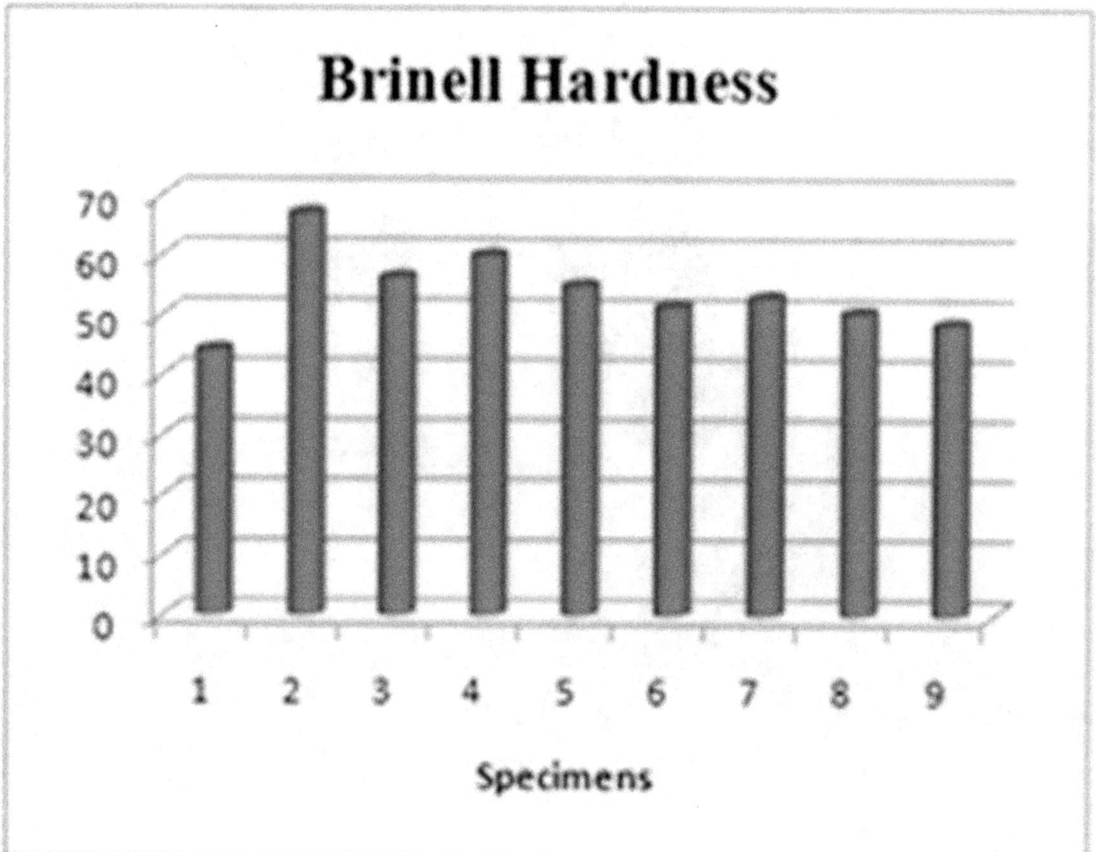

Figure 6. Impact testing machine

Table 7. Impact test results

Specimens	1	2	3	4	5	6	7	8	9
Izod Impact strength (Nm)	2	3	4	5	4	3	8	6	4

Figure 7. Impact test results

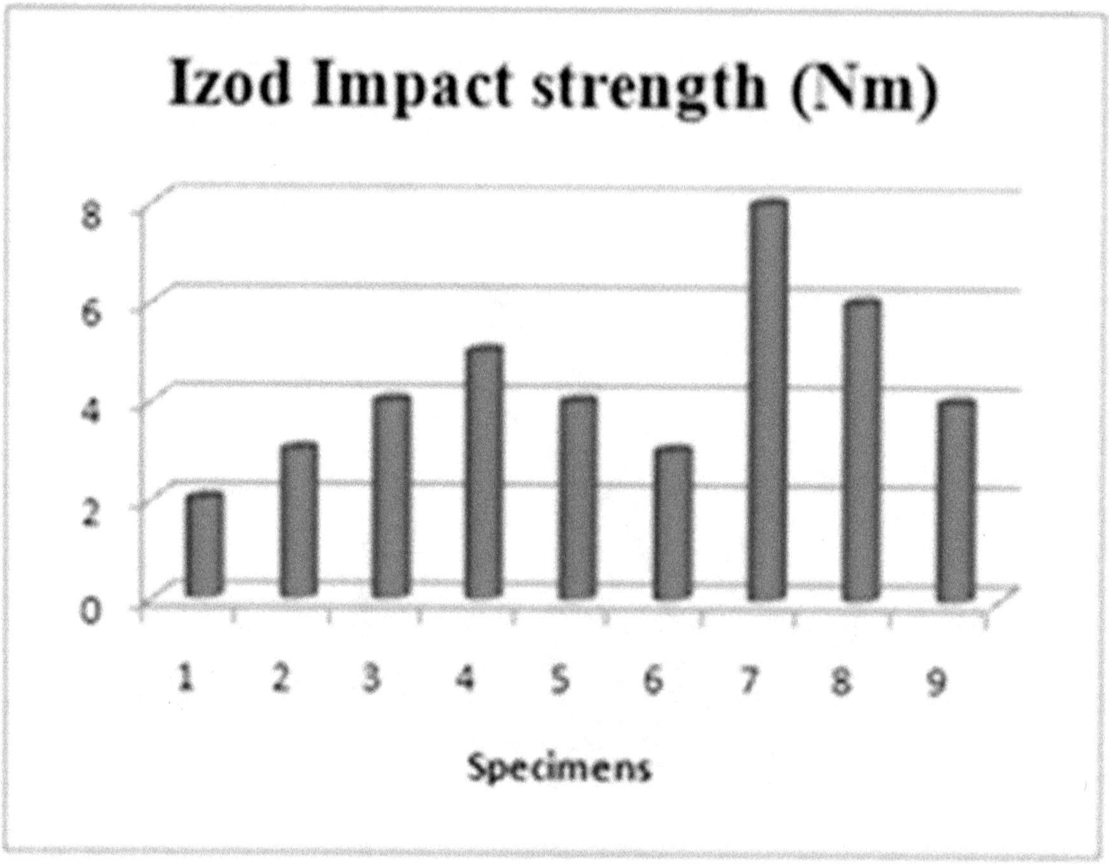

$$Brinell\ Hardness\ Number = \frac{2P}{\pi D\left[D - \sqrt{D^2 - d^2}\right]} \tag{1}$$

where,

P is the load applied on indenter (250 kgf)

D is the diameter of steel ball indenter (5 mm)

d is the diameter of ball impression created in mm.

The Brinell hardness test is conducted in all the nine specimens and the results are mentioned below in the form of Table 6. A graph is drawn taking different compositions on X-axis and the corresponding hardness values on Y-axis. It is denoted as Fig. 5.

From the hardness test results, it is clear that, the hardness values of composite specimens are higher than that of individual alloy. But the hardness values increase with increase in the percentage of B_4C addition. The hardness values are high, when the percentage of B_4C is high. It reduces with the addition of RHA with B_4C, but still higher than that of individual aluminium alloy.

Impact Test

Impact strength is the capacity of a material to withstand blows without fracture. It is also called as 'hammering effect'. The impact test is done as per Standard test methods for notched bar tensile strain Impact test method ASTM E23 in Impact testing machine shown in Fig. 6.

In Izod impact test, a notch of 2 mm depth and 45° inclination is cut in the specimen measuring 75x10x10 mm. The specimen is fixed vertically like a cantilever beam and is struck by a single blow from the suddenly released hammer at exactly 28 mm above the notch in the testing machine. The energy absorbed in breaking the specimen can be measured from the scale provided in the machine (Saravanan & Senthilkumar, 2014).

After initial preparations, all the nine specimens are subjected to Izod impact test and the results are shown below in the form of Table 7 and Fig. 7:

From the above results, it is certain that, the impact strength increases with increase in percentage of reinforcement. The composite specimens are having high impact strength when compared to individual aluminium alloy. But, RHA contributes more to the improvement of impact strength as the impact energy increases with increase in the weight % of RHA and is higher than that of composites having only B_4C reinforcement.

Corrosion Test

The corrosion resistance of the composites when compared to individual alloy is studied through mass loss and corresponding corrosion rate measurements by immersion corrosion test (Babu et al., 2014). In this test, exclusive samples are prepared by cutting separate pieces from the various combinations of composite specimens to the size of 10x10x10 mm, grinded carefully with coarse and fine emery with grit sizes ranging from 320 to 1200 and then polished using alumina cloth according to ASTM G31 Standard practice for lab immersion corrosion testing of metals. They are then degreased, washed in acetone and made to dry in hot air oven for about 2 days. Finally, they are dipped in acid and base solutions of 1N and kept immersed for about 24 hours. Then they are taken out and the loss in their weight and corrosion rate of the sample are calculated using the formulae given in Eq. (2) and Eq. (3) respectively:

$$ML = m_i - m_f \tag{2}$$

where,

M_L is the mass loss (g),
m_i is the initial weight before immersion (g) and
m_f is the final weight after immersion (g).

$$C.R = \frac{K * ML}{\rho * A * T} \tag{3}$$

where,

C.R is the corrosion rate (mm per year)

Figure 8. Immersion corrosion test

K is the conversion constant to convert hours to years (24*365 = 8760)
M_L is the mass loss (g),
ρ is the density of the composite (initial mass/volume) (g/mm^3),
A is the effective area of the sample (length x breadth of the facing side) (mm^2),
T is time of exposure (hours).
Fig. 8 shows the experimental setup for immersion corrosion test.

Table 8. Corrosion test results

Specimens	m_i		m_f		W_L		C.R	
	g		g		g		mmpy	
	A	B	A	B	A	B	A	B
1	2.7	2.4	2	0.8	0.7	1.6	1021	2333
2	1.9	1.8	1.7	1.2	0.2	0.6	291.6	874.8
3	2.2	2.2	1.9	1.4	0.3	0.8	437.4	1166
4	3.4	3.4	2.9	2	0.5	1.4	729	2041
5	2.8	2.2	2.2	1.1	0.6	1.1	874.8	1604
6	2.3	2.4	2	1.2	0.3	1.2	437.4	1744
7	3	2.5	2.8	1.7	0.2	0.8	291.6	1166
8	2.3	2.2	1.8	0.8	0.5	1.4	729	2041
9	2.8	2.6	2.2	1.1	0.6	1.5	874.8	2187

Figure 9. Corrosion test results

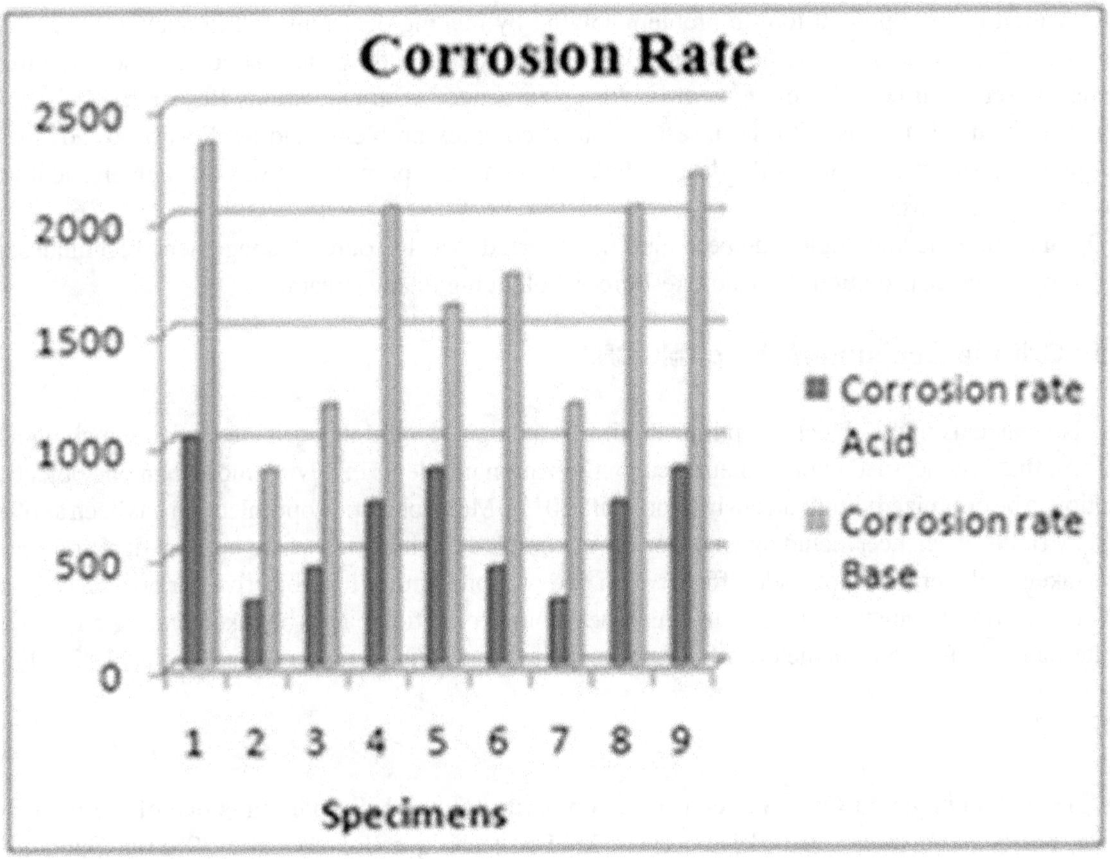

Before the commencement of the corrosion test, acidic and basic solutions of 1 N or 1 Normality are prepared. So, to say, normality is the measurement of the solution in which the weight equivalent to the molecular weight of the compound is dissolved in 1000 ml of water. A strong acid, HCl (36.46 ml in 1 l of water) and a strong base, NaOH (40 g in 1 l of water) is used for this purpose.

Corrosion test is conducted in both acidic and alkaline media to understand response of composite specimens and resistance offered to both the environments. The results are represented in Table 8 and Fig. 9 indicated below:

From the results, it is evident that the corrosion rate of the composite samples immersed in acid and base solutions are far more lesser than that of individual aluminium alloy. It also infers that the base is more corrosive to aluminium based alloys and composites in nature than acid.

Initially, the specimen having highest percentage of B_4C as reinforcement is having lowest corrosion rates. Later, in 10% reinforcement, the composite which is having B_4C as a minor constituent than RHA is giving lower corrosion rates than B_4C as a major contributor. In contrary, in 5% reinforcement, the composite which is having RHA as a minor constituent than B_4C is giving lower corrosion rates than RHA as a major contributor.

OPTIMIZATION

Optimization is an important tool in problem solving by making decisions and in analyzing physical systems. In mathematical terms, an optimization is a method of finding the best possible solution from among the set of all feasible solutions. It is closely associated with Operations Research, a relatively broader domain, that deals with the investigation of complex problems and application of advanced analytical and statistical tools for the design and development of proper solutions through efficient and economical decisions.

Optimization methodologies can be sometimes referred to as Resource Management Techniques, as they provide optimal solutions to handle resources, belonging to any stream.

Multi Criteria Decision Making (MCDM)

It is also called as multi-objective optimization. It is an area of multiple attribute decision making and analysis, that is concerned with mathematical optimization problems involving more than one objective function to be optimized simultaneously (Kou et al., 2012). Multi-objective optimization has been applied in many fields of science, including engineering, economics and logistics where optimal decisions need to be taken in the presence of trade-offs between two or more conflicting objectives or criteria (Hwang & Yoon, 1981). It depicts a situation in which best course of action has to be taken among a given set of alternatives which are characterized by contrasting or contradictive properties (Peng et al., 2011).

TOPSIS

TOPSIS or Technique for Order Preference by Similarity to the Ideal Solution is one of the more accurate, practical, versatile and widely accepted MCDM technique (Roszkowska, 2005) in Operations Research in which the best possible solution is selected among different alternatives through ranking of Closeness index or Closeness coefficient by having least distance from positive ideal or ideal best solution (Opricovic & Tzeng, 2004) (Shivakoti et al.,2017). The 6-step algorithm for ranking alternatives by TOPSIS is as follows (Mahmoodzadeh et al., 2007):

Step 1: Creation of an evaluation matrix represented in Eq. (4) comprising of m alternatives or combinations and n criteria or objectives, in which the intersection of each alternative and criteria is denoted as x_{ij} given below:

$$\begin{pmatrix} x_{11} & x_{12} & \cdots & x_{1n} \\ x_{21} & x_{22} & \cdots & x_{2n} \\ \vdots & \vdots & \cdots & \vdots \\ x_{m1} & x_{m2} & \cdots & x_{mn} \end{pmatrix} \tag{4}$$

Step 2: Balancing of the given numbers and their matrix, since the numbers represent different values with different measuring units and different multiplication factors (Elsayed et al., 2017). Hence, the matrix must be normalized using the normalization method shown as Eq. (5) below:

$$r_{ij} = \frac{x_{ij}}{\sqrt{\sum_{i=1}^{m} x_{ij}^{2}}} \tag{5}$$

where,

i = 1,2,…,m

j = 1,2,…,n.

The normalized matrix R is represented as Eq. (6):

$$\begin{pmatrix} r_{11} & r_{12} & \cdots & r_{1n} \\ r_{21} & r_{22} & \cdots & r_{2n} \\ \vdots & \vdots & \cdots & \vdots \\ r_{m1} & r_{m2} & \cdots & r_{mn} \end{pmatrix} \tag{6}$$

Step 3: Each criterion must be assigned a weight factor (Olson, 2004). This weightage is to be determined objectively so that the sum of all weight factors must be equal to 1.

A weighted matrix is to be created as per Eq. (7), in which:

$$a_{ij} = w_j * r_{ij} \tag{7}$$

The resultant matrix will be like the one represented in Eq. (8):

Table 9. Criteria and their values (Decision matrix)

Specimens	Tensile strength	Hardness	Impact strength	Acidic corrosion rate	Basic corrosion rate
1	125	44.02	2	1021	2333
2	129	66.81	3	291.6	874.8
3	80	56.2	4	437.4	1166
4	92	59.86	5	729	2041
5	97	54.75	4	874.8	1604
Specimens	Tensile strength	Hardness	Impact strength	Acidic corrosion rate	Basic corrosion rate
6	107	51.25	3	437.4	1744
7	102	52.79	8	291.6	1166
8	85	50.28	6	729	2041
9	89	48.41	4	874.8	2187

$$\begin{pmatrix} a_{11} & a_{12} & \cdots & a_{1n} \\ a_{21} & a_{22} & \cdots & a_{2n} \\ \vdots & \vdots & \cdots & \vdots \\ a_{m1} & a_{m2} & \cdots & a_{mn} \end{pmatrix} \tag{8}$$

Step 4: Determination of the positive ideal solution (A_j^+) and the negative ideal solution (A_j^-) of the alternatives. The coordinates for A_j^+ of positive ideal solution $A^+ = (A_1^+, A_2^+, \ldots A_n^+)$ are chosen using the formula given as Eq. (9):

$$A_i^+ = \begin{cases} \max a_{ij} & for\ j = 1,2,...,k \\ \min a_{ij} & for\ j = k+1,...,n \end{cases} \tag{9}$$

The coordinates for A_j^- of negative ideal solution $A^- = (A_1^- A_2^- \ldots A_n^-)$ are chosen using the formula given as Eq. (10):

$$A_i^- = \begin{cases} \max a_{ij} & for\ j = 1,2,...,k \\ \min a_{ij} & for\ j = k+1,...,n \end{cases} \tag{10}$$

Step 5: Calculation of the distance between the target alternative *i* and the negative ideal solution A^- is given by Eq. (11):

$$S^- = \sqrt{\sum_{j=1}^{n}((a_{ij} - A_j^-)^2)} \tag{11}$$

Table 10. Weightage given to criteria

S.No.	Criteria	Type	Objective type	Weightage
1	Tensile strength	Beneficiary (Larger-the-better)	Maximization	0.20
2	Hardness	Beneficiary (Larger-the-better)	Maximization	0.20
3	Impact strength	Beneficiary (Larger-the-better)	Maximization	0.20
4	Acidic corrosion rate	Cost or non-beneficiary (Smaller-the-better)	Minimization	0.20
5	Basic corrosion rate	Cost or non-beneficiary (Smaller-the-better)	Minimization	0.20

Similarly, the distance between the target alternative i and the positive ideal solution A^+ is given by Eq. (12):

$$S^+ = \sqrt{\sum_{j=1}^{n}((a_{ij} - A_j^{+})^2)} \qquad (12)$$

Step 6: The relative distance of the points from the ideal solution known as the closeness index (Bhutia & Phipon, 2012) is calculated using the Eq. (13):

$$CI = \frac{S^-}{S^+ + S^-} \qquad (13)$$

Finally, ranking of closeness index in the descending order is done with the highest valued alternative being the most ideal one (Ren et al., 2007).

Table 11. Normalized decision matrix

Tensile strength	Hardness	Impact strength	Acidic corrosion rate	Basic corrosion rate
0.409	0.271	0.143	0.499	0.444
0.422	0.411	0.215	0.142	0.166
0.262	0.346	0.286	0.214	0.222
0.301	0.368	0.358	0.356	0.388
0.317	0.337	0.286	0.427	0.305
0.35	0.315	0.215	0.214	0.332
0.334	0.325	0.573	0.142	0.222
0.278	0.309	0.43	0.356	0.388
0.291	0.298	0.286	0.427	0.416

Table 12. Weight normalized decision matrix

Tensile strength	Hardness	Impact strength	Acidic corrosion rate	Basic corrosion rate
0.082	0.054	0.029	0.1	0.089
0.084	0.082	0.043	0.028	0.033
0.052	0.069	0.057	0.043	0.044
0.06	0.074	0.072	0.071	0.078
0.063	0.067	0.057	0.085	0.061
0.07	0.063	0.043	0.043	0.066
0.067	0.065	0.115	0.028	0.044
0.056	0.062	0.086	0.071	0.078
0.058	0.06	0.057	0.085	0.083

Table 13. Positive ideal solution matrix

Tensile strength	Hardness	Impact strength	Acidic corrosion rate	Basic corrosion rate
0.084	0.082	0.115	0.028	0.033
0.084	0.082	0.115	0.028	0.033
0.084	0.082	0.115	0.028	0.033
0.084	0.082	0.115	0.028	0.033
0.084	0.082	0.115	0.028	0.033
0.084	0.082	0.115	0.028	0.033
0.084	0.082	0.115	0.028	0.033
0.084	0.082	0.115	0.028	0.033
0.084	0.082	0.115	0.028	0.033

Table 14. Negative ideal solution matrix

Tensile strength	Hardness	Impact strength	Acidic corrosion rate	Basic corrosion rate
0.052	0.054	0.029	0.1	0.089
0.052	0.054	0.029	0.1	0.089
0.052	0.054	0.029	0.1	0.089
0.052	0.054	0.029	0.1	0.089
0.052	0.054	0.029	0.1	0.089
0.052	0.054	0.029	0.1	0.089
0.052	0.054	0.029	0.1	0.089
Tensile strength	Hardness	Impact strength	Acidic corrosion rate	Basic corrosion rate
0.052	0.054	0.029	0.1	0.089
0.052	0.054	0.029	0.1	0.089

Table 15. S^+ and S^- of TOPSIS

S^+	S^-
0.128	0.0294
0.072	0.1009
0.069	0.0792
0.079	0.0568
0.089	0.0457
0.084	0.066
0.027	0.1215
0.077	0.0655
0.101	0.0335

RANKING BASED OPTIMIZATION OF ABOVE PRESENTED WORK

As far as the above conducted work involving evaluation of mechanical properties is concerned, the ranking based optimization using TOPSIS is carried out for the identification of best proportion or composition of composites from the various combinations (Wu et al., 2010).

Let us consider, the nine different combinations of composite specimens as alternatives. The different mechanical and chemical properties such as tensile strength, hardness, impact strength and corrosion rate respectively obtained as a result of conducting corresponding tests are considered as criteria.

Here, higher values are expected for tensile strength, hardness and impact strength, whereas lower values are desirable for acidic and basic corrosion rates. The decision matrix is represented as Table 9.

As all the properties are considered equally important for the effective selection of optimal composite combination, equal weight is given to the properties/ criteria as shown in Table 10. The sum of all weights must be equal to 1.

Table 16. Closeness index

CI
0.187
0.642
0.504
0.361
0.291
0.419
0.773
0.417
0.213

Table 17. Rank order based on closeness index

CI	Rank	Specimens	Tensile strength	Hardness	Impact strength	Acidic corrosion rate	Basic corrosion rate
0.187	9	1	125	44.02	2	1021	2333
0.642	2	2	129	66.81	3	291.6	874.8
0.504	3	3	80	56.2	4	437.4	1166
0.361	6	4	92	59.86	5	729	2041
0.291	7	5	97	54.75	4	874.8	1604
0.419	4	6	107	51.25	3	437.4	1744
0.773	1	7	102	52.79	8	291.6	1166
0.417	5	8	85	50.28	6	729	2041
0.213	8	9	89	48.41	4	874.8	2187

OPTIMIZATION BASED ON PROCEDURAL STEPS OF TOPSIS

The formulae given in the step by step procedure of TOPSIS is applied to predict the best composite combination for evaluating and optimizing the properties in hybrid composites with different ratio of reinforcements through ranking.

Normalized Decision matrix (R) represented as Table 11

Weight Normalized Decision matrix represented as table

The element or value in each cell of the normalized decision matrix table is multiplied by the above-mentioned weightage (0.2) to form the weight normalized decision matrix table represented as Table 12.

Creation of positive and negative ideal solution matrix tables

The positive and negative ideal solution matrix is framed by taking the largest value and smallest value from each criterion/ property and apply it to the entire column (Deng et al., 2000) as shown in Tables 13 & 14 respectively.

Creation of column sets S^+ and S^- for the distances from the target alternative to positive ideal solution A^+ and from the target alternative to negative ideal solution A^- presented in Table 15.

Calculation of Closeness Index (CI)

The closeness index or closeness coefficient is the shortest distance from the ideal solution. It is calculated from the formula, $CI = \frac{S^-}{S^- + S^+}$ and shown below as Table 16.

Table 18. Top most ranked composite specimen

Rank	Specimen	Tensile strength	Hardness	Impact strength	Acidic corrosion rate	Basic corrosion rate
1	**7**	**102**	**52.79**	**8**	**291.6**	**1166**

Ranking of alternatives/combinations based on closeness index given as Table 17.

Optimized or Top Ranked Proportion of Composites

Normally, optimized value is not the largest value, but the most economical and efficient value. It is neither the highest nor the lowest, but optimally intermediate, when contrasting criteria having expectation of opposite behavior/characteristics are involved (Deng et al., 2000). The top ranked or optimized proportion is represented as Table 18 below:

From the results obtained, the optimum composite combination is 95% Al 6061+ 3% B_4C+ 2% RHA.

CONCLUSION

From the above experimental work, we have the following conclusions:

- Hybrid AMCs are having profound applications in various process and product industries.
- Stir casting is the most suitable and economical method of fabricating MMCs.
- Boron Carbide, with its density even lesser than that of aluminium alloy is having good miscibility in AMC.
- By utilizing waste by products from agriculture like rice husk ash as reinforcement in AMCs, we can convert the waste into wealth, thereby preserving ecology.
- Reinforcements like B_4C and RHA can give enhanced mechanical properties when added with Al 6061.
- Superior mechanical properties can be obtained due to the even and uniform distribution of reinforcement particles inside the metallic matrix, without forming clusters.
- MCDM plays an inevitable role for solving complex problems and capacity planning in production, industrial, logistics and supply chain management.
- TOPSIS is one of the most powerful, robust and most practical MCDM technique for optimizing the process parameters.

REFERENCES

Babu, B., Karthikeyan, S. K., & Adithya, V. (2014). Experimental Investigation and Analysis of Corrosion and Hardness using Aluminium Composites. *International Journal of Latest Trends in Engineering and Technology*.

Bhandare, R. G., & Sonawane, P. M. (2013). Preparation of aluminium matrix composite by using stir casting method. *International Journal of Engineering and Advanced Technology*, *3*(3), 61–65.

Bhutia, P. W., & Phipon, R. (2012). Application of AHP and TOPSIS method for supplier selection problem. *IOSR Journal of Engineering*, *2*(10), 43–50. doi:10.9790/3021-021034350

Bodunrin, M. O., Alaneme, K. K., & Chown, L. H. (2015). Aluminium matrix hybrid composites: A review of reinforcement philosophies; mechanical, corrosion and tribological characteristics. *Journal of Materials Research and Technology*, *4*(4), 434–445. doi:10.1016/j.jmrt.2015.05.003

Deng, H., Yeh, C. H., & Willis, R. J. (2000). Inter-company comparison using modified TOPSIS with objective weights. *Computers & Operations Research*, *27*(10), 963–973. doi:10.1016/S0305-0548(99)00069-6

Dhinakaran, S., & Moorthy, T. V. (2014). Fabrication and Characteristic of Boron Carbide Particulate Reinforced Aluminum Metal Matrix Composites. *International Journal of Engineering Research Science*, *3*(7), 1051–1078.

Diyaley, S., Shilal, P., Shivakoti, I., Ghadai, R. K., & Kalita, K. (2017). PSI and TOPSIS based selection of process parameters in WEDM. *PeriodicaPolytechnica Mechanical Engineering*, *61*(4), 255–260.

Elsayed, E. A., Dawood, A. S., & Karthikeyan, R. (2017). Evaluating alternatives through the application of TOPSIS method with entropy weight. *Int. J. Eng. Trends Technol*, *46*(2), 60–66. doi:10.14445/22315381/IJETT-V46P211

Fatile, O. B., Akinruli, J. I., & Amori, A. A. (2014). Microstructure and mechanical behaviour of stir-cast Al-Mg-Sl alloy matrix hybrid composite reinforced with corn cob ash and silicon carbide. *International Journal of Engineering and Technology Innovation*, *4*(4), 251.

Gopal Krishna, U. B., Sreenivas Rao, K. V., & Vasudeva, B. (2013). Effect of boron carbide reinforcement on aluminium matrix composites. *International Journal of Metallurgical & Materials Science and Engineering (IJMMSE)*, *3*, 41–48.

Gupta, V., & Takhi, S. (2015). Effects of rice husk ash particulates on the mechanical and tribological properties of the aluminum metal composite reinforced with aluminum oxide. *International Journal for Scientific Research and Development*, *3*(3), 2321–0613.

Hwang, C. L., & Yoon, K. (1981). *Multiple Attribute Decision Making–Methods and Applications. Springer-Verlag*.

Kou, G., Lu, Y., Peng, Y., & Shi, Y. (2012). Evaluation of classification algorithms using MCDM and rank correlation. *International Journal of Information Technology & Decision Making*, *11*(01), 197–225. doi:10.1142/S0219622012500095

Krishna, S. M., Shridhar, T. N., & Krishnamurthy, L. (2015). Research significance, applications and fabrication of hybrid metal matrix composites. *International Journal of Innovative Science. Engineering & Technology*, *2*(5), 227–237.

Kumar, A., Mohanta, K., Kumar, D., & Prakash, O. (2012). *Properties and industrial applications of rice husk: a review*. Academic Press.

Kumar, S., Sangwan, P., Dhankhar, R. M. V., & Bidra, S. (2013). Utilization of rice husk and their ash: A review. *Res. J. Chem. Env. Sci*, *1*(5), 126–129.

Lancaster, L., Lung, M. H., & Sujan, D. (2013, January). Utilization of agro-industrial waste in metal matrix composites: towards sustainability. In Proceedings of World Academy of Science, Engineering and Technology (No. 73, p. 1136). World Academy of Science, Engineering and Technology (WASET).

Mahmoodzadeh, S., Shahrabi, J., Pariazar, M., & Zaeri, M. S. (2007). Project selection by using fuzzy AHP and TOPSIS technique. *World Academy of Science, Engineering and Technology*, *30*, 333–338.

Mali, A., & Sonawane, S. A. (2014). Review on Effect of Hybrid Reinforcement on Mechanical Behavior of Aluminium Matrix Composite. *International Journal of Engine Research*, *3*(5).

Olson, D. L. (2004). Comparison of weights in TOPSIS models. *Mathematical and Computer Modelling*, *40*(7-8), 721–727. doi:10.1016/j.mcm.2004.10.003

Opricovic, S., & Tzeng, G. H. (2004). Compromise solution by MCDM methods: A comparative analysis of VIKOR and TOPSIS. *European Journal of Operational Research*, *156*(2), 445–455. doi:10.1016/S0377-2217(03)00020-1

Peng, Y., Kou, G., Wang, G., & Shi, Y. (2011). FAMCDM: A fusion approach of MCDM methods to rank multiclass classification algorithms. *Omega*, *39*(6), 677–689. doi:10.1016/j.omega.2011.01.009

Peters, M., Kumpfert, J., Ward, C. H., & Leyens, C. (2003). Titanium alloys for aerospace applications. *Advanced Engineering Materials*, *5*(6), 419–427. doi:10.1002/adem.200310095

Rao, S. R., & Padmanabhan, G. (2012). Fabrication and mechanical properties of aluminium-boron carbide composites. *International Journal of Materials and Biomaterials Applications*, *2*(3), 15-18.

Raviteja, T., Radhika, N., & Raghu, R. (2014). Fabrication and mechanical properties of stir cast Al-Si12Cu/B4C composites. *International Journal of Research in Engineering and Technology*, *3*(07), 343–346. doi:10.15623/ijret.2014.0307058

Ren, L., Zhang, Y., Wang, Y., & Sun, Z. (2007). Comparative analysis of a novel M-TOPSIS method and TOPSIS. *Applied Mathematics Research Express*, •••, 2007.

Roszkowska, E. (2011). Multi-criteria decision making models by applying the TOPSIS method to crisp and interval data. *Multiple Criteria Decision Making/University of Economics in Katowice*, *6*, 200-230.

Saravanan, S. D., & Senthilkumar, M. (2014). Mechanical Behavior of Aluminum (AlSi10Mg)-RHA Composite. *IACSIT International Journal of Engineering and Technology*, *5*(6), 4834–4840.

Shankar, G., Jayashree, P. K., Shetty, R., Kini, A., & Sharma, S. S. (2013). Individual and combined effect of reinforcements on stir cast aluminium metal matrix composites-a review. *International Journal of Current Engineering and Technology*, *3*(3), 922–934.

Shivakoti, I., Pradhan, B. B., Diyaley, S., Ghadai, R. K., & Kalita, K. (2017). Fuzzy TOPSIS-based selection of laser beam micro-marking process parameters. *Arabian Journal for Science and Engineering*, *42*(11), 4825–4831. doi:10.100713369-017-2673-1

Subrahmanyam, A. P. S. V. R., Narsaraju, G., & Rao, B. S. (2015). Effect of rice husk ash and fly ash reinforcements on microstructure and mechanical properties of aluminium alloy (AlSi10Mg) matrix composites. *International Journal of Advanced Science and Technology*, *76*, 1–8. doi:10.14257/ijast.2015.76.01

Usman, A. M., Raji, A., Hassan, M. A., & Waziri, N. H. (2014). A comparative study on the properties of Al-7% Si-Rice husk ash and Al-7% Si-Bagasse ash composites produced by stir casting. *International Journal of Engineering Science*, *3*(8), 1–7.

Vengatesh, D., & Chandramohan, V. (2014). Aluminium alloy metal matrix composite: Survey paper. *International Journal of Engineering Research and General Sciences*, *2*(6), 792–796.

Wu, C. S., Lin, C. T., & Lee, C. (2010). Optimal marketing strategy: A decision-making with ANP and TOPSIS. *International Journal of Production Economics*, *127*(1), 190–196. doi:10.1016/j.ijpe.2010.05.013

Zavadskas, E. K., Turskis, Z., & Kildienė, S. (2014). State of art surveys of overviews on MCDM/MADM methods. *Technological and Economic Development of Economy*, *20*(1), 165–179. doi:10.3846/20294913.2014.892037

ADDITIONAL READING

Duckstein, L., & Opricovic, S. (1980). Multi objective optimization in river basin development. *Water Resources Research*, *16*(1), 14–20. doi:10.1029/WR016i001p00014

Gowdy, G. H., Reddy, M. G., Sreenivasulu, B., & Ravuri, M. (2014). Multi Objective Optimization of Process Parameters in WEDM during Machining of SS304. *Procedia Materials Science*, *5*, 1408–1416. doi:10.1016/j.mspro.2014.07.459

Kuo, Y., Yang, T., & Huang, G. W. (2008). The use of grey relational analysis in solving multiple attribute decision-making problems. *Computers & Industrial Engineering*, *55*(1), 80–93. doi:10.1016/j.cie.2007.12.002

Mian, S. A., & Christine, N. D. (1999). Decision-making over the project life cycle: An analytical hierarchy approach. *Project Management Journal*, *30*(1), 40–52. doi:10.1177/875697289903000106

Opricovic, S. (1998). *Multicriteria optimization of civil engineering systems*. Faculty of Civil Engineering.

Pavlicic, D. (2001). Normalization affects the results of MADM methods. *Yugoslav Journal of Operations Research*, *11*(2), 251–265.

Triantaphyllou, E. (2000). *Multi-Criteria Decision Making Methods: A Comparative Study*. Kluwer Academic Publishers. doi:10.1007/978-1-4757-3157-6

Zeleny, M. (1982). *Multiple Criteria Decision Making*. Mc-Graw-Hill.

KEY TERMS AND DEFINITIONS

Criteria: Criteria is the set of properties or attributes of the material. (they may be either beneficiary or non-beneficiary).

Decision Matrix: Decision matrix is defined as the matrix table constructed with various alternatives or specimens as rows and criteria as columns.

Formability: Formability is defined as the ability of a material to be made into any shape, size or dimension.

Impact Strength: Impact strength is the property of a material to withstand sudden heavy load and absorb energy before getting fractured.

Machinability: Machinability is defined as the degree of ease in removing parts from a material in the form of chips, powders, or flakes.

Stir Casting: Stir casting is one of the most popular and widely used method in which material formation (mainly Metal Alloys and Metal Matrix Composites) is done by melting metals and casting them into suitable shapes and sizes by pouring them into cavities. It is also called as liquid metallurgy.

Tensile Strength: Tensile strength is the property of a ductile material to be elongated till failure occurs.

Chapter 7
A Comparative Assessment of TOPSIS Variants in Multi-Attribute Optimization of EDM Process Parameters

Ankit Kumar Singh
Indian Institute of Science, Bangalore, India

Srinivasan Murugan
https://orcid.org/0000-0001-6326-911X
Dhofar University, Oman

Dhruva Kumar
Sikkim Manipal Institute of Technology, Sikkim Manipal University, Majhitar, India

Saurabh Sharma
Sikkim Manipal Institute of Technology, Sikkim Manipal University, Majhitar, India

ABSTRACT

The machining aspects of EDM sustains the discharge energy and discharge duration. This generated energy causes debris removal and evaporation of material. Therefore, in this chapter, an investigation is carried out to perform optimization of input process variables through the EDM process of ASTM A681 steel metal wielding the modified weighted TOPSIS MCDM technique. The study consists of process variables like discharge current, discharge duration, and discharge energy for the output response measurement of material removal rate (MRR) and average surface roughness. The multi-criteria decision approach is based on entropy, SDV, and mean-weighted TOPSIS and finds an output response related to the closeness coefficient between the feasible solution and the ideal solution. From the result, it is observed that the rank produced by mean weighted TOPSIS and SDV-TOPSIS are almost the same. But the rank produced by entropy-TOPSIS is slightly different as compared to the other method. However, the first, second, and third ranks are the same for all three techniques.

DOI: 10.4018/978-1-7998-7206-1.ch007

INTRODUCTION

Increase in demand for quality finishing and energy-efficient machining process has been a point of contention since the past few decades. Extensive energy consumption and rise in loss of materials in machining have coerced the manufacturing system, to find a suitable way to optimize the resources and preserve the environment from losses (Gostimirovic, et al., 2018) (Mahajan, et al., 2018) (Singh, et al., 2018). With the expansion in the research effort and worldwide demand to improve the energy-efficient machining, an optimized EDM process has been suggested to fathom this challenge. Electrical Discharge Machining is a popular and robust non-traditional machining approach being intensively used for modern day's problems. This machine is used to cut intricate shapes and grooving pattern irrespective of the material hardness i.e. like super alloys & alloys. EDM is a non-contact metal removing approach which works on the principle of thermo-electrical erosion, controlled by high-frequency pulse generator in a dielectric medium. This generator develops a huge potential (Volts) between tool and material filled with a dielectric, causing the metal to vaporize and remove. This act of huge potential eliminates the mechanical stress, chatter and vibrations in the work-piece (Ganesh, et al. 2020).

The growth of high potential difference leads to a rise in temperature and formation of plasma across the tool & workpiece gap, mostly hang on rate of generation and distribution of discharge energy throughout the machining region. This generated energy entirely works on the amount of discharge current and discharge duration, while the distribution of energy depends on the physical and chemical characteristics of the discharge area. This characteristic of energy efficiency in manufacturing is a critical subject in the industry but no studies have been done in non-conventional machining process till yet. Marin et al. experimentally investigated the thermal properties of the discharged energy and its influence on MRR, gap, SR and recast layer using a copper electrode. Analysis and experimental research showed that methodical selection of apt variables of discharge energy is favourable in EDM machining (Gostimirovic, et al., 2012). Lin et al. studied the EDM features related with electrical discharge energy using cemented tungsten carbide K10 & P10 graded tool. The machining variables were assorted to inspect the effect of discharge energy on machining aspects like MRR, EWR, heat affected layers' surface cracks and surface roughness. The experiment results show that the MRR rises with growth in discharge energy. Also, surface roughness and surface cracks increase with a high level of discharge energy. Lin et al. suggested a development of semi-empirical model for the MRR for various materials by using dimensional analysis. The associated input process variables were screened with design of experiment in the machining operation (Lin, et al., 2008). Sanchez et al. (Sanchez, et al., 2009) made a computer simulation to theoretically estimate a MRR and surface finish. The model was built on the numerical computation of temperature fields within the workpiece. The crucial barrier was the difficultly in estimation of energy dependent variables in the discharge zone. Rebelo et al. (Rebelo, et al., 1998) carried in-depth study of the impact of EDM discharge energy on the surface integrity of martensitic steels. In his work significant featured output responses were evaluated such as roughness of the surface, the metallurgical structure, the residual stress state and the surface crack network of the near-surface layers. Gadalla et al. (Gadalla & Tsai, 1989) estimated the effect of cobalt and tungsten carbide content on performance measurement by cemented tungsten carbide tool. There performed data discloses that increasing the cobalt content stabilizes the EDM machining and also increases the MRR rate. Giridharan et al. (Giridharan & Samuel, 2015) studied energy conservation as an important aspect of the EDM process. In his study, the importance of discharge energy on the performance of WEDM turning process namely MRR, surface finish, recast layer and surface crack was analysed. They found that with an increase in discharge energy the

MRR, surface roughness & thickness of the recast layer increases. Fenggou and Dayong (Fenggou & Dayong, 2004) studied the importance of artificial intelligence in EDM performance and conducted their experiment extensively. In their work, a highly efficient and intelligent optimization system was made using artificial neural networks to optimized genetic algorithm. Keshkin et al. (Keskin, Halkaci, & Kizil, 2006) studied and performed experiment to estimate variables influencing surface roughness. A significant and robust equation was obtained to measure surface roughness using power, pulse time, and spark time variables. Wen Li and Sami kara (Li & Kara, 2015) studied this effect and developed an empirical approach to characterise the energy efficiency of EDM. Their model suggested a contrary correlation between specific energy consumption and material removal rate. Due to the stochastic essence of EDM process, the analytical and semi-empirical MRR models cannot forecast the true MRR. Therefore, combined empirical models were used to estimate the consumption of energy in EDM with an exactness of about 90%.

Even though overhead literature survey ensures that a significant and diverse quality of work has been done to optimize the EDM energy-efficient process using Genetic Algorithm. But there is further research scope and advancement left. Hence, in this article, an effort has been made to choose and compare an optimal methodology i.e., Modified weighted TOPSIS optimization- SDV, Mean & Entropy to obtain a pragmatic combination of variables of discharge current, discharge duration, and discharge energy. TOPSIS is a highly acceptable and generic slant to identify a viable alternative in a constructed decision matrix i.e., nearest to the best solution and farthest from the worst solution in a multi-dimensional performance array. Entropy and SDV weights are competent in improving the TOPSIS performance through decision/weight matrix via discrete probability distribution and pairwise relative comparison matrix. These weighted distributions avoid biasing at each level i.e. experimental/personal and converges the data with consistency and a high degree of accuracy when combined with general TOPSIS steps. Modified weighted TOPSIS optimization has a large sum of applications in resource supply chain management, waste management, design and engineering, mechanical manufacturing systems and logistics, etc. The advantage of Modified TOPSIS is because of its coherence and aptness in decision making as it maintains the same number of criteria steps and instantly reviews different approaches or perhaps to stand on its own as a single decision-making tool regardless of problem matrix.

MATERIALS AND METHODS

In this experimentation, a manganese-alloyed air-hardening tool steel ASTM A681 of composition (0.9%C, 2%Mn, and 0.2%V) was selected. An electrolytic copper tool with 99.9% purity of 20x10mm cross-section and refined petroleum was selected for the machining process. The machining condition i.e. discharge current, discharge duration and discharge energy is shown in table -1. The discharge current I_e = 1/50 A (current density 0.5/25 A/cm^2) while the discharge duration varied between t_e = 1/100μs. The discharge voltage is constant for all trails is 20V. The measured performance characteristics were material removal rate (MRR) and surface roughness (SR). The surface roughness was evaluated using roughness meter via Mahr Perthen Perthometer, S5P, Germany as in reference (Gostimirovic, et al., 2018).

Table 1. Machining Process Parameters (Gostimirovic, et al., 2018)

Sl. No.	Discharge Current	Discharge Duration	Discharge Energy	MRR	Surface roughness
1	1	1	20	0.86	1.8
2	1	2	40	1.28	1.9
3	1	5	100	2.35	2.1
4	1	7	140	1.97	2.3
5	5	1	100	3.22	3.9
6	5	2	200	4.16	4.2
7	5	5	500	6.47	5.1
8	5	7	700	4.31	5.1
9	9	2	360	7.71	8.2
10	9	5	900	14.89	8.8
11	9	7	1260	9.49	9
12	9	10	1800	5.24	9.8
13	13	2	520	6.13	9.2
14	13	5	1300	18.71	9.4
15	13	7	1820	10.71	9.7
16	13	10	2600	6.62	10.3
17	20	5	2000	24.49	10.2
18	20	7	2800	31.82	10.4
19	20	10	4000	26.7	10.8
20	20	20	8000	17.11	11.2
21	30	7	4200	39.92	10.8
22	30	10	6000	53.3	11.3
23	30	20	12000	56.82	11.8
24	30	50	30000	40.48	12.5
25	50	10	10000	46.36	11.8
26	50	20	20000	66.83	12.5
27	50	50	50000	72.92	13.2
28	50	100	100000	60.6	13.4
28	50	100	100000	60.6	13.4

Multi-Criteria Decision Making

Modified entropy-TOPSIS, SDV-TOPSIS and general TOPSIS are carried out in this study. These methods were utilized because of their robust nature, high accuracy, versatility, and less computational cost to predict responses with the least data set.

TOPSIS

The technique for order preference by similarity to the ideal solution (TOPSIS) is a reliable, generic and widely accepted MCDM approach in operation and production research. In this concept, a best alternative is searched assumed on the closeness coefficient from the ideal best & worst solution (Diyaley, et al., 2017). The generic aim behind this method is to gauge the alternatives on the Euclidian distance scale, i.e. the least span from ideal best and farthest from ideal worst solution. The alternatives are then ordered based on their rank & performance.

The steps for TOPSIS approach are as follows—

Step 1. Decision matrix design and assumption of the weight matrix.

Let $D=x_{ij}$ be a decision matrix, where $x_{ij} \in \mathbb{R}$.

$$D = \begin{bmatrix} x_{11} & x_{12} & \cdots & x_{1n} \\ x_{21} & x_{22} & \cdots & x_{2n} \\ \vdots & \vdots & \ddots & \vdots \\ x_{m1} & x_{m2} & \cdots & x_{mn} \end{bmatrix} \tag{1}$$

Weight vector may be expressed as $W=[w_1 \ldots w_n]$ where $\sum_{j=1}^{n}(w_1 \quad \ldots \quad w_n)=1$

Step 2. Construction of the normalized decision matrix n_{ij} of each criterion using equation (2),

$$n_{ij} = \frac{x_{ij}}{\sqrt{\sum_{i=1}^{m} x_{ij}^2}} \tag{2}$$

Step 3. Determination of weighted normalized matrix using equation (3),

$$N_{ij}=w_j*n_{ij} \text{ for } i\in[1,m] \text{ and } j\in[1,n] \tag{3}$$

Step 4. Estimation of the ideal positive (*best*) and ideal negative (*worst*) solutions using equation (4) and (5) respectively.

$$A_j^+ = \begin{cases} \max.N_{ij} \mid \mathrm{j} \in \mathrm{B} \\ \min.N_{ij} \mid \mathrm{j} \in \mathrm{C} \end{cases} \tag{4}$$

$$A_j^+ = \begin{cases} \min.N_{ij} \mid \text{j} \in \text{B} \\ \max.N_{ij} \mid \text{j} \in \text{C} \end{cases} \quad (5)$$

Where B is a vector of benefit function and C is the vector of the cost function, for $i \in 1,m$ and $j \in 1,n$.

Step 5. Determination of the separation measurement and relative closeness coefficient. In TOPSIS, the difference of each response from ideal positive (*best*) solution is given by equation (6).

$$S_i^+ = \sqrt{\sum_{j=1}^{n} \left(N_{ij} - A_j^+\right)^2} \text{ for } i \in [1,m] \text{ and } j \in [1,n] \quad (6)$$

Similarly, the difference between each response from the ideal negative (*worst*) solution is given by equation (7).

$$S_i^- = \sqrt{\sum_{j=1}^{n} \left(N_{ij} - A_j^-\right)^2} \text{ for } i \in [1,m] \text{ and } j \in [1,n] \quad (7)$$

The corresponding closeness coefficient (CC_i) of the ith alternative is calculated using equation (8).

$$CC_i = \frac{S_i^-}{S_i^+ + S_i^-} \quad (8)$$

Where $0 \leq CC_i \leq 1$, $i \in [1,m]$

Step 6. The final step is to rank the alternatives in decreasing order of closeness coefficient.

WEIGHT ALLOCATION STRATEGY

Mean Weight Method

The TOPSIS decision matrix is based on equal weight distribution to responses (Phipon, et al., 2020). Equal weights are assigned to give equal importance and influence of each parameter on selecting the optimal set. For the current criteria i.e. *MRR & SR*, the weight vector is expressed as

W= [0.5 0.5]

Standard Deviation (SDV) Method

Standard deviation builds an unbiased & unprejudiced assignment of weights (Phipon, et al., 2020). It improves the MCDM and lowers the subjective weight stress. The constitutional steps are finalized in SDV through the following equations to change the varied scales and norms into structured & quantifiable units to figure out their tangible weights.

$$B_{ij} = \frac{x_{ij} - \min(x)_{ij}}{\max(x)_{ij} - \min(x)_{ij}} \tag{9}$$

$$SDV_J = \sqrt{\frac{\sum_{i=1}^{m}\left(B_{ij} - B_j\right)^2}{m}} \tag{10}$$

Where B_j is the average of the values for the ith measure, where j=1,2,3.
The weight vector is given by equation (11) as,

$$W_j = \frac{SDV_j}{\sum_{j=1}^{n} SDV_j} \tag{11}$$

For the current problem, the SDV weight vector is

W= [0.492 0.508]

Entropy Method

Entropy-TOPSIS is a new practical and speedy approach to predict the best alternative (Phipon, et al., 2020). Entropy weighted removes subjective influence and uses objective decision to derive suitable weight vectors to compute the finite rank. The entropy weight function is assumed on the discrete probability distribution.

$$e_j = \frac{-1}{\ln(m)} \sum_{i=1}^{m} n_{ij} \ln\left(n_{ij}\right) \tag{12}$$

The degree of diversity (d) possessed by each criterion is evaluated as,

$$d_j = 1 - e_j,\ j = 1,2,3 \tag{13}$$

And the weight objective for each criterion is given by

$$W_j = \frac{d_i}{\sum_{i=1}^{n} d_i} \tag{14}$$

For the current problem, the entropy weight vector is

W= [0.5612 0.4338]

RESULTS AND DISCUSSION

Weight Calculation

In the present work the weight allocation has been done by using three different methods i.e. mean weight method, standard deviation method and entropy method. In mean weight method equal weight 0.5 has been allocated to both the response parameters. The weight calculation by standard deviation method is done by using three different equations and it is given in equation (9), (10) and (11). The weight allocated for the MRR and surface roughness are 0.492 and 0.508, respectively. Similarly, the weight by entropy method has been calculated by using equation (12), 13) and (14). The weight allocated for the MRR and surface roughness are 0.5612 and 0.4338, respectively by the entropy method.

MCDM Using TOPSIS

In the present work the TOPSIS method is used to find the best alternative by considering the different weight calculated by various method. The detailed steps of TOPSIS method have been described in the methodology section. The TOPSIS method is based on the Euclidean distance of each alternative from both negative and positive ideal solution (NIS and PIS). By considering the various weight, the distance to the PIS (PID)and distance to the NIS (NID) have been calculated and it is shown in Figure. (1), (2) and (3). By considering the PID and NID the closeness coefficient (CC) has been calculated by using all three different weight calculation method and it is shown in Figure. (4). From the graph it is observed that the CC values for both mean and standard methods are almost same due to the approximately equal weight. However, the CC values for the entropy methods are bit different. Based on the CC values the ranks have been computed for the three different weight calculation method (Diyaley, et al., 2017) (Phipon, et al., 2020). The rank using TOPSIS for all the three weight calculation methods are presented in Figure 5. In all the three cases the 1st 2nd, and 3rd rank is same and represented by expt. no 26, expt. no 27 and expt. no 23, respectively. As the CC values of both SDV method and mean weight method are almost equal, so there is no variation of rank between them. For both mean weight and SDV TOPSIS method experiment no 16 is giving the last rank, whereas expt. no 1 is giving the last rank for the entropy-TOPSIS method. From table no 1 it is noticed that the expt. 26, 27 and 23 have the higher current and discharge time and that leads to give a better performance with respect to MRR and surface roughness.

Figure 1. PID and NID for mean weight TOPSIS

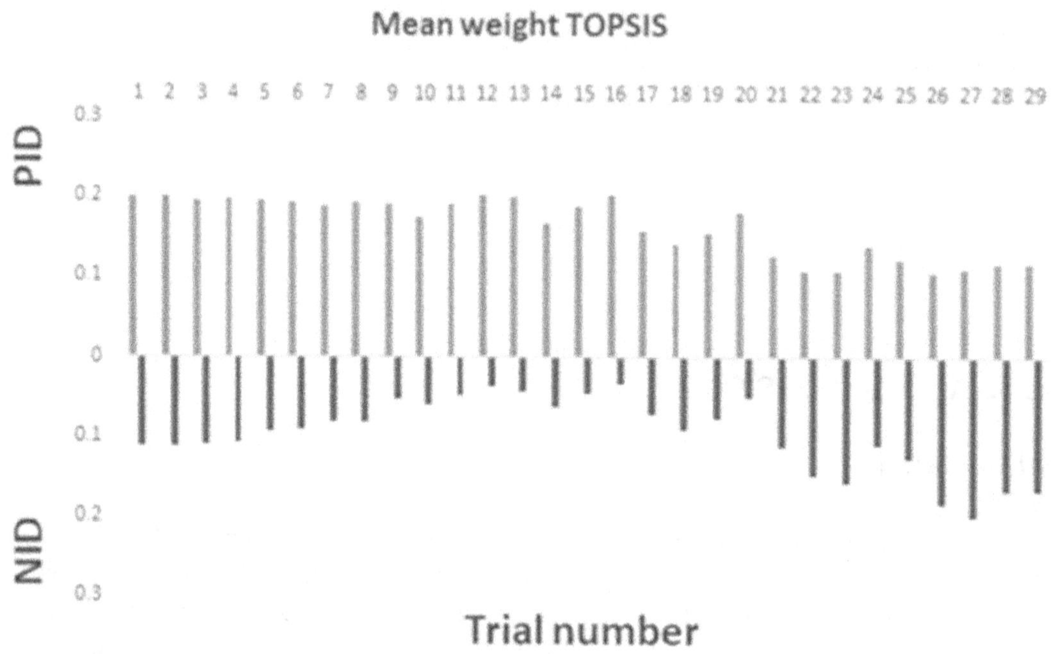

Figure 2. PID and NID for SDV weight TOPSIS

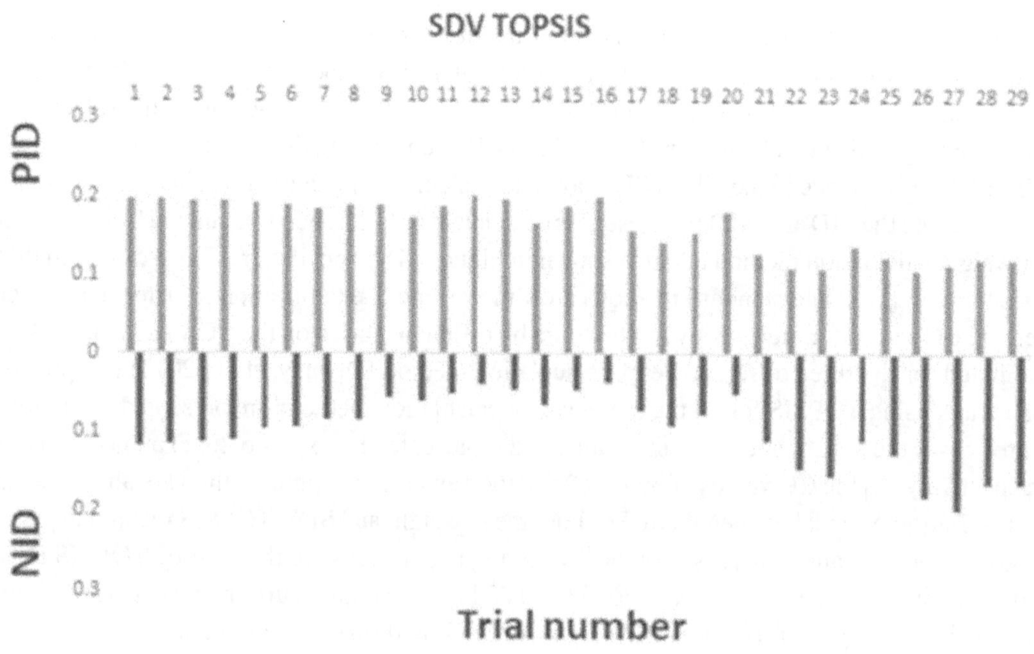

Figure 3. PID and NID for entropy weight TOPSIS

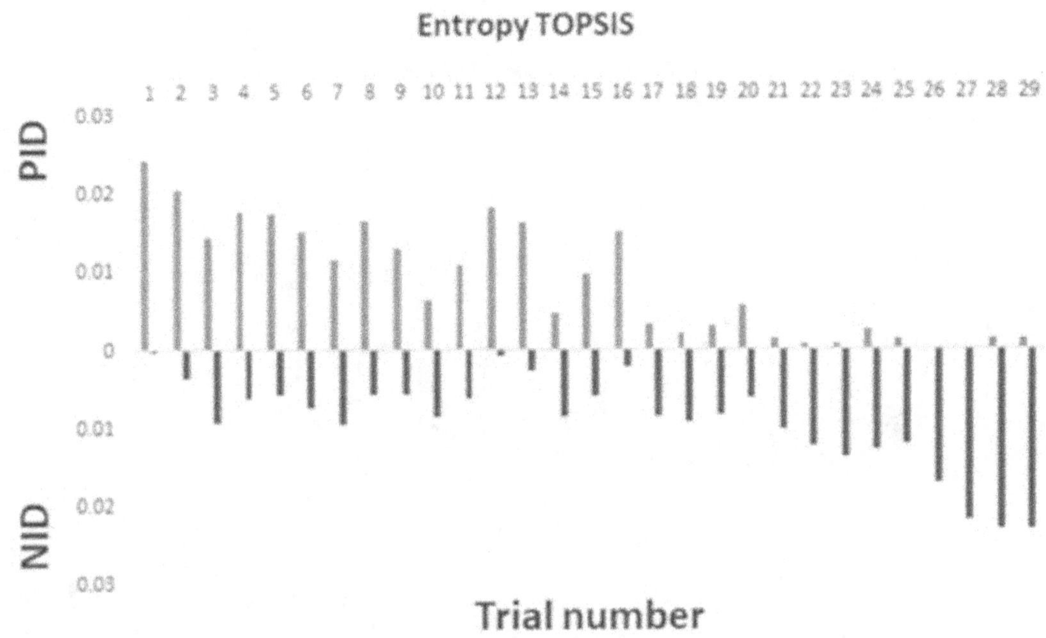

Figure 4. Comparison of closeness coefficients

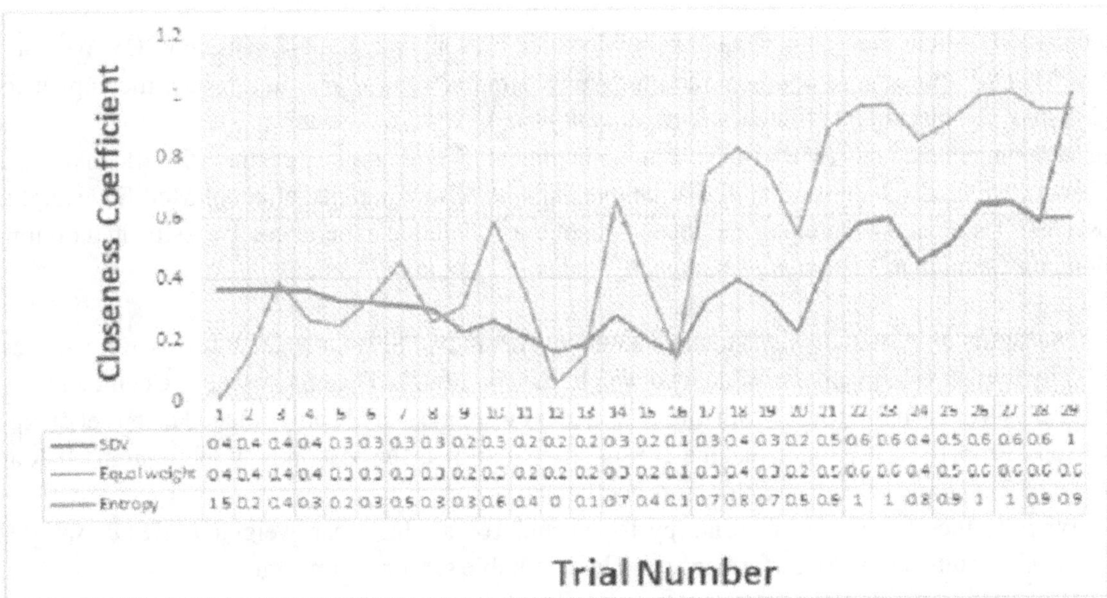

Figure 5. Rank as calculated by the three TOPSIS variants

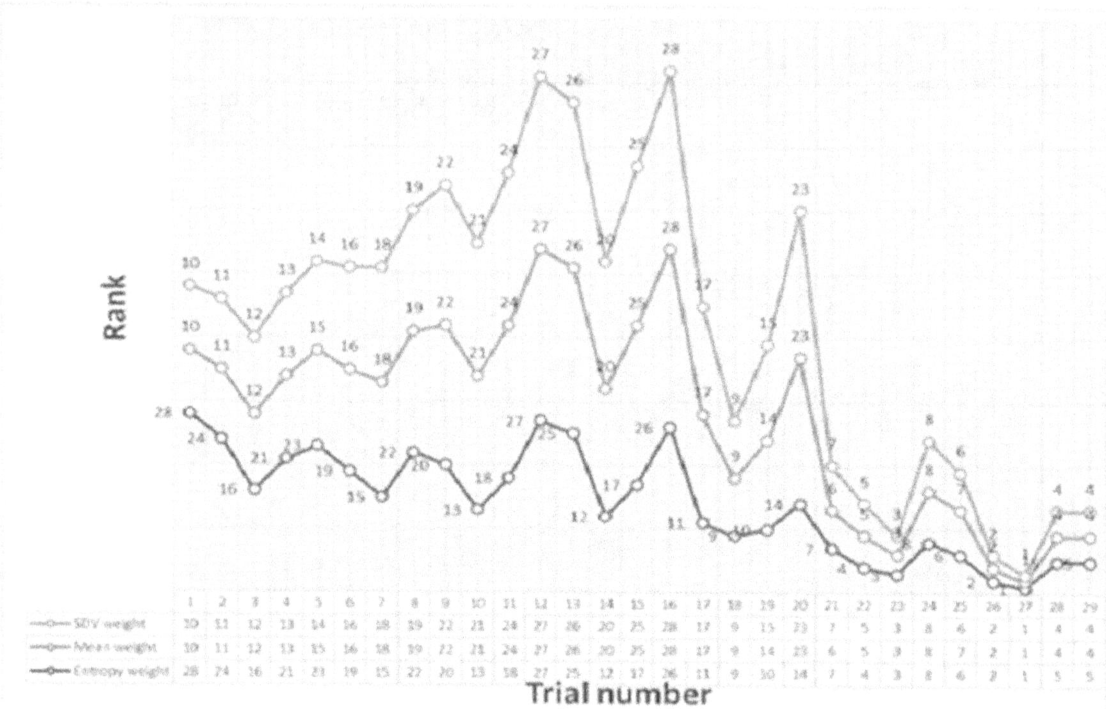

CONCLUSION

The presented article shows a general Topsis, Modified Entropy Weighted-Topsis and SDV Weighted-Topsis MCDM approach to optimize the machining parameters in EDM. In a probe, the copper tool used to study the machining efficiency & effectiveness in ASTM A681 material. The recorded responses were Material removal rate (MRR) and Surface Roughness (SR). The study portrays the effectiveness of discharge current, discharge energy and discharge duration with combined effect of Modified weighted Topsis MCDM in EDM. Based on the above data of the concluded experiment and estimated optimize sample, we found that-

- The mean weighted Topsis, Modified Entropy-Topsis and SDV-Weighted Topsis were successfully implemented with a well-defined weight matrix, PID, NID, and Closeness-Coefficient.
- Analysis of each Responses with different method were measured and estimated using set of equations. From these, we found that the 27th set of input array is the best optimal set for high MRR and low SR on basis of rank and closeness coefficient.
- Whereas the 1th array set for Entropy-Topsis and 16th set for mean weighted & SDV weighted Topsis results the worst performance and least suitable set for optimization.
- From performance characteristic, both SDV & mean topsis yields same output.

REFERENCES

Diyaley, S., Shilal, P., Shivakoti, I., Ghadai, R. K., & Kalita, K. (2017). PSI and TOPSIS based selection of process parameters in WEDM. *Periodica Polytechnica Mechanical Engineering*, *61*(4), 255–260. doi:10.3311/PPme.10431

Fenggou, C., & Dayong, Y. (2004). The study of high efficiency and intelligent optimization system in EDM sinking process. *Journal of Materials Processing Technology*, *149*(1-3), 83–87. doi:10.1016/j.jmatprotec.2003.10.059

Gadalla, A. M., & Tsai, W. (1989). Machining of WC-Co composites. *Material and Manufacturing Process*, *4*(3), 411–423. doi:10.1080/10426918908956301

Ganesh, N., Ghadai, R. K., Bhoi, A. K., Kalita, K., & Gao, X. Z. (2020). An Intelligent Predictive Model-Based Multi-Response Optimization of EDM Process. *Computer Modeling in Engineering & Sciences*, *124*(2), 459–476. doi:10.32604/cmes.2020.09645

Giridharan, A., & Samuel, G. L. (2016). Analysis on the effect of discharge energy on machining characteristics of wire electrical discharge turning process. *Proceedings of the Institution of Mechanical Engineers. Part B, Journal of Engineering Manufacture*, *230*(11), 2064–2081. doi:10.1177/0954405415615732

Gostimirovic, M., Kovac, P., Sekulic, M., & Skoric, B. (2012). Influence of discharge energy on machining characteristics in EDM. *Journal of Mechanical Science and Technology, 26*(1), 173-179.

Gostimirovic, M., Pucovsky, V., Sekulic, M., Radovanovic, M., & Madic, M. (2018). Evolutionary multi-objective optimization of energy efficiency in electrical discharge machining. *Journal of Mechanical Science and Technology*, *32*(10), 4775–4785. doi:10.100712206-018-0925-y

Keskin, Y., Halkacı, H. S., & Kizil, M. (2006). An experimental study for determination of the effects of machining parameters on surface roughness in electrical discharge machining (EDM). *International Journal of Advanced Manufacturing Technology*, *28*(11-12), 1118–1121. doi:10.100700170-004-2478-8

Li, W., & Kara, S. (2015). Characterising energy efficiency of electrical discharge machining (EDM) processes. *Procedia Cirp*, *29*, 263–268. doi:10.1016/j.procir.2015.01.039

Lin, Y. C., Chen, Y. F., Lin, C. T., & Tzeng, H. J. (2008). Electrical discharge machining (EDM) characteristics associated with electrical discharge energy on machining of cemented tungsten carbide. *Materials and Manufacturing Processes*, *23*(4), 391–399. doi:10.1080/10426910801938577

Mahajan, R., Krishna, H., Singh, A. K., & Ghadai, R. K. (2018, June). A Review on Copper and its alloys used as electrode in EDM. *IOP Conference Series. Materials Science and Engineering*, *377*(1).

Phipon, R., Shivakoti, I., Kalita, K., Kibria, G., Sharma, A., & Ghadai, R. K. (2020). Laser beam micro engraving on silicon carbide. *Materials and Manufacturing Processes*, 1–11.

Rebelo, J. C., Dias, A. M., Kremer, D., & Lebrun, J. L. (1998). Influence of EDM pulse energy on the surface integrity of martensitic steels. *Journal of Materials Processing Technology*, *84*(1-3), 90–96. doi:10.1016/S0924-0136(98)00082-X

Sánchez, J. A., Izquierdo, B., Ortega, N., Pombo, I., Plaza, S., & Cabanes, I. (2009). Computer simulation of performance of electrical discharge machining operations. *International Journal of Computer Integrated Manufacturing*, *22*(8), 799–811. doi:10.1080/09511920902741125

Singh, A. K., Mahajan, R., Tiwari, A., Kumar, D., & Ghadai, R. K. (2018). Effect of dielectric on electrical discharge machining: A review. *Journal of Marine Science and Engineering*, *377*, 012184.

Chapter 8
An Integrated CRITIC-MARCOS Technique for Analysing the Performance of Steel Industry

Rishi Dwivedi
Xavier Institute of Social Service, India

Kanika Prasad
National Institute of Technology, Jamshedpur, India

Prashant Kumar Jha
Xavier Institute of Social Service, India

Shamvavi Singh
Xavier Institute of Social Service, India

ABSTRACT

Steel has played critical role in the expansionist ambition of successful aristocrats and laying the foundation of enviable empires. Modern economies are built on strong infrastructure, communication, and transport networks. Industries, like automobiles, consumer durables, real estate, cannot be imagined without steel. Its synonymy with growth of an economy can be gauged by the fact that matrices, like per capital steel consumption and its contribution to the GDP, are considered as parameters of economic development. It is one of the most tracked industries by the investors, analysts, and financial institutions. In this chapter, a novel multi-criteria decision-making tool while integrating measurement alternatives and ranking according to compromise solution (MARCOS) and criteria importance through intercriteria correlation (CRITIC) methods is developed for evaluation of steel organisations which are constituents of BSE 200 index. The results derived from the implementation of integrated MARCOS-CRITIC model aid diverse stakeholders in making informed investment decisions.

DOI: 10.4018/978-1-7998-7206-1.ch008

INTRODUCTION

Steel has a major influence on the vitality of human life and is equally crucial for the evolution of any modern economy. It is one of the most important, multi-functional and versatile product for the modern world. Steel finds a wide variety of application ranging from construction, industrial machinery to consumer products. Steel industry plays a decisive role in the development process of a country. It basically provides a base for other industries, like transportation, manufacturing, automobile, construction etc. as steels' products are used as the input for these industries. The level of per capita consumption of steel is treated as an important index for level of socio-economic development and standard of living of the people in any country. It is one of the most important industries in India too and plays an important role in strengthening the economy. As per a report published by Indian Steel Association the Indian steel industry contributes nearly 2% to the country's gross domestic product (GDP) with an output multiplier of 1.4 on GDP and employment multiplier of 6.8 in year 2019. The production and consumption of steel has increased in the recent years. According to a report of India Brand Equity Foundation, India's steel production capacity has expanded to 137.975 million tons in financial year 2019 and is likely to play a major role in global economy. India replaced Japan to become the world's second largest steel producer in 2019, with crude steel production of 111.2 million tons. India's finished steel consumption grew at a CAGR of 7.5% during 2008-2019 to reach 97.54 million tons. Thus, the importance of steel industry in Indian economy is conspicuous as it has major impact on the GDP of the country and is equally vital for the development of other industries and the mankind. Therefore, it becomes imperative to understand the various factors germane to steel industry which impact its performance and efficiency. The performance evaluation of steel industry can be carried out using multiple parameters deploying different financial ratios, such as profitability, liquidity, solvency, debt-coverage etc. But, each ratio has its own importance. For example, profitability ratios represent operational performance of a company, i.e. how profitable company is. It also represents how profitably owner's funds have been utilized in the company. On the other hand, debt-coverage ratio is a measurement of ability of a company to meet the debt obligations over the course of one year. It is noticed that different parameters throw up different performance score for the same organization. It implies that an enterprise ranked first on current ratio may be having different rank based on debt equity ratio. Therefore, it becomes difficult to determine which organization is doing better if all parameters are taken into consideration. Moreover, in order to measure the performance of the company in a comprehensive way all important financial ratios need to be included in performance evaluation system. Thus, to solve this problem a multi-criteria decision making (MCDM) tool is needed. Therefore, this paper for the very first time implements a novel integrated technique based on Measurement Alternatives and Ranking according to COmpromise Solution (MARCOS) and CRiteria Importance through Intercriteria Correlation (CRITIC) methods to analyze the performance of the steel companies listed in benchmark BSE 200 index. CRITIC method finds objective criteria weights by eliminating decision makers' effect on the decision-making process. MARCOS method provides the ranking of alternatives, and helps managers and all other decision makers to address the problem easily and solve it by applying the ranking of alternatives. This integrated MARCOS-CRITIC method will help the policy makers, stakeholders, financial analyst and investors to know which steel organization is performing superior on all parameters. The proposed model is very flexible and simple.

LITERATURE REVIEW

Dvorkin and Toscano (2003) analyzed the finite element models for simulating the service performance of the tubular steel products used in the oil industry. Singh et al. (2007) presented a method for development of composite sustainability performance index (CSPI) that addressed the sustainable performance of steel industries along all the three pillars of sustainability - economic, environmental and societal. Liu et al. (2011) collected, classified and evaluated the disclosed information of listed steel companies and also analyzed the level of their environmental disclosure. Rout et al. (2013) studied the effective implementation of ISO 50000 in two different steel plants and examined the performance of the system in organization. Ma et al. (2014) proposed circular economy efficiency composite index (CEECI) values to estimate circular economy performance of Wu'an Iron and Steel Group (WISG). Wu et al. (2017) proposed two-stage data envelopment analysis (DEA) model for efficiency evaluation of sustainable manufacturing. The model was applied to wastewater recycling and reusing in the main iron and steel makers in China. Nkonyana et al. (2019) employed predictive analysis to solve the complex challenges faced in steel industry.

Diakoulaki et al. (1995) proposed a method for determination of objective weights which was based on the quantification of two fundamentals notions of MCDM, the contrast intensity and the conflicting character of the evaluation criteria. Madic and Radovanović (2015) introduced an integrated MCDM method employing range of value (ROV) and CRITIC techniques for solving the non-traditional machining processes selection problems. Adalı and Işık (2017) designed a combined CRITIC-multi attribute utility theory method in contract manufacturer selection problem. Ghorabaee et al. (2018) developed a new hybrid MCDM approach to deal with the evaluation process in the fuzzy environment. Fuzzy extensions of the step-wise weight assessment ratio analysis (SWARA) and CRITIC methods for determining subjective and objective weights of criteria were proposed. Tuş and Adalı (2019) applied the new combined decision-making approach based on CRITIC and weighted aggregated sum product assessment (WASPAS) methods for the time and attendance software selection problem of the private hospital. Peng et al. (2020) evaluated 5G industry based on CRITIC method with score function.

Stević et al. (2020) proposed a novel MCDM model called MARCOS technique for sustainable supplier selection in private healthcare industry. Stanković et al. (2020) developed a new fuzzy MARCOS model for traffic risk assessment. Stević and Brković (2020) designed a methodology based on MARCOS technique to evaluate human resource of a transport organization. Chakraborty et al. (2020) developed D-MARCOS approach to choose the best performing supplier in a leading Indian iron and steel making industry based on seven selective evaluation criteria and opinions of three decision makers. Badi and Pamucar (2020) implemented hybrid Grey theory-MARCOS methods for decision-making regarding the selection of suppliers in the Libyan Iron and Steel Company (LISCO) to help it compete.

From the reviewed literature it can be understood that a good amount of research work has already been carried out by the past researchers on the financial performance evaluation of various industries using different MCDM tools. Moreover, a new MCDM method, MARCOS model is successfully applied in some industries for supplier selection and human resource assessment. Therefore, this paper implements an almost unexplored MCDM technique i.e. MARCOS method for performance evaluation of Indian steel organizations on the basis of financial parameters. These enterprises are selected from a benchmarked index, i.e. BSE 200. A focus group is formed to select key financial ratios based on which the steel organizations in BSE 200 are evaluated. CRITIC method is applied at the first stage to compute the relative weights or priority of the 22 identified financial parameters. Afterwards, MARCOS method is implemented to find out the comparative efficiency of the selected organizations. The results derived from the implementation of integrated CRITIC-MARCOS technique can help the assorted stakeholders to understand the long-term sustenance of the identified steel organizations.

CRITIC METHOD

Criteria weights are affected as much from characteristics of the criteria as from subjective point of view of the decision makers. Such subjective weighting of the criteria is usually shaped by the decision makers experience, knowledge and perception of the problem. However this leads to doubt about reliability of the results. To overcome such problems, objective weighting approaches are used. CRITIC method finds objective criteria weights by eliminating decision makers' effect on the decision making process. At the same time CRITIC technique does not separate the criteria as beneficial and non-beneficial. Besides, it tries to reveal the intensity of the contrast in the structure of the decision making problem. To determine the criteria contrast, the standard deviation of normalized criterion values by columns and the correlation coefficients of all pairs of columns are used. Moreover, CRITIC technique belongs to the class of correlation methods and is based on the analytical examination of decision matrix to determine the information contained in the criteria by which the alternatives are evaluated.

CRITIC method needs a few application steps, but initially it is assumed that there is a set of m feasible alternatives, $A_i(i=1,2,3,\ldots,m)$ and n evaluation criteria $C_j(j=1,2,3,\ldots,n)$ in the problem. The application of CRITIC method is presented in the following steps:

Step 1: The initial decision matrix $X = \left[x_{ij}\right]_{mxn}$ is formed in this stage. It shows the performance of different alternatives with respect to various criteria, where x_{ij} is the performance measure of ith alternative with respect to j^{th} criterion, m is the total number of alternatives and n is the total number of criteria.

Step 2: Decision matrix is normalized using the following equation:

$$X_{ij} = \frac{x_{ij} - x_{ij}^{min}}{x_{ij}^{max} - x_{ij}^{min}} \tag{1}$$

Whereas, $x_{ij}^{max} = \max\left(x_{ij}, i = 1,\ldots..,m\right)$ and $x_{ij}^{min} = \min\left(x_{ij}, i = 1,\ldots..,m\right)$

X_{ij} is the normalized performance value of ith alternative on j^{th} criterion. Here, it should be noted that normalization does not take into account the type of criteria.

Step 3: The criteria weights are determined by considering both standard deviation of the criterion and its correlation between other criteria. The standard deviation of normalized criterion values by columns and the correlation coefficients of all pairs of columns are used to determine the criteria contrast. The weight of the jth criterion (W_j) is obtained as:

$$W_{j=} \frac{C_j}{\sum_{i=1}^{n} C_j} \tag{2}$$

Whereas C_j is the quantity of information contained in j^{th} criterion determined as:

$$C_j = \sigma_j \sum_{i=1}^{m} (1 - r_{ij}) \tag{3}$$

Where σ_j standard deviation of the jth criterion and r_{ij} is the correlation coefficient between j^{th} and i^{th} criteria. For a criterion the high standard deviation and low correlation with the other criteria mean that given criterion weight is high. Because the higher value of C_j implies a greater amount of information that is obtained from the given criterion.

Analyzing the Factors Impacting the Performance of Steel Organisation Using CRITIC Method

There are numerous benchmark indices in India on which shares of assorted companies are traded. Here, BSE 200 is selected as the benchmark index from where steel organizations are derived for ultimate evaluation. It implies that steel companies listed on BSE 200 are considered as the alternatives in this research work. These are the steel organization for which sustainability assessment is carried out. Next, two executives from steel enterprises and two experts from the related field are selected to form a focus group. The evaluation criteria chosen in this research work is identified in consultation with this focus group. Past researches are also referred to select the evaluation parameters. Financial factors, like valuation, profitability, liquidity, solvency, management efficiency, debt-coverage and profit and loss account ratios are considered here for examining the performance of selected steel enterprises. The requisite data of the steel enterprises listed on BSE 200 are compiled from the websites, such as www.moneycontrol.com and www.money.rediff.com for the financial year 2018-19. A brief description of all the selected parameters is mentioned below:

1. **Earnings per Share (EPS) (in Rs):** EPS serves as an indicator of company's profitability and is a major factor determining company's stock price. A higher EPS indicates more value because investors will pay more for a company with higher profits.
2. **Sales (in Rs):** A sale is basically a transaction done between two or more parties, typically a buyer and a seller, in which goods or services are exchanged for money or other assets.
3. **Face Value (in Rs):** Face value is a financial term used to describe the nominal value of a security, as stated by its issuer and it is the value printed on the face of a stock, bond, or other financial instrument or document.
4. **Return on Average Equity (ROAE):** It is a financial ratio that measures the profitability of a company in relation to the average shareholders' equity. It is important for analysts to confirm high ROAE measures with other return ratios to ensure a growing ROAE is due to growing sales and improved productivity instead of growing debt.
5. **Operating Profit Per Share (in Rs):** It is a profitability ratio that reflects the operating profit of a company produces from its operation divided by the outstanding shares of its common stock. Higher the operating profit per share, the more profitable it is considered.
6. **Net Operating Profit Per Share (in Rs):** It is the ratio of the profit of a company which is left after subtracting for cost of goods sold, operating expenses, interest and taxes to the outstanding

shares of its common stock. Higher the value of net operating profit per share, more beneficial it is to the organization.

7. **Operating Profit Margin (%):** It measures what percentage of total revenue is made up of operating income. This ratio is important for both creditors and investors because it helps to show how strong and profitable a company's operations are.
8. **Gross Profit Margin (%):** It indicates how much profit a company makes after paying off its cost of goods sold. The higher the profit margin, the more efficient a company is.
9. **Net Profit Margin (%):** It indicates how much net income a company makes with total sales achieved. A higher net profit means that a company is more efficient at converting sales into actual profit.
10. **Return on Capital Employed (ROCE) (%):** ROCE helps to understand how well a company is generating profits from its capital. A higher ROCE shows higher percentage of the company's value can ultimately be returned as profit to stockholders.
11. **Return on Net Worth (RONW) (%):** RONW indicates the efficiency of an organization's management. Its increasing percentage indicates the prudent use of shareholder's money and reflects higher efficiency.
12. **Return on Long Term Fund (%):** It provides a test of profitability related to the sources of long term funds. It reflects the earning capacity of the firm.
13. **Current Ratio:** The current ratio is a commonly used to measure of short-run solvency, the ability of the firm to meet its short-term liabilities. A low ratio indicates that a firm may not be able to pay its future obligations in time.
14. **Quick Ratio:** The quick or acid-test ratio is a more rigorous test of short-run solvency than the current ratio. It gives the relationship between the liquid assets and current liabilities. The higher the ratio, better the company's liquidity and financial health is.
15. **Debt-Equity (D/E) Ratio:** D/E ratio is the ratio of a company's to total liabilities by its shareholder equity. It is used to evaluate a company's financial leverage. Low D/E ratio shows a lower amount of financing by debt via lenders, versus funding through equity via shareholders.
16. **Interest Cover:** It measures the ability of a company to pay the interest on its outstanding debt. Higher interest coverage ratio indicates stronger financial health and shows organization is not vulnerable to volatile interest rates.
17. **Inventory Turnover Ratio:** It is the ratio showing how many times a company has sold and replaced inventory during a given period.
18. **Debtors' Turnover Ratio:** It is also called as receivables turnover ratio and shows how quickly the credit sales are converted into the cash. This ratio measures the efficiency of an organization in managing and collecting the credit issued to customers. Higher the debtors' turnover ratio better is the credit management of the organization.
19. **Investments Turnover Ratio:** It determines the revenues produced by a business to its debt and equity. High investment turnover ratio means that the company is efficient in its use of resources. This results in a higher value of the shareholders' investment.
20. **Total Assets Turnover Ratio:** This ratio measures the value of a company's sales relative to the value of its assets. Higher asset turnover ratio indicates enterprise's efficiency in generating the revenue from its assets.
21. **Number of Days in Working Capital (NWC) (in Days):** NWC is the very important indicator of efficient working capital management. Lower the days of working capital, better is the efficiency

of working capital. It basically expresses how much of net operating working capital is invested in acquiring the sales.

22. **Material Cost Composition (MCC) (in Rs):** MCC is the cost of materials used to manufacture a product and these costs are acquired by material resources for the necessary business to provide the services.

To determine the weightage or priority of identified parameters, initial decision matrix is formed while incorporating six steel companies listed on BSE 200 index as alternatives. The name of the steel enterprises listed on BSE 200 index is Tata Steel, Hindalco, JSW Steel, Jindal Steel and Power, NMDC Limited and SAIL. After initial decision matrix is formed while enlisting the six enterprises as alternatives and identified 21 parameters as criteria, normalization of initial decision matrix is carried out using Equation (1). Next, while employing Equations 2 and 3, weights of different identified parameters for evaluating steel organization are estimated as shown in Table 1.

MARCOS MODEL

MARCOS method is a powerful and robust tool for optimizing multiple goals. Compared with other methods, this method is simple, effective and easy to sort and optimize the process. MARCOS refreshes the MCDM domain by introducing an algorithm for analyzing the relationship between alternatives and reference points. The MARCOS method integrates three starting points to provide a robust decision; defining reference points (ideal and anti-ideal values), determining the relationship between alternatives and ideal/anti-ideal values, and defining the utility degree of alternatives in relation to the ideal and anti-ideal solution. Also, the results obtained by the MARCOS method are more reasonable due to the fusion of the results of the ratio approach and reference point sorting approach. In this section, the algorithm of MARCOS method is presented. The MARCOS method is based on defining the relationship between alternatives and reference values (ideal and anti-ideal alternatives). On the basis of the defined relationships, the utility functions of alternatives are determined, and compromise ranking is made in relation to ideal and anti-ideal solutions. Decision preferences are defined on the basis of utility functions. Utility functions represent the position of an alternative with regard to an ideal and anti-ideal solution. The best alternative is the one that is closest to the ideal and at the same time furthest from the anti-ideal reference point. The MARCOS method is performed through the following steps:

Step 1: Formation of an initial decision-making matrix. Multi-criteria models include the definition of a set of n criteria and m alternatives. In the case of group decision-making, a set of r experts should be formed to evaluate alternatives according to the criteria. In the case of group decision-making, expert evaluation matrices are aggregated into an initial group decision-making matrix.

Step 2: Formation of an extended initial matrix. In this step, the extension of the initial matrix is performed by defining the ideal (AI) and anti-ideal (AAI) solution.

Table 1. Weights of different identified parameters

Parameter	Weight
EPS (C1)	0.04
Sales (C2)	0.06
Face value (C3)	0.07
ROAE (C4)	0.04
Operating profit per share (C5)	0.05
Net operating profit per share (C6)	0.05
Operating profit margin (C7)	0.03
Gross profit margin (C8)	0.03
Net profit margin (C9)	0.04
ROCE (C10)	0.03
RONW (C11)	0.04
Return on long term fund (C12)	0.03
Current ratio (C13)	0.04
Quick ratio (C14)	0.04
D/E ratio (C15)	0.07
Interest cover (C16)	0.04
Inventory turnover ratio (C17)	0.04
Debtors turnover ratio (C18)	0.05
Investment turnover ratio (C19)	0.04
Total assets turnover ratio (C20)	0.06
NWC (C21)	0.04
Material cost composition (C22)	0.07

$$X = \begin{array}{c} \\ AAI \\ A_1 \\ A_2 \\ \cdots \\ A_m \\ AI \end{array} \begin{array}{c} \begin{array}{cccc} C_1 & C_2 & \cdots & C_n \end{array} \\ \begin{bmatrix} x_{aa1} & x_{aa2} & \cdots & x_{aan} \\ x_{11} & x_{12} & \cdots & x_{1n} \\ x_{21} & x_{22} & \cdots & x_{2n} \\ \cdots & \cdots & \cdots & \cdots \\ x_{m1} & x_{m2} & \cdots & x_{mn} \\ x_{ai1} & x_{ai2} & \cdots & x_{ain} \end{bmatrix} \end{array} \tag{4}$$

The anti-ideal solution (AAI) is the worst alternative while the ideal solution (AI) is an alternative with the best characteristic. Depending on the nature of the criteria, AAI and AI are defined by applying equations (5) and (6):

$$AAI = \min_i x_{ij} \text{ if } j \in B \text{ and } \max_i x_{ij} \text{ if } j \in C \tag{5}$$

$AI = \max_i x_{ij}$ if $j \in B$ and $\min_i x_{ij}$ if $j \in C$ (6)

Where, "B" represents a benefit group of criteria, while "C" represents a group of cost criteria.

Step 3: Normalization of the extended initial matrix (X). The elements of the normalized matrix

$N=[n_{ij}]mxn$ are obtained by applying equations (7) and (8):

$$n_{ij} = \frac{x_{ai}}{x_{ij}} \text{ if } j \in C \quad (7)$$

$$n_{ij} = \frac{x_{ij}}{x_{ai}} \text{ if } j \in B \quad (8)$$

Where elements x_{ij} and x_{ai} represent the elements of the matrix X.

Step 4: Determination of the weighted matrix $V=[v_{ij}]mxn$. The weighted matrix V is obtained by multiplying the normalized matrix N with the weight coefficients of the criterion w_j, equation (9).

$$v_{ij} = n_{ij} \times w_j \quad (9)$$

Step 5: Calculation of the utility degree of alternatives K_i. By applying equations (10) and (11), the utility degrees of an alternative in relation to the anti-ideal and ideal solution are calculated.

$$K_i^- = \frac{S_i}{S_{aai}} \quad (10)$$

$$K_i^+ = \frac{S_i}{S_{ai}} \quad (11)$$

Where $S_i(i=1, 2. .. m)$ represents the sum of the elements of the weighted matrix V, equation (12).

$$Si = \sum_{i=1}^{n} V_{ij} \quad (12)$$

Step 6: Determination of the utility function of alternatives $f(K_i)$. The utility function is the compromise of the observed alternative in relation to the ideal and anti-ideal solution. The utility function of alternatives is defined by equation (13).

$$f(K_i)=\frac{K_i^+ + K_i^-}{1+\frac{1-f(K_i^+)}{f(K_i^+)}+\frac{1-f(K_i^-)}{f(K_i^-)}} \tag{13}$$

Where $f(K_i^-)$ represents the utility function in relation to the anti-ideal solution, while $f(K_i^+)$ represents the utility function in relation to the ideal solution. Utility functions in relation to the ideal and anti-ideal solution are determined by applying equations (14) and (15)

$$f(K_i^-)=\frac{K_i^+}{K_i^+ + K_i^-} \tag{14}$$

$$f(K_i^+)=\frac{K_i^-}{K_{i+}^+ K_i^-} \tag{15}$$

Step 7: Ranking the alternatives. Ranking of the alternatives is based on the final values of utility functions. It is desirable that an alternative has the highest possible value of the utility function.

Application of Integrated CRITIC-MARCOS Method for Analyzing the Performance of Steel Organizations

Here the steel organizations are selected as the companies listed on Sensex 200 as mentioned earlier. The steps of the MARCOS method as discussed above in the previous section are implemented in order to find out the ranking of the alternatives. Firstly the extended decision matrix is formed while using Equations 4, 5 and 6. Equations 7 and 8 are then employed to normalize the extended decision matrix. Subsequently, the weighted-normalized matrix is calculated using the Equation 9. Equation 10, 11 and 12 are applied in the next step to compute the utility degree of alternatives. The last stage of application of MARCOS method is estimation of utility function which is calculated while using equations 13, 14 and 15. The ranking of the alternatives are carried out on the basis of the obtained utility function score $f(K_i)$ as shown in Table 2.

Table 2. Ranking of the steel organization

Steel Organization	$f(K_i)$	RANK
Tata Steel	0.081	2
Hindalco	0.064	3
JSW Steel	0.020	6
Jindal Steel and Power	0.041	4
NMDC Limited	1.038	1
SAIL	0.039	5

Application of MARCOS-CRITIC model offers ranking as depicted in Table 2. Evaluation based on the 22 parameters pertaining to the chosen companies, NMDC limited comes out on the top with highest utility $f(K_i)$ value of 1.038 while TATA STEEL ranks distant second with $f(K_i)$ value of 0.081. HINDALCO, JSW Steel, Jindal Steel and Power and SAIL occupy 3rd, 4th, 5th and 6th spot in the ranking based on $f(K_i)$ value. From the ranking derived in Table 2, it is observed that NMDC and TATA STEEL are the top two companies; in that order, in terms of performance as captured by $f(K_i)$ value.

FUTURE RESEARCH DIRECTIONS

The future scope of this current research work may include implementation of this integrated CRITIC-MARCOS method for the assessment all the steel organizations in India with respect to financial parameters in order to identify the superior performing steel enterprise of entire nation. This will also assist the various stakeholders to understand the significance of various parameters in terms of the sustainable growth which in turn will further help them to achieve long term sustainability.

CONCLUSION

Steel industry and its value to the economy and civilization cannot be appreciated enough. The industry has carried the weight of majority of strides humankind have made since centuries. It is very difficult to imagine the vast expanse of infrastructure, rapid rise of automobile sector in the last century, railroad extension, telecom coverage, utility provision etc without steel in the picture. Very few materials can boast of both its timelessness and indispensability as steel can. Not only it is difficult to find argument against focus on steel industry for its role in many future innovations and technology, but also why this is one of the most tracked industry on the radar of investors. Investors and analysts have developed many tools over the years to evaluate the relative attractiveness of companies among its peers in the industry. Many parameters have been developed thus far to enable analysts taking an investment call; or recommending for clients. Some parameters have yielded good visibility, and some have turned out to be not so useful. In this work a novel approach, called MARCOS-CRITIC model, has been applied for evaluation of parameters consistent with their respective importance. The application of MARCOS-CRITIC model reduces the biases proffered by parameters taken individually or in combination. The companies belonging to steel industry have been selected from BSE 200 index. The outcome of the model discussed is very interesting and it is likely to be very useful for analysts, researchers and investors.

REFERENCES

Adalı, E. A., & Işık, A. T. (2017). CRITIC and MAUT methods for the contract manufacturer selection problem. *European Journal of Multidisciplinary Studies*, *2*(5), 93–101. doi:10.26417/ejms.v5i1.p93-101

Badi, I., & Pamucar, D. (2020). Supplier selection for steelmaking company by using combined Grey-MARCOS methods. *Decision Making: Applications in Management and Engineering*, *3*(2), 37–48.

Chakraborty, S., Chattopadhyay, R., & Chakraborty, S. (2020). An integrated D-MARCOS method for supplier selection in an iron and steel industry. *Decision Making: Applications in Management and Engineering*, *3*(2), 49–69.

Diakoulaki, D., Mavrotas, G., & Papayannakis, L. (1995). Determining objective weights in multiple criteria problems: The critic method. *Computers & Operations Research*, *22*(7), 763–770. doi:10.1016/0305-0548(94)00059-H

Dvorkin, E. N., & Toscano, R. G. (2003). Finite element models in the steel industry: Part II: Analyses of tubular products performance. *Computers & Structures*, *81*(8-11), 575–594. doi:10.1016/S0045-7949(02)00403-0

Ghorabaee, M. K., Amiri, M., Zavadskas, E. K., & Antucheviciene, J. (2018). A new hybrid fuzzy MCDM approach for evaluation of construction equipment with sustainability considerations. *Archives of Civil and Mechanical Engineering*, *18*(1), 32–49. doi:10.1016/j.acme.2017.04.011

Liu, Z. G., Liu, T. T., McConkey, B. G., & Li, X. (2011). Empirical analysis on environmental disclosure and environmental performance level of listed steel companies. *Energy Procedia*, *5*, 2211–2218. doi:10.1016/j.egypro.2011.03.382

Ma, S. H., Wen, Z. G., Chen, J. N., & Wen, Z. C. (2014). Mode of circular economy in China's iron and steel industry: A case study in Wu'an city. *Journal of Cleaner Production*, *64*, 505–512. doi:10.1016/j.jclepro.2013.10.008

Madic, M., & Radovanović, M. (2015). Ranking of some most commonly used nontraditional machining processes using ROV and CRITIC methods. *UPB Sci. Bull., Series D*, *77*(2), 193–204.

Nkonyana, T., Sun, Y., Twala, B., & Dogo, E. (2019). Performance evaluation of data mining techniques in steel manufacturing industry. *Procedia Manufacturing*, *35*, 623–628. doi:10.1016/j.promfg.2019.06.004

Peng, X., Zhang, X., & Luo, Z. (2020). Pythagorean fuzzy MCDM method based on CoCoSo and CRITIC with score function for 5G industry evaluation. *Artificial Intelligence Review*, *53*(5), 3813–3847. doi:10.100710462-019-09780-x

Rout, M. K., Badgayan, N. D., Pattnaik, S. K., & Rout, M. K. (2013). An Empirical Study on Energy Management Standard (ISO 50000): Effectiveness Performance Evaluation for Steel Plants and Comparison with ISO 9000, ISO 14000, ISO 18000 & ISO. *International Journal of Engineering Research and Technology*, *2*, 2278–0181.

Singh, R. K., Murty, H. R., Gupta, S. K., & Dikshit, A. K. (2007). Development of composite sustainability performance index for steel industry. *Ecological Indicators*, *7*(3), 565–588. doi:10.1016/j.ecolind.2006.06.004

Stanković, M., Stević, Ž., Das, D. K., Subotić, M., & Pamučar, D. (2020). A new fuzzy MARCOS method for road traffic risk analysis. *Mathematics*, *8*(3), 457. doi:10.3390/math8030457

Stević, Ž., & Brković, N. (2020). A Novel Integrated FUCOM-MARCOS Model for Evaluation of Human Resources in a Transport Company. *Logistics*, *4*(1), 4. doi:10.3390/logistics4010004

Stević, Ž., Pamučar, D., Puška, A., & Chatterjee, P. (2020). Sustainable supplier selection in healthcare industries using a new MCDM method: Measurement of alternatives and ranking according to COmpromise solution (MARCOS). *Computers & Industrial Engineering*, *140*, 106231. doi:10.1016/j.cie.2019.106231

Tuş, A., & Adalı, E. A. (2019). The new combination with CRITIC and WASPAS methods for the time and attendance software selection problem. *Opsearch*, *56*(2), 528–538. doi:10.100712597-019-00371-6

Wu, H., Lv, K., Liang, L., & Hu, H. (2017). Measuring performance of sustainable manufacturing with recyclable wastes: A case from China's iron and steel industry. *Omega*, *66*, 38–47. doi:10.1016/j.omega.2016.01.009

Chapter 9
Data-Driven Genetic Programming-Based Symbolic Regression Metamodels for EDM Process

Kanak Kalita
https://orcid.org/0000-0001-9289-9495
Vel Tech Rangarajan Dr. Sagunthala R&D Institute of Science and Technology, India

Ranjan Kumar Ghadai
Sikkim Manipal Institute of Technology, Sikkim Manipal University, Majhitar, India

Dinesh S. Shinde
SVKM's NMIMS, Mukesh Patel School of Technology Management and Engineering, Shirpur, India

Xiao-Zhi Gao
School of Computing, University of Eastern Finland, Finland

ABSTRACT

In this research, a data-driven approach to metamodeling of manufacturing/machining processes is developed. Instead of the conventionally used second-order polynomial regression metamodels, a non-predefined form-free approach is discussed. The highly adaptive metamodeling strategy, called symbolic regression, is carried out by using genetic programming. A central composite design based experimental dataset on electric discharge machining is used as the training and the testing data. Four different process parameters namely (voltage, pulse on time, pulse off time, and current) are used as the independent parameters to quantify three different responses (material removal rate, electrode wear rate, and surface roughness). The performance of the metamodels are evaluated by using various statistical metrics like R^2, MAE, MSE. The performance of the metamodels on the training and testing data is found to be adequate for all the responses.

DOI: 10.4018/978-1-7998-7206-1.ch009

INTRODUCTION

When the conventional machining processes are unable to process the materials, non-traditional machining like electric discharge machining (EDM) are used. EDM produces high accuracy and is applicable for any conductive material. EDM is a non-traditional machining process (NTMP) in which material removal takes place with the help of thermal energy generated spark generated through electricity flow between the electrode tool and workpiece under the envelope of dielectric fluid. Similar to some NTMPs like laser cutting process; it does not require force to remove material. EDM is popular in tool and mold making. EDM can be applied to machine any conductive material, even the hardest of materials such as tungsten carbide. It processes the material without force which makes it suitable for fragile materials too. Turning, cutting, drilling, milling, etc. operations can be performed using EDM. Applied voltage, current, pulse on and off time, tool gap etc. are the controlling parameters of EDM process and material removal rate (MRR), surface roughness, overcut, tool wear rate, electrode wear rate etc. are the machining output characteristics. The process parameters, their interaction and influence on the response are the main concerns of the researchers working in the field of machining of materials using EDM. Some researchers studied the effect of the process controlling parameters of the EDM process and measured the 3D surface topography for the surface quality of machined surface and stated that applied current in the most significant parameter (Ramasawmy & Blunt, 2004). The EDM drilling process parameters of Inconel 718 materials were studied by Kuppan et al. (Kuppan, Rajadurai, & Narayanan, 2008) for selecting the best combination of process parameters aiming maximum MRR and minimum surface roughness. Researchers also studied the effect of change in the electrode conditions in EDM for better outcomes. Pandey et al. (Srivastava & Pandey, 2012) applied the ultrasonic waves to cool the electrode during EDM of high-speed steel and reported that current machining system improved the surface integrity of the workpiece compared to conventional EDM.

During recent developments in the manufacturing industries, the market need is to generate the product with high quality for less time and cost. Achievement of this aim is possible through optimization of the various machining process parameters. Optimization has become a vital engineering tool which gives the best utilization of given resources for maximum possible outcomes. Different techniques used in this context are simulated annealing, particle swarm optimization, ant colony optimization, artificial bee colony etc. It is found in the literature that GA is one of the most preferred optimizers in the optimization of the machining process parameters (Yusup, Zain, & Hashim, 2012). It is based on the survival of fittest principle of genetics, which starts with a basic pool of alternative called population and head towards the best solution. However, to mathematically optimize the process an approximation function of the machining/manufacturing process is needed. This is because in general, no closed-form equations are available for expressing the machining/manufacturing process as a function of the process parameters. In this regard, researchers have relied on metamodeling approaches to form approximate relations between the independent process parameters and the dependent responses. Polynomial regression is perhaps the most widely used metamodeling strategy in machining/manufacturing processes. Metamodels based on evolutionary programming techniques are becoming very popular for optimization of parameters of the machining process. Genetic programming (GP) metamodel can be used to developed relationships between the input process parameters for superior output in case of various engineering applications. This approach significantly reduces computational time and cost (Kalita, Mukhopadhyay, Dey, & Haldar, 2019). Critical engineering problems like turbulent flow in pipe bends can also be solved using the

GP metamodel-based evaluation with understanding the influence of different parameters affecting the system (Narayanan, Joshi, Dutta, & Kalita, 2019).

Zhang et al. (Zhang, et al., 2013) optimize the process parameter of medium speed wire-EDM process in machining SKD11 materials using genetic programming (GP) integrated with back-propagation neural network and reported that the GP approach evolves better results for optimization of material removal rate (MRR) and surface roughness. Garg et al. (Garg & Lam, 2016) utilized the genetic programming to model the multi-response of the EDM process for environmental (energy consumption and dielectric fluid consumption) and manufacturing aspects (tool wear ratio) and concluded that the current approach can establish a relationship between the process parameters for multiple objectives. Marin et al. (Gostimirovic, Rodic, Kovac, Pucovsky, & Savkovic, 2014) applied the GP for modelling of the EDM process parameter for the surface roughness considering process parameters such as pulse duration and current, whereas Salman et al. (Salman & Kayacan, 2008) considered similar study with GP considering EDM process parameters viz., applied current and voltage, time of pulse on, time of pulse off and stated that the GP methodology is most suitable for optimization environment such EDM process parameters and yields best solution compared to other techniques. Somashekhar et al. (Somashekhar, Ramachandran, & Mathew, 2010) used the GP modelling of materials removal rate during micro-EDM process combined with artificial neural network and concluded that the optimization using GP can predict the combination of best process parameters in most precise machining situation like micro-EDM. The machining properties such as MRR, rate of tool wear and the surface characteristics of the powder metallurgy product machined using EDM were studied in the literature, considering variation in rotational speed of the tool, applied current, time of pulse and pressure of the dielectric fluid and the results are used to established non-linear regression models for said machining properties using non-linear Goal programming based on GP to optimise the machining conditions (Kanagarajan, Karthikeyan, Palanikumar, & Davim, 2009). The input process parameters of EDM viz., pulse current, time of pulse and duty factor and output machining characteristics MRR and surface roughness were considered and modelling is established between them, the performance of which had been verified with the help of few real experimental studies. Single and multi-objective optimisation problems have been considered and solved using GP for determination of a combination of process parameters (Maji, Pratihar, & Patra, 2010). Bose et al. (Bose & Pain, 2020) selected best suitable EDM process parameters using GP approach considering process parameters such as applied voltage, current and pulse timing and process outcomes such as MRR and tool wear in machining high chromium high carbon tool steel using titanium nitride coated copper electrode tool and reported that the GP is the effective tool in the critical task like the selection of process parameter combination in EDM. Mandal et al. (Mandal, Pal, & Saha, 2007) attempted to model and optimization of the complicated EDM process using data-driven metamodel programming viz., backpropagation GP algorithm integrated with artificial neural network considering output properties MRR and surface roughness and process control parameters such as input voltage and current, pulse time, etc. Tzeng et al. (Tzeng & Chen, 2013) optimize the EDM process parameters for MRR, the ratio of electrode wear and surface roughness in machining SKD61 using genetic programming with remarks that current GP approach gives better outcomes in terms prediction and validated experimental results than other methods. Padhee et al. (Padhee, Nayak, Panda, Dhal, & Mahapatra, 2012) did parametric optimization of powder mixed EDM process using genetic programming with an objective of minimum surface roughness and maximum MRR during machining EN31 steels with copper electrodes.

Wire-EDM process is used to machine hard material. The controlling parameters of wire-EDM such as wire tension and wire travel were also prime factors in the machining of hard materials. The GP based fuzzy logic controller can be applied to analyse the closed-loop wire tension controlling system with faster response lesser errors (Yan & Fang, 2008). Sharma et al. (Sharma, Khanna, & Gupta, 2015) utilize wire-EDM process with a brass wire as an electrode tool in machining high strength low alloy steel considering controlling process parameters pulse on and off time, the gap between electrode and workpiece, current and tension in the wire for machining overcut produced and reported that GP based overcut model can be used to optimize the control factors with very less computational efforts, time and cost. In their study, Shajan et al. (Kuriakose & Shunmugam, 2005) used a multiple-response regression model to establish a relation between input parameters (Pulse time, applied voltage and current) and output variables (machining speed and surface roughness) using GP in the Wire-EDM process with remarks that the current approach is significant in the optimization process.

From the literature review, it is clear that understanding the effect of process parameters is of paramount importance. In this regard, it is often seen that researchers have tapped the tremendous potential of approximation functions or metamodels that act as predictive models or surrogates to the actual experiment. By carefully building metamodels from a handful of experimental data the economic and computational cost involved in understanding the machining process can be drastically reduced. In this paper, a data-driven and form-free intelligent approach to metamodeling by using the evolutionary power of genetic programming is discussed. By applying a variety of statistical accuracy metrics, the utility and the efficacy of the GP based symbolic regression metamodels is established.

METHODOLOGY

Experimental Details

The experimental data used in this research is taken from Soundhar et al. (Soundhar, Zubar, Sultan, & Kandasamy). In the experiment, a newly developed titanium alloy (Ti-13Zr-13Nb (TZN)) prepared by powder metallurgy process is used for the machining through the EDM process. The powder of Titanium (Ti), Niobium (Nb) and Zirconium (Zr) are used to develop the new Titanium alloy which has a wide industrial application. Hydride dihydride process is used to produce the powder. Graphite and kerosene are used as EDM tool and dielectric respectively for the machining of the newly developed Ti alloy. Various input parameters such as voltage, pulse on time, pulse off time, and current are considered for conducting the experiments. Face centred central composite design has been used for the experimental design. Material removal rate (MRR) and electrode wear rate (EWR) and surface roughness (SR) are the response parameters for the machining of the Ti alloy using EDM. Both MRR and EWR are being calculated by the ratio of weight difference of workpiece and tool with the machining time.

Genetic Programming

Genetic Programming (GP) is a powerful computational intelligence technique that can perform symbolic regression to build explanatory models based on the provided training datasets. It is a highly automated process that requires no manual intervention once the algorithm is started. On the contrary, a polynomial regression carried out by RSM may require the use of additional statistical tests like ANOVA to determine the significance of the polynomial terms in the model. The GP, on the other hand, self-prunes the insignificant terms because of the inherent evolutionary traits.

GP starts with a randomly generated population of candidate solutions called individuals or GP trees. This population is called as generation zero. Each GP tree is made up of two key ingredients — functions and terminals. Mathematical operators like +, -, *, / etc.; trigonometric functions like sin, cos etc., exponential functions, logarithmic function etc. form the functions. Terminals, on the other hand, are comprised of constants and variables. Fitness for each GP tree of the population is calculated. The fitness is calculated using some predefined metrics, like mean squared error or coefficient of determination. In the case of using mean squared error as the metric lower values are considered better. Similarly, in the case of the coefficient of determination, the higher the values, the better. Next, the population in generation zero is improved by using three key genetic operators called selection, crossover and mutation. An additional genetic property called elitism is also used in the present work. Crossover is the process of randomly grafting chosen parts from one GP to another GP tree. This is done by associating a higher probability of selection for the crossover of higher fitness GP trees. A mutation is used to create new GP trees for the new population by randomly altering a small part or node of the selected GP tree. Elitism is the act of copying a small proportion of the fittest GP trees, unchanged into the next generation. This improved population makes up the subsequent generation. This process is repeated until a predefined accuracy on the performance measurement scale or the maximum number of generations is reached.

Metamodel Validation Metrics

The metamodels are evaluated by using certain residue-based statistical parameters. The residue may be defined as the difference between the experimental and the predicted responses. A dependent parameter or output response (y) may be expressed as a function of the independent parameters $\left(x's\right)$ in the following generic form.

$$y = f\left(x_1, x_2, x_3, \ldots\ldots x_k\right) + \varepsilon \tag{1}$$

In eqn. (1), f denotes the approximate response surface, ε is the normally distributed statistical error and k is the maximum number of independent parameters. If the metamodel predicted values are denoted as ($\widehat{y_i}$), then the residues ($\varepsilon i_)$ may be computed as,

$$\varepsilon_i = y_i - \widehat{y_i} \tag{2}$$

The sum of squared error in predictions (SSE) i.e. the sum of squared residues may be calculated as,

$$SSE = \sum_{i=1}^{n} \varepsilon_i^{\,2} \tag{3}$$

where n is the number of sample points.

Similarly, the TSS or total sum of squares may be calculated as,

$$TSS = \sum_{i=1}^{n} \left(y_i - \bar{y}\right)^2 \tag{4}$$

where $\bar{y}$ is the mean of the response y.

The coefficient of determination R^2 can be calculated as

$$R^2 = 1 - \frac{SSE}{TSS} \tag{5}$$

MAE (mean-absolute-error) provides an absolute measure of prediction error in metamodels. MAE is calculated as

$$MAE = \frac{\sum_{i=1}^{n} \left| y_i - \hat{y}_i \right|}{n} \tag{6}$$

Similarly, MSE (mean-squared-error) is calculated as

$$MSE = \frac{\sum_{i=1}^{n} \left(y_i - \hat{y}_i \right)^2}{n} \tag{7}$$

RESULTS AND DISCUSSION

Figure 1 shows the scatter plots of process parameters and output responses. Figure 2 shows the box plots representing the spread of the process parameters and responses. In the case of the process parameters, the distribution is such that there are no data below Q1 and beyond Q3. On the other hand, the spread of the response variables is somewhat normal. A couple of outliers are seen in the MRR and in EWR.

Material Removal Rate

Based on the normalized training data presented in Table 1, the following symbolic regression metamodel for material removal rate is developed. A separate testing data is presented in Table 2.

$$MRR = \left(\frac{\left(\beta_0.I + \beta_1.P_{off}\right) P_{on}.\beta_2}{\beta_3.I + \beta_4.P_{on}} + \beta_5 \right) \tag{8}$$

It is seen from the above equation that the process parameter voltage (V) is auto-pruned by the GP algorithm in expressing the MRR as a function of the process parameters. The values of the coefficients in eqn. 8 are presented in Table 3.

Figure 1. Scatter plots of process parameters and output responses involved in the study

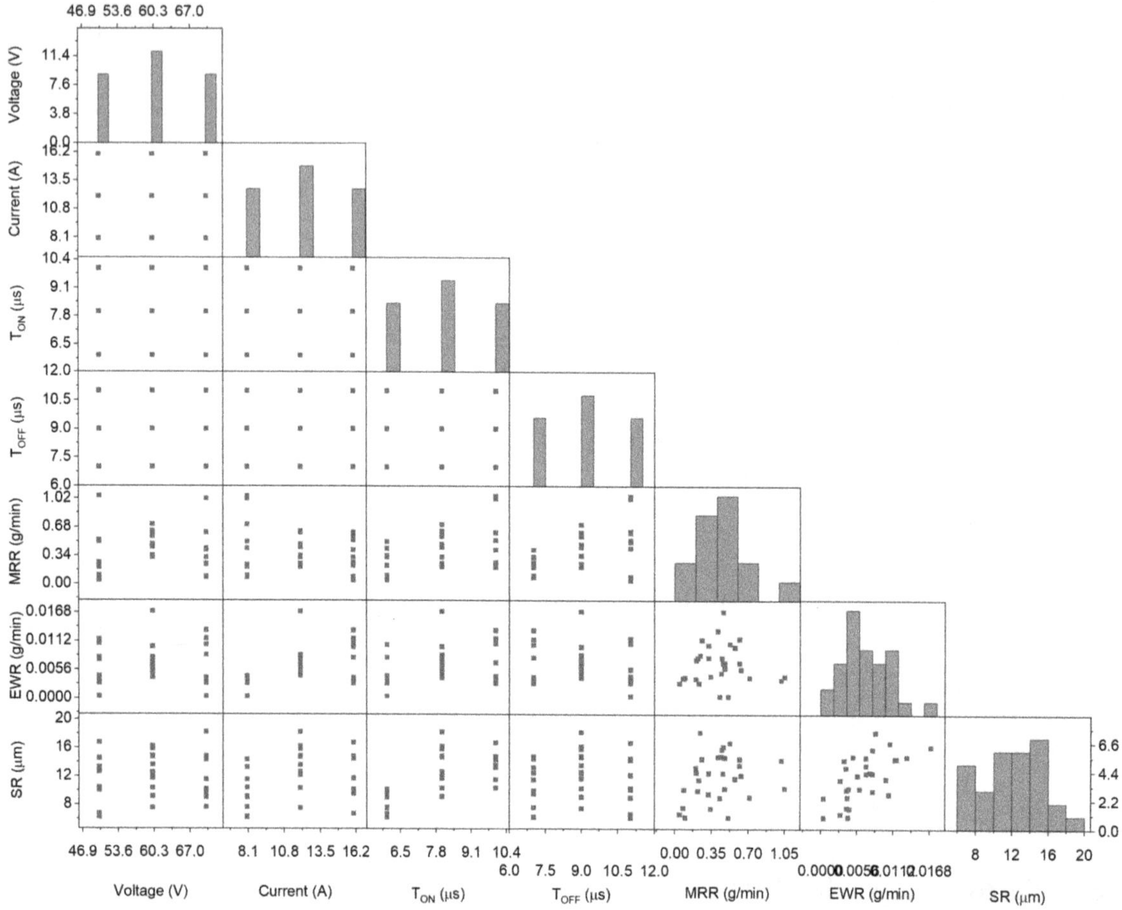

The predictions of the GP metamodel for MRR is compared with the training and testing experimental data in Figure 3. It should be noted that closer the data points are to the identity (diagonal) line in Figure 3, better is the estimation of the metamodel. While data points below the identity line suggest underprediction by GP metamodel, data points above the identity line suggest overprediction. Datapoints on the identity line suggest ideal estimation i.e. the GP metamodel predicts the MRR with 100% accuracy. It is clear from Figure 3., that the MRR prediction capability of the GP metamodel is similar for training and testing data. This is further confirmed by the residue in training and testing data for the MRR GP metamodel shown in Figure 4. The relative error for the MRR GP metamodel on training and testing data is shown in Figure 5. The accuracy and the prediction performance of the MRR GP metamodel is further evaluated by using R^2, MAE, MSE and reported in Table 4. The relative statistical impact (or contribution) of the process parameters on the MRR GP metamodel is evaluated and presented as pie charts in Figure 6. It is seen that P_{off} has the highest statistical impact in the MRR GP metamodel, followed by P_{on} and current. For both the training data and overall data the relative statistical impact (or contribution) of the process parameters on the MRR GP metamodel's predictions is seen to be consistent.

Figure 2. Normalized range of process parameters and output responses

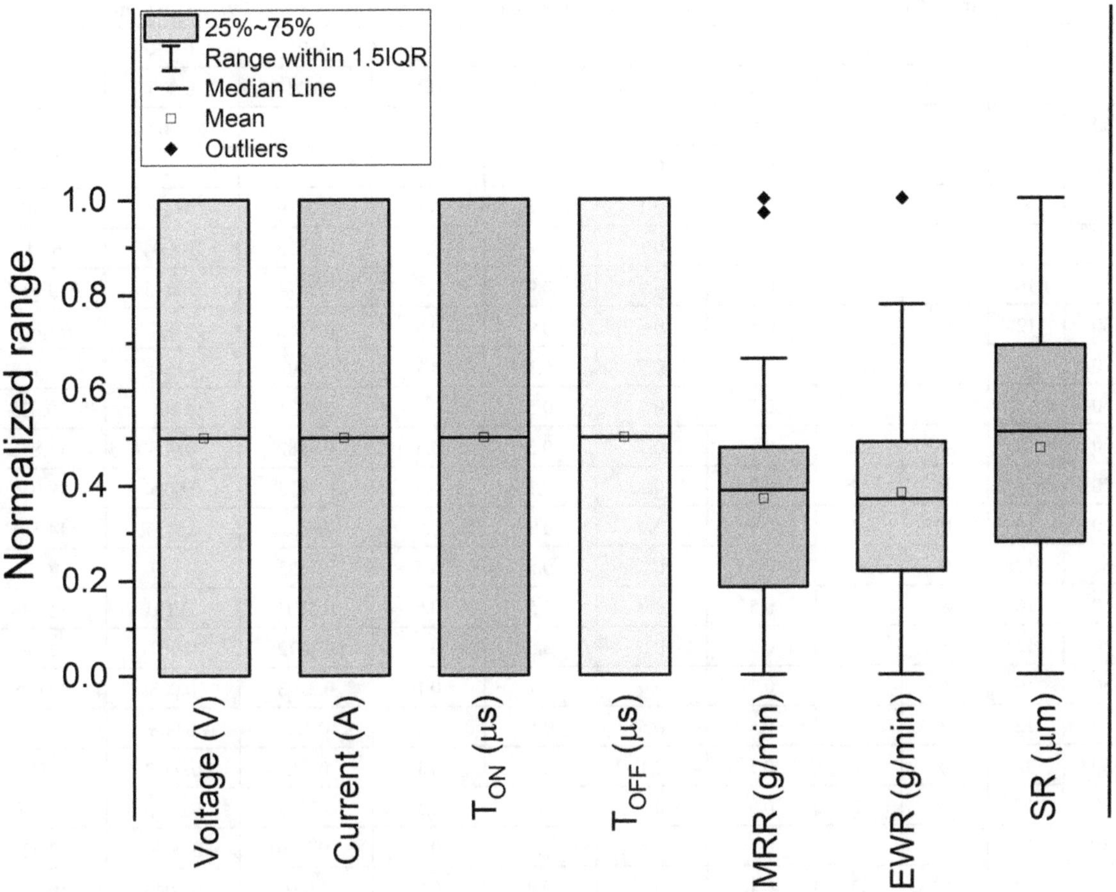

Tool Wear Rate

Using the experimental training data for EWR presented in Table 1, the following relation for EWR is developed using GP based symbolic regression.

$$EWR = \left(\beta_0.I - \beta_1.P_{on} + \frac{\beta_2.P_{off}}{P_{on}} + \frac{\beta_3.I}{P_{on}} + \beta_4 \right) \quad (9)$$

Table 1. Experimental data used for training GP metamodels (Soundhar, Zubar, Sultan, & Kandasamy)

Process parameters				Process parameters (coded)				Output responses		
V	I	P_{on}	P_{off}	V	I	P_{on}	P_{off}	MRR	EWR	SR
70	8	10	7	0.9	0.1	0.9	0.1	0.23	0.003	11.558
60	12	8	9	0.5	0.5	0.5	0.5	0.473	0.007	14.717
60	12	8	9	0.5	0.5	0.5	0.5	0.441	0.005	14.867
70	8	6	7	0.9	0.1	0.1	0.1	0.0789	0.004	7.647
50	8	6	11	0.1	0.1	0.1	0.9	0.5075	0.0004	6.245
50	16	10	7	0.1	0.9	0.9	0.1	0.2574	0.0115	14.514
60	12	10	9	0.5	0.5	0.9	0.5	0.6162	0.007	13.608
70	16	6	11	0.9	0.9	0.1	0.9	0.086	0.004	10.168
60	12	8	9	0.5	0.5	0.5	0.5	0.4731	0.006	10.325
60	12	8	9	0.5	0.5	0.5	0.5	0.4482	0.008	15.851
70	8	6	11	0.9	0.1	0.1	0.9	0.4272	0.0004	9.04
70	16	10	11	0.9	0.9	0.9	0.9	0.616	0.0117	14.514
60	12	8	9	0.5	0.5	0.5	0.5	0.4623	0.017	16.24
60	16	8	9	0.5	0.9	0.5	0.5	0.5707	0.0101	11.728
60	12	8	9	0.5	0.5	0.5	0.5	0.4572	0.007	12.485
50	16	6	7	0.1	0.9	0.1	0.1	0.2193	0.008	10.008
60	12	8	7	0.5	0.5	0.5	0.1	0.322	0.008	12.512
50	8	6	7	0.1	0.1	0.1	0.1	0.099	0.0042	6.301
70	16	6	7	0.9	0.9	0.1	0.1	0.3206	0.0105	9.577
50	12	8	9	0.1	0.5	0.5	0.5	0.205	0.0076	12.629
50	8	10	11	0.1	0.1	0.9	0.9	1.051	0.0043	10.389
60	12	8	11	0.5	0.5	0.5	0.9	0.6305	0.0057	12.196
60	12	6	9	0.5	0.5	0.1	0.5	0.34	0.0044	7.545
50	16	6	11	0.1	0.9	0.1	0.9	0.0448	0.003	6.753
50	8	10	7	0.1	0.1	0.9	0.1	0.2004	0.0039	13.289

Table 2. Experimental data used for testing GP metamodels (Soundhar, Zubar, Sultan, & Kandasamy)

Process parameters				Process parameters (coded)				Output responses		
V	I	P_{on}	P_{off}	V	I	P_{on}	P_{off}	MRR	EWR	SR
60	8	8	9	0.5	0.1	0.5	0.5	0.7129	0.0041	9.149
70	12	8	9	0.9	0.5	0.5	0.5	0.2412	0.0085	18.214
50	16	10	11	0.1	0.9	0.9	0.9	0.525	0.0107	16.758
70	16	10	7	0.9	0.9	0.9	0.1	0.4086	0.0133	14.814
70	8	10	11	0.9	0.1	0.9	0.9	1.0208	0.0036	14.322

Table 3. Coefficients of genetic programming metamodels

MRR	EWR	SR
β_0=1.4574	β_0=0.009832	β_0=0.027337
β_1=2.3545	β_1=0.001952	β_1=0.311068
β_2=0.6797	β_2=0.000589	β_2=0.173889
β_3=2.5934	β_3=0.000471	β_3=0.534502
β_4=1.3646	β_4=0.004906	β_4=0.83864
β_5=0.1007	-	β_5=0.015114
-	-	β_6=0.01158
-	-	β_7=0.015325
-	-	β_8=0.062816
		β_8=28.419

Figure 3. Experimental vs.GP predicted MRR for training and testing data

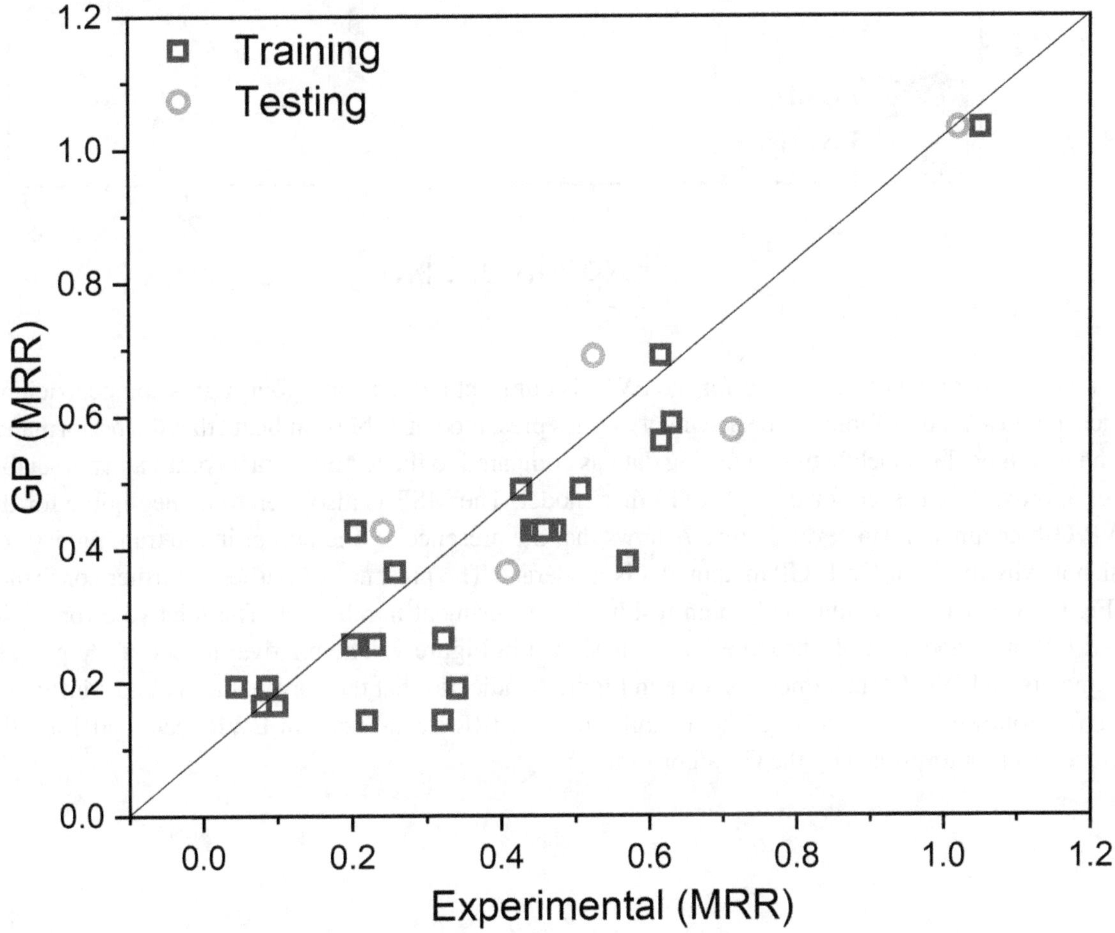

Figure 4. Residuals of GP MRR metamodel for training and testing data

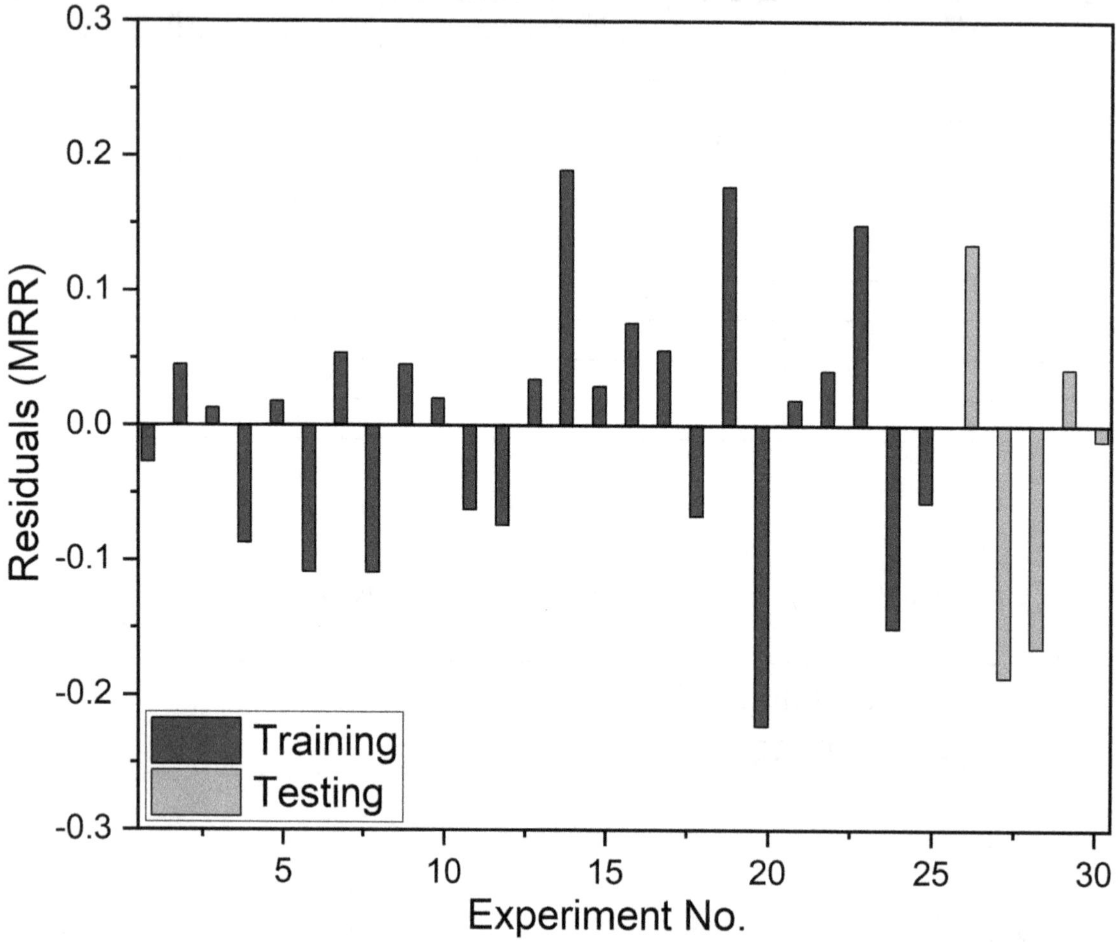

It is seen that the GP expression for the EWR is compact and has only four regression coefficients which are presented in Table 3. The accuracy metrics presented in Table 4 indicate that the performance of the metamodel is much better on testing data as compared to the testing data. Overall an appreciable R^2 of approx. 73% is seen for the EWR GP metamodel. The MSE is also seen to be negligible for the EWR GP metamodel. However, Figure 7 shows that the presence of one outlier in the training dataset, which is why the R^2 of EWR GP metamodel is moderate. The presence of outlier is further confirmed in Figure 8 which shows unusually high residue for experiment number 13. The relative error of the EWR GP metamodel for all the experiments is shown in Figure 9. The relative impact of the process parameters on EWR GP metamodel shown in Figure 6 indicates that the current has the most influence on EWR followed by P_{on} and P_{off}. Much similar to the MRR metamodel, in EWR metamodel too the voltage (V) is auto-pruned by the GP algorithm.

Table 4. Performance of genetic programming metamodels

Metric	Training	Testing	Overall
	MRR		
R^2	0.810007	0.790693	0.821045
MAE	0.077307	0.107930	0.082412
MSE	0.009276	0.016418	0.010467
	EWR		
R^2	0.673483	0.969353	0.72811
MAE	0.001108	0.000605	0.001025
MSE	0.002094	0.000891	3.79E-06
	SR		
R^2	0.810618	0.205201	0.710085
MAE	0.993028	2.860651	1.304296
MSE	1.639477	10.283959	3.080224

Surface Roughness

Eqn. 10 is an approximate relation build using GP symbolic regression approach to quantify the surface roughness as a function of I, V, P_{on} and P_{off}.

$$SR = \left(\beta_0.V - \frac{\beta_1.V}{P_{off}} - \frac{\beta_2}{I} - \frac{\beta_3}{P_{on}} + \frac{\left(\beta_4 + \beta_5.I\right)}{\left(\beta_6.P_{off} - \beta_{7.}V - \beta_8\right)} + \beta_9 \right) \tag{10}$$

The values of the coefficients in the above equation are presented in Table 3. Figure 10 shows tt the SR GP metamodel has good estimation on the training data but on the testing data the performance is somewhat poor. Figure 11 shows that for 60% of the testing dataset the residues are relatively high. This is further confirmed by the relative error plot in Fig, 12. The R^2 of the SR GP metamodel are 81%, 20% and 71% for training, testing and overall data relatively. Similarly, the MSE is seen to be much higher for the testing as compared to the training (% approx.) data. The pie plots in Figure 6 indicate that the P_{on} has the highest relative influence on the SR GP metamodel as per the training data evaluation.

CONCLUSION

In this research, the effectiveness of genetically optimized symbolic regression metamodel is evaluated. Three different output responses relating to the EDM process are considered. The metamodels are evaluated based on several metrics like R^2, MAE, MSE for both training and testing data. It is seen that the GP trained symbolic regression metamodels do not need the use of additional statistical techniques like Box-Cox transformation or analysis of variance. Based on the training data the GP prunes insignificant terms from the symbolic regression metamodels. It is seen that for all the three metamodels for the various responses the overall prediction performance in terms of coefficient of determination is more than 70%. Further, all the GP metamodels are compact, easy to interpret and has much lesser terms than a traditionally used full-scale second-order polynomial regression metamodel. The non-predefined form free ability of the GP based symbolic regression approach is another advantage as it can accommodate higher levels of nonlinearity in a compact form.

Figure 5. Relative error of GP MRR metamodel for training and testing data

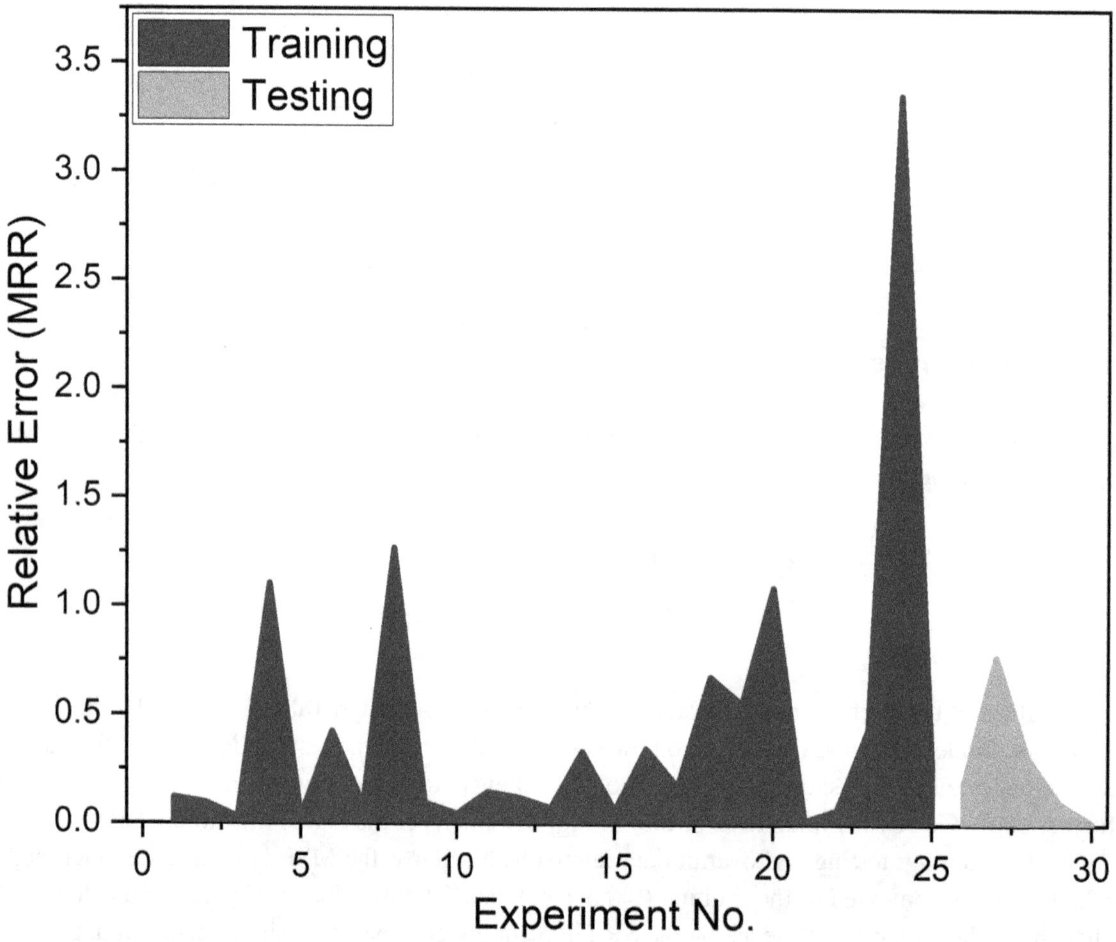

Figure 6. Relative impact of the process parameters on GP metamodels for training and overall data

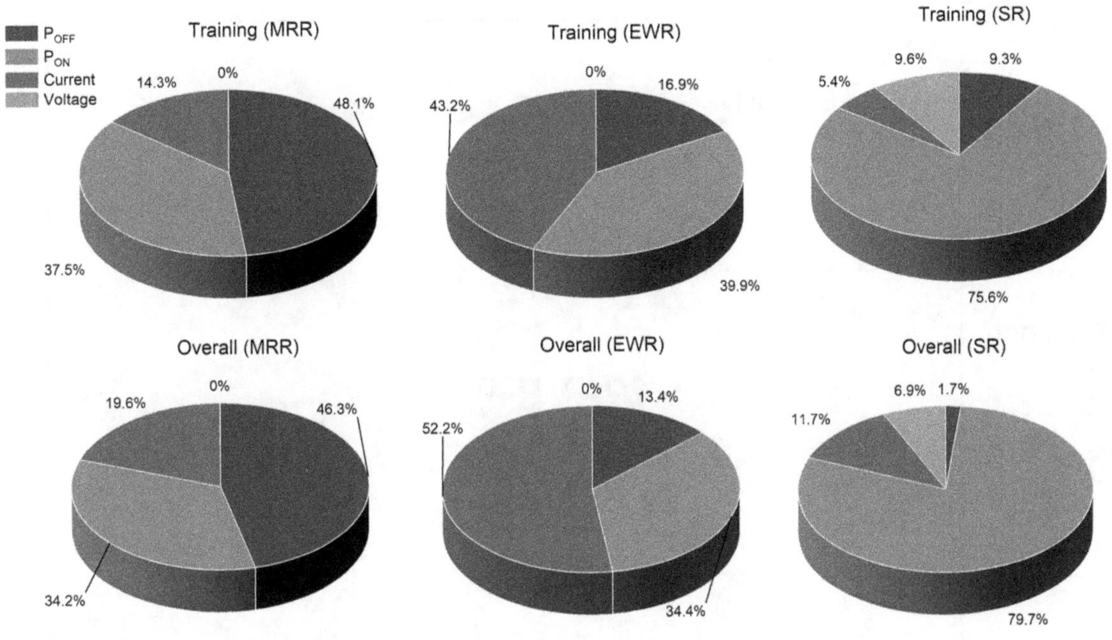

Figure 7. Experimental vs.GP predicted EWR for training and testing data

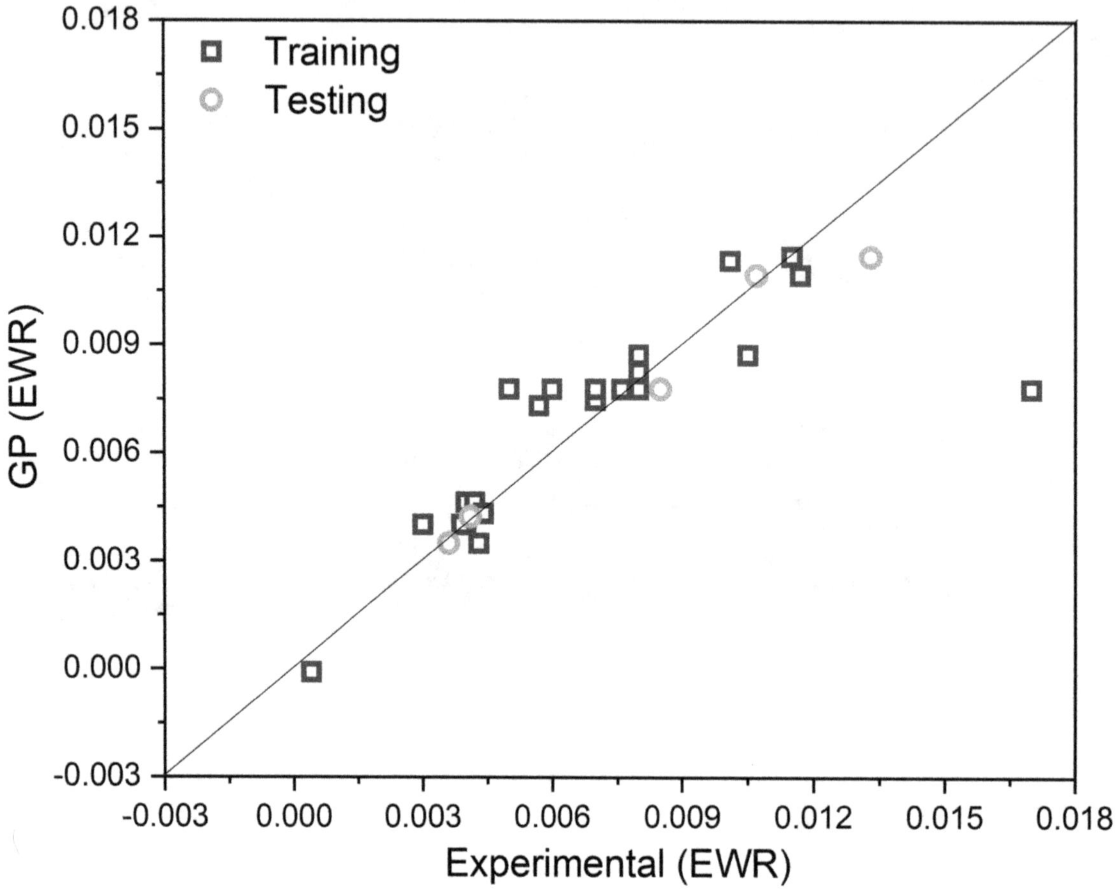

Figure 8. Residuals of GP EWR metamodel for training and testing data

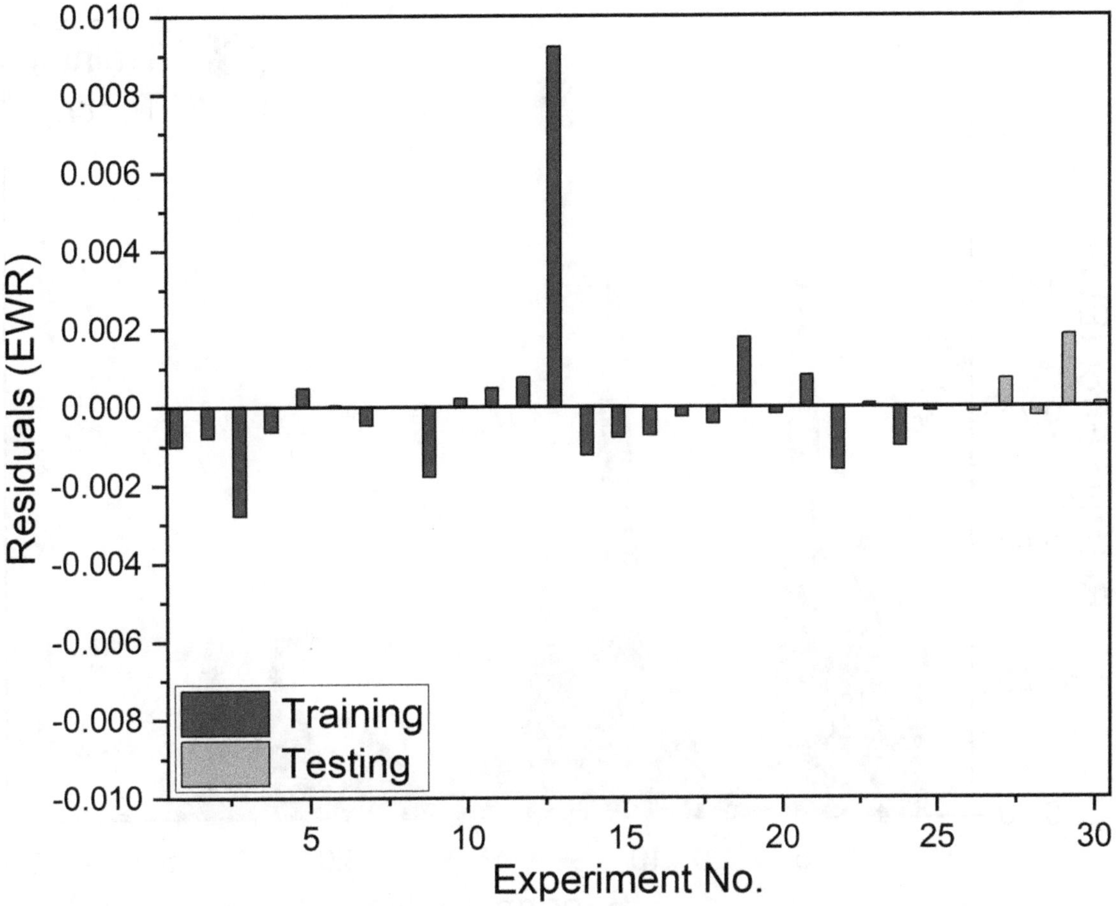

Figure 9. Relative error of GP EWR metamodel for training and testing data

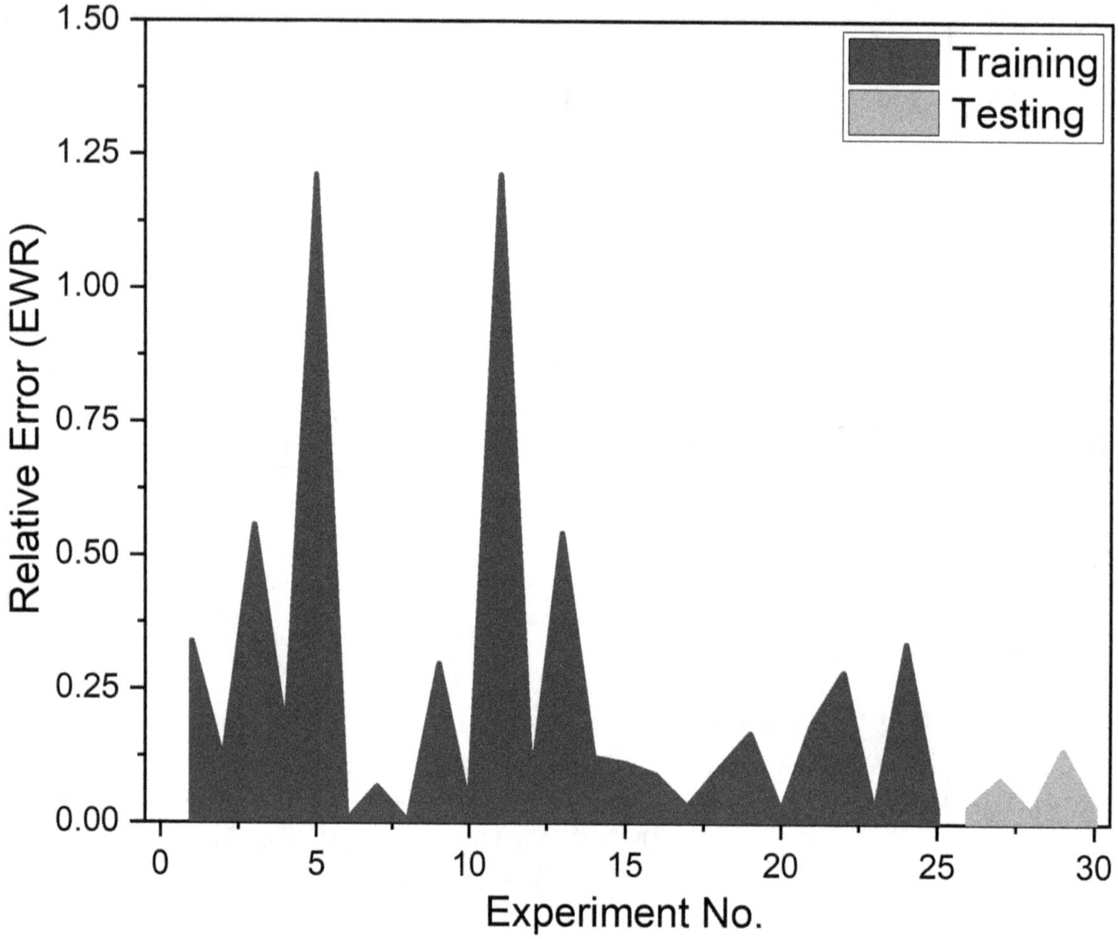

Figure 10. Experimental vs.GP predicted SR for training and testing data

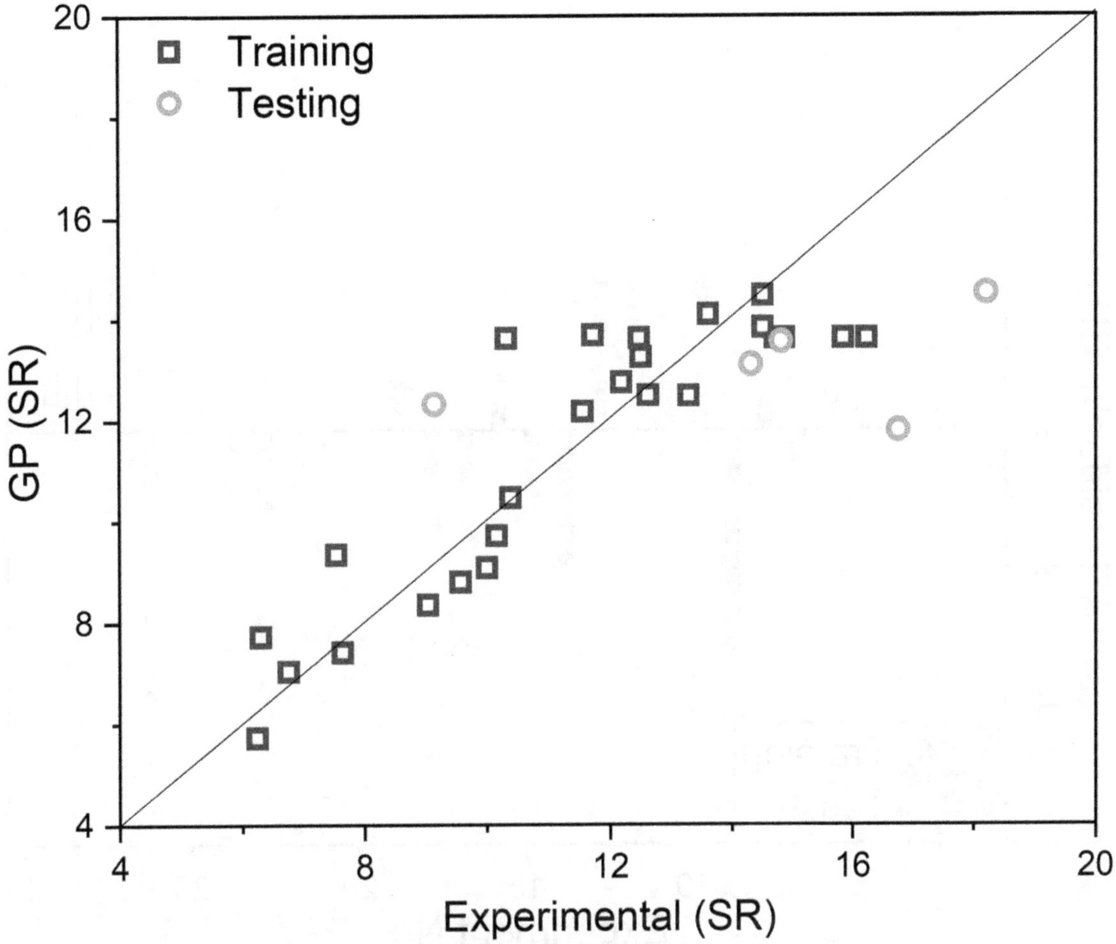

Figure 11. Residuals of GP SR metamodel for training and testing data

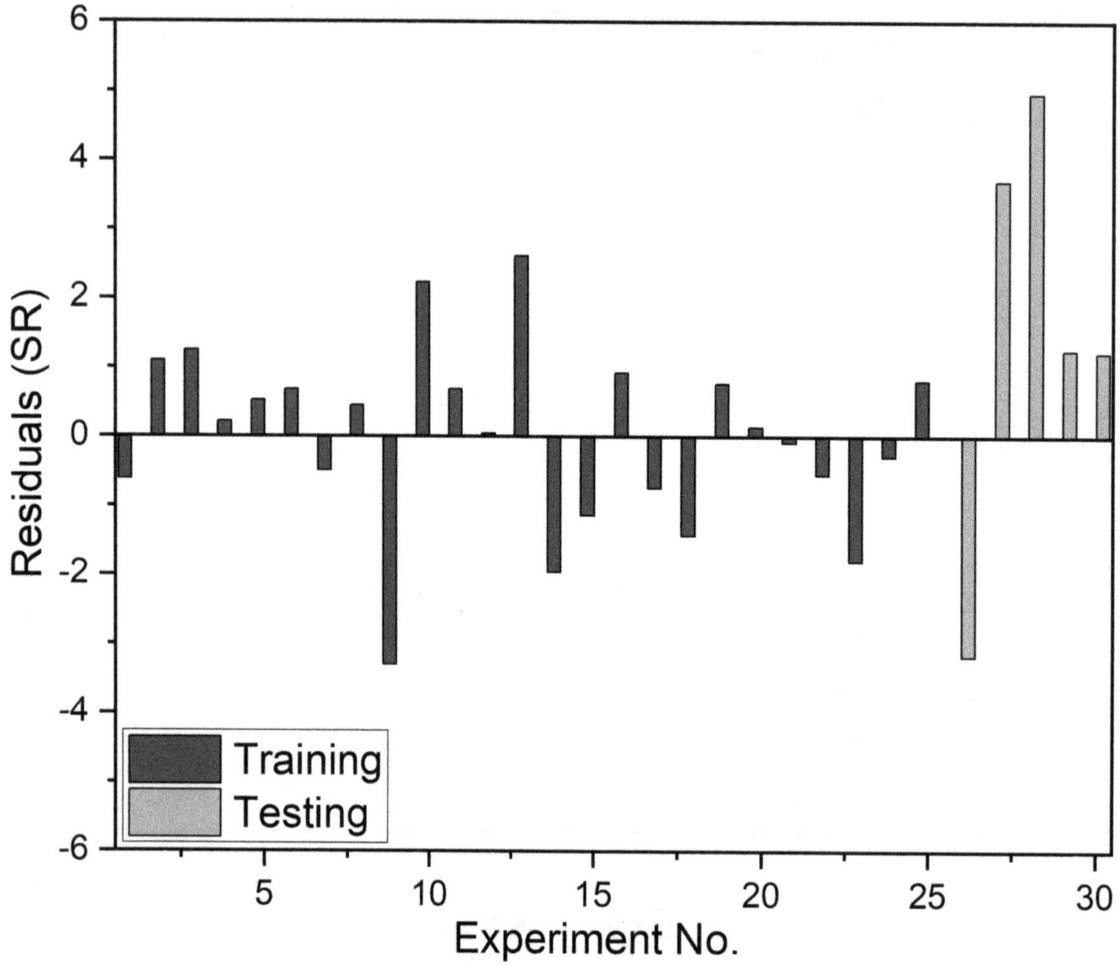

Figure 12. Relative error of GP SR metamodel for training and testing data

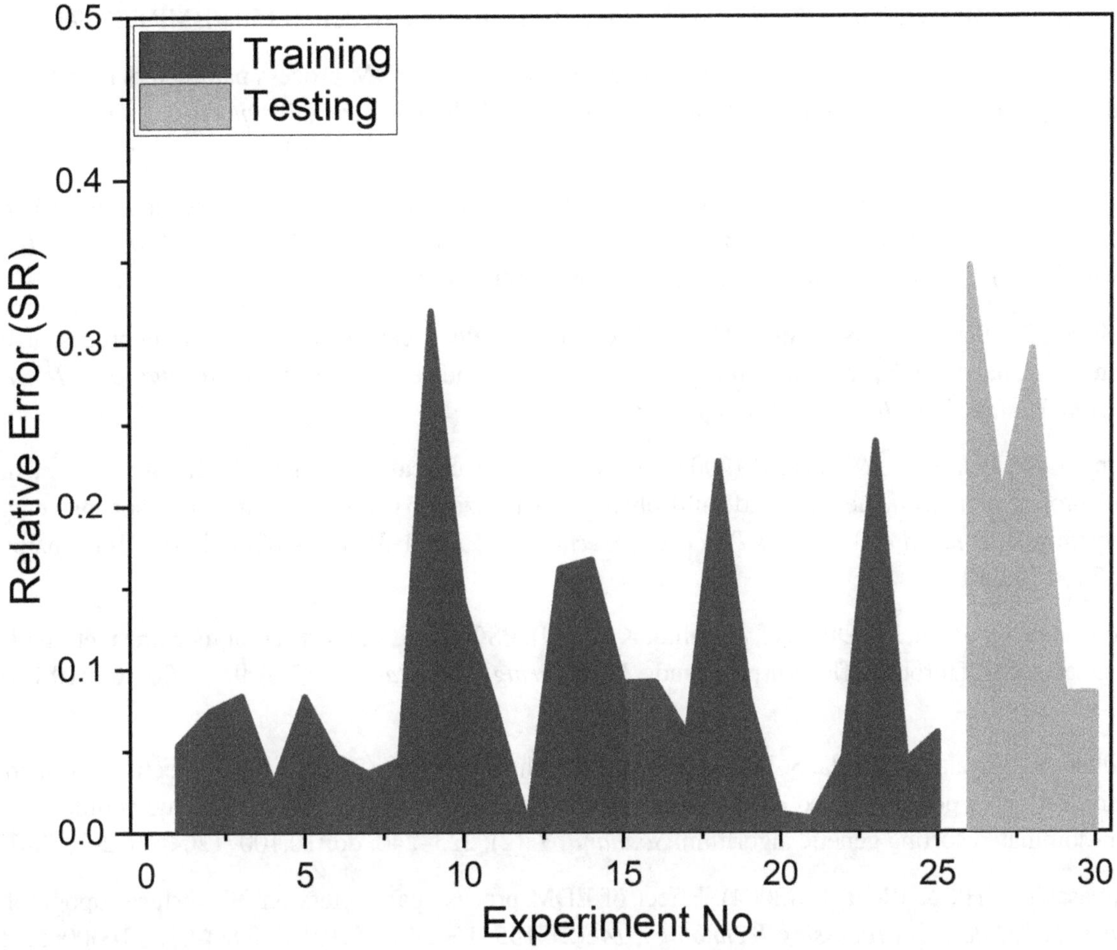

REFERENCES

Bose, G. K., & Pain, P. (2020). 9 Optimization of EDM process through evolutionary computing and fuzzy MCDM techniques. *Soft Computing: Techniques in Engineering Sciences*, *1*, 165.

Garg, A., & Lam, J. S. (2016). Modeling multiple-response environmental and manufacturing characteristics of EDM process. *Journal of Cleaner Production*, *137*, 1588–1601. doi:10.1016/j.jclepro.2016.04.070

Gostimirovic, M., Rodic, D., Kovac, P., Pucovsky, V., & Savkovic, B. (2014). Application of neuro-fuzzy systems and genetic programming for modelling surface roughness in electrical discharge machining. *Annals of the Faculty of Engineering Hunedoara*, *12*, 137.

Kalita, K., Mukhopadhyay, T., Dey, P., & Haldar, S. (2019). Genetic programming-assisted multi-scale optimization for multi-objective dynamic performance of laminated composites: The advantage of more elementary-level analyses. *Neural Computing & Applications*, 1–25.

Kanagarajan, D., Karthikeyan, R., Palanikumar, K., & Davim, J. P. (2009). Application of goal programming technique for electro discharge machining (EDM) characteristics of cemented carbide (WC/Co). *International Journal of Materials & Product Technology, 35*(1/2), 216–227. doi:10.1504/IJMPT.2009.025228

Kuppan, P., Rajadurai, A., & Narayanan, S. (2008). Influence of EDM process parameters in deep hole drilling of Inconel 718. *International Journal of Advanced Manufacturing Technology, 38*(1-2), 74–84. doi:10.100700170-007-1084-y

Kuriakose, S., & Shunmugam, M. S. (2005). Multi-objective optimization of wire-electro discharge machining process by non-dominated sorting genetic algorithm. *Journal of Materials Processing Technology, 170*(1-2), 133–141. doi:10.1016/j.jmatprotec.2005.04.105

Maji, K., Pratihar, D. K., & Patra, S. (2010). Modelling of electrical discharge machining process using regression analysis, adaptive neuro-fuzzy inference system and genetic algorithm. *International Journal of Data Mining. Modelling and Management, 2*, 75–94.

Mandal, D., Pal, S. K., & Saha, P. (2007). Modeling of electrical discharge machining process using back propagation neural network and multi-objective optimization using non-dominating sorting genetic algorithm-II. *Journal of Materials Processing Technology, 186*(1-3), 154–162. doi:10.1016/j.jmatprotec.2006.12.030

Narayanan, G., Joshi, M., Dutta, P., & Kalita, K. (2019). PSO-tuned support vector machine metamodels for assessment of turbulent flows in pipe bends. *Engineering Computations, 37*(3), 981–1001. doi:10.1108/EC-05-2019-0244

Padhee, S., Nayak, N., Panda, S. K., Dhal, P. R., & Mahapatra, S. S. (2012). Multi-objective parametric optimization of powder mixed electro-discharge machining using response surface methodology and non-dominated sorting genetic algorithm. *Sadhana, 37*(2), 223–240. doi:10.100712046-012-0078-0

Ramasawmy, H., & Blunt, L. (2004). Effect of EDM process parameters on 3D surface topography. *Journal of Materials Processing Technology, 148*(2), 155–164. doi:10.1016/S0924-0136(03)00652-6

Salman, Ö., & Kayacan, M. C. (2008). Evolutionary programming method for modeling the EDM parameters for roughness. *Journal of Materials Processing Technology, 200*, 347–355.

Sharma, N., Khanna, R., & Gupta, R. D. (2015). WEDM process variables investigation for HSLA by response surface methodology and genetic algorithm. *Engineering Science and Technology, an International Journal, 18*, 171–177.

Somashekhar, K. P., Ramachandran, N., & Mathew, J. (2010). Optimization of material removal rate in micro-EDM using artificial neural network and genetic algorithms. *Materials and Manufacturing Processes, 25*(6), 467–475. doi:10.1080/10426910903365760

Soundhar, A., Zubar, H., Sultan, M., & Kandasamy, J. (n.d.). Dataset on optimization of EDM machining parameters by using central composite design. *Data in Brief, 23*, 103671.

Srivastava, V., & Pandey, P. M. (2012). Effect of process parameters on the performance of EDM process with ultrasonic assisted cryogenically cooled electrode. *Journal of Manufacturing Processes, 14*(3), 393–402. doi:10.1016/j.jmapro.2012.05.001

Tzeng, C.-J., & Chen, R.-Y. (2013). Optimization of electric discharge machining process using the response surface methodology and genetic algorithm approach. *International Journal of Precision Engineering and Manufacturing*, *14*(5), 709–717. doi:10.100712541-013-0095-x

Yan, M.-T., & Fang, C.-C. (2008). Application of genetic algorithm-based fuzzy logic control in wire transport system of wire-EDM machine. *Journal of Materials Processing Technology*, *205*(1-3), 128–137. doi:10.1016/j.jmatprotec.2007.11.091

Yusup, N., Zain, A. M., & Hashim, S. Z. (2012). Evolutionary techniques in optimizing machining parameters: Review and recent applications (2007–2011). *Expert Systems with Applications*, *39*(10), 9909–9927. doi:10.1016/j.eswa.2012.02.109

Zhang, G., Zhang, Z., Guo, J., Ming, W., Li, M., & Huang, Y. (2013). Modeling and optimization of medium-speed WEDM process parameters for machining SKD11. *Materials and Manufacturing Processes*, *28*(10), 1124–1132. doi:10.1080/10426914.2013.773024

ADDITIONAL READING

Ganesh, N., Ghadai, R. K., Bhoi, A. K., Kalita, K., & Gao, X. Z. (2020). An Intelligent Predictive Model-Based Multi-Response Optimization of EDM Process. *Computer Modeling in Engineering & Sciences*, *124*(2), 459–476. doi:10.32604/cmes.2020.09645

Ghadai, R. K., & Kalita, K. (2020). Accurate estimation of DLC thin film hardness using genetic programming. *International Journal of Materials Research*, *111*(6), 453–462. doi:10.3139/146.111911

Gostimirovic, M., Pucovsky, V., Sekulic, M., Rodic, D., & Pejic, V. (2019). Evolutionary optimization of jet lag in the abrasive water jet machining. *International Journal of Advanced Manufacturing Technology*, *101*(9-12), 3131–3141. doi:10.100700170-018-3181-5

Hayajneh, M. T., & Abdellahia, M. (2019). Prediction Performance of End-Milling Process by Gene Expression Programming. *Jordan Journal of Mechanical & Industrial Engineering*, *13*(2).

Panda, B., Shankhwar, K., Garg, A., & Savalani, M. M. (2019). Evaluation of genetic programming-based models for simulating bead dimensions in wire and arc additive manufacturing. *Journal of Intelligent Manufacturing*, *30*(2), 809–820. doi:10.100710845-016-1282-2

Sharma, N., Kumar, K., Raj, T., & Kumar, V. (2019). Porosity exploration of SMA by Taguchi, regression analysis and genetic programming. *Journal of Intelligent Manufacturing*, *30*(1), 139–146. doi:10.100710845-016-1236-8

Yunus, M., & Alsoufi, M. S. (2019). Mathematical modeling of multiple quality characteristics of a laser microdrilling process used in Al7075/SiCp metal matrix composite using genetic programming. *Modelling and Simulation in Engineering*.

KEY TERMS AND DEFINITIONS

Predictive Model: Predictive models (also referred to as metamodels) are approximate functions of the actual function. They are inexpensive and can be used as surrogates to the actual functions.

Process Parameter: The various independent variables involved in a physical process or phenomenon may be referred to as process parameters.

Process Response: The measured output (i.e., dependent variable) of a process is known as its response. Example—material removal rate, surface roughness, etc.

Regression: It is a statistical approach to express the process response as a function of the process parameters.

Chapter 10
Machine Learning-Based Predictive Modelling of Dry Electric Discharge Machining Process

Kanak Kalita
https://orcid.org/0000-0001-9289-9495
Vel Tech Rangarajan Dr. Sagunthala R and D Institute of Science and Technology, India

Dinesh S. Shinde
SVKM's NMIMS, Mukesh Patel School of Technology Management and Engineering, Shirpur, India

Ranjan Kumar Ghadai
Sikkim Manipal Institute of Technology, Sikkim Manipal University, Majhitar, India

ABSTRACT

The conventional methods like linear or polynomial regression, despite their overwhelming accuracy on training data, often fail to achieve the same accuracy on independent test data. In this research, a comparative study of three different machine learning techniques (linear regression, random forest regression, and AdaBoost) is carried out to build predictive models for dry electric discharge machining process. Six different process parameters namely voltage gap, discharge current, pulse-on-time, duty factor, air inlet pressure, and spindle speed are considered to predict the material removal rate. Statistical tests on independent test data show that despite linear regression's considerable accuracy on training data, it fails to achieve the same on independent test data. Random forest regression is seen to have the best performance among the three predictive models.

DOI: 10.4018/978-1-7998-7206-1.ch010

INTRODUCTION

Electrical Discharge Machining (EDM) is a non-traditional machining process used to process electrically conductive materials with the help of material erosion by spark bombardment on the surface. Melting and vaporization of material from the surface of part produces replicate of the tool on the part surface. A little gap is present between the tool and workpiece. The dielectric fluid is used to take away the material debris produced due to material removal from the surface and also cools the workpiece and tool. The electrolyte disposal after machining, fire hazard during handling, etc. are the major issue with the conventional EDM process. Dry EDM uses an envelope of gases like compressed air, argon, helium, nitrogen, oxygen, etc. in place of dielectric fluid, which makes it green machining process. The friction of the workpiece and tool due to dielectric, lesser heat affected zone, lesser residual stresses, any direction used for machining due to absence of dielectric tank, etc. makes dry EDM superior to conventional EDM. Dry EDM has the controlling parameters similar to conventional EDM such as applied voltage, current, pulse on and off time and frequency, the gap between tool and workpiece etc. Apart from these, types and conditions of gas used, the flow rate of gas etc. are the controlling process parameters, with typical output such as material removal rate (MRR), surface finish, overcut, etc. Apart from using gaseous medium solely in machining, combined dielectric fluid and gas can also be used in the EDM process and methodology is named as near dry-EDM process which combines advantages of the conventional and dry-EDM process (Kao, Tao, & Shih, 2007). In dry-EDM use of flat face electrode or tool leads to uncontrolled plasma extension hence the material removal rate reduces, also the material removed debris carrying becomes ineffective, which can be overcome by the use of different slotted electrodes (Puthumana & Joshi, 2011). Gholipoor et al. (Gholipoor, Baseri, & Shabgard, 2015) have compared the performance of near dry machining with the conventional and dry-EDM process in the drilling of SPK steels considering toll wear ratio, MRR and surface roughness and concluded that surface quality obtained by near-dry machining is better than other EDM processes. Also, the near-dry machining gives lower surface crack on the machined surface. Dhakar et al. (Dhakar & Dvivedi, 2016) evaluated the process parameters (i.e. applied current, gas pressure, duty factor, and lift of tool) in case of near dry machining of high-speed steels for MRR, tool wear rate and surface roughness.

Researchers are interested to study the process parameters of dry-EDM for different applications. Saha et al (Saha & Choudhury, 2009) studied the influence of the dry-EDM parameters on machining of mild steel such as gap voltage, discharge current, pulse-on time, duty factor, air pressure and spindle speed considering output response as MRR, surface roughness and tool wear rate and reported that applied current is the most significant factor followed by duty factor, air pressure and machining speed. Masanori Kunleda et al. (Kunleda, et al., 2003) investigated dry-EDM milling process parameters of steel specimen for MRR and observed that when the applied electric power goes beyond certain values the MRR increases drastically over the surface because of the chemical reaction between the oxygen gas and object material. Islam et al. (Islam, Li, & Ko, 2017) studied the process parameters of deburring of drilled holes in carbon fibre reinforced plastic (CFRP) using dry-EDM process with remarks that dry EDM with oxygen gas produces better machining than using air. Also, the dry EDM is more productive than that of conventional EDM. Jia Tao et al. (Tao, Shih, & Ni, 2008) studied experimentally the process parameters of dry and near dry-EDM mailing for maximum MRR and minimum surface finish.

Optimization of dry-EDM process parameters for best output is the prime concern of analysts. Pragadish et al. (Pragadish & Kumar, 2016) optimized the process parameters of dry-EDM such as applied current, ON time of pulse, applied voltage, gas pressure and cutting speed for maximum MRR and minimum surface finish with remarks that the applied current is the most significant process parameter.

To improve the characteristics of dry-EDM researchers utilized different methodologies. Joshi et al. (Joshi, Govindan, Malshe, & Rajurkar, 2011) used pulsating magnetic field during dry-electric discharge machining and reported that the current approach enhanced the productivity by 30% with no tool wear relative to the process without the magnetic field, whereas Masanori et al. (Kunieda, Takaya, & Nakano, 2004) utilized piezoelectric actuators to improve the performance of dry-EDM process for MRR with the conclusion that the issue of power short circuit in dry-EDM can be overcome using current piezo-electric actuators. Normally the electrode or tool in dry-EDM is made stationary during machining, but Vineet et al. (Yadav, Kumar, & Dvivedi, 2019) machined the hard high-speed steels using dry-EDM with rotating tool considering MRR, surface roughness and overcut as responding outcome and reported that tool rotation speed significantly affects the outcomes. Reza et al. (Teimouri & Baseri, 2013) had given magnetic field assistance to the machining tool and ultrasonic vibrations to the workpiece during dry-EDM machining and develop the mathematical model of input parameters for MRR, surface roughness, electrode wear rate, and overcut with remarks that the performance of dry-EDM is drastically improved with this approach of machining. Rendi et al. (Kurniawan, et al., 2017) presented model of process parameters (i.e. capacitance, pulse on time and amplitude of vibration) for burr removal rate in the ultrasonic-assisted dry-EDM process of carbon fibre reinforced plastic (CFRP).

Predictive modelling is the engineering tool used to select the process parameter combination to increase the machining outcomes. To achieve the desired outcome, the modelling approach by the optimization of the process tool such as machine learning can be obtained. The experimental study can validate the predictive model. The modelling capabilities of artificial intelligence techniques for predicting the outcome in the machining process improves the production characteristics (Mirkoohi, Bocchini, & Liang, 2019) (Kant & Sangwan, 2015). Jogendra Kumar et al. (Kumar, Verma, & Mondal, 2020) implemented the predictive modelling of parameters for their optimization in the drilling of polymer nano-composites and stated that the current study predicts the output of the developed model for the best set of process parameters, which was validated by an experimental confirmatory test, which have shown good validation with the actual data Martinez-Alvarado et al. (Martínez-Alvarado, et al., 2018) presented a predictive model for EDM process parameters such as applied voltage, applied current, pulse frequency, and type of electrode for machining high strength steels, which is experimentally verified. The dry-EDM process being green machining preferred over the conventional EDM process. As it involves relatively a greater number of controlling parameters, their optimal selection through modelling and prediction involves improved machining characteristics (Teimouri & Baseri, 2012). Kumar investigated a predictive model of process parameter in dry-wire EDM for monel alloy and reported that the approach gives a path to the optimal selection of the process parameters for better surface finish, durability and reliability. Nishant et al. (Singh, Singh, Kumar, & Sharma, 2019) did a comparative study between statistical and soft-computing based predictive modelling during dry-EDM of D3 die steel using helium gas as the dielectric for maximum MRR and minimum surface roughness considering applied current, ON time of the pulse, duty cycle, rotations of the electrode and helium gas pressure as process parameters.

From the literature, it is seen that the use of very advanced machine learning algorithms in EDM process modelling is rare. No studies on the use of AdaBoost and random forest regression are found for process modelling of non-traditional machining operations. Thus, a novel attempt has been made in this study to compare the efficacy of three machine learning models viz. linear regression, random forest regression and Adaboost in predictive modelling of an EDM process.

MATERIALS AND METHODS

Experimental Details

The experimental data for the present analysis are taken from Saha et al (Saha & Choudhury, 2009). The experiments are conducted by using a numerically controlled Die sinking Electro discharge machining equipment. EN32 mild steel and copper are considered as a work material and tool for carrying out the experiments using the EDM process. The shape of the workpiece is a cuboid having dimension 75mm x 20 mm x 5 mm. A thin-walled tubular-shaped copper tool with a central hole is used for the experiment. Due to the presence of a hole in the copper tool a replica of tool formed on the work surface. For machining the EN32 mild steel using EDM, high-pressure air is considered as dielectric and it is supplied to the discharge gap through the copper tool. The experimental design is based on the central composite design (CCD) by considering input parameters like gap voltage (V_g), spindle rotational speed (N), pulse-on time (T_{on}), discharge current (I_d), duty factor (D) and air inlet pressure (P). The response parameter is the material removal rate (MRR), which is calculated by considering the ratio of weight difference and machining time.

Machine Learning Algorithms

Linear Regression

The linear regression is used to predict the value of a dependent variable, which is strongly correlated to another independent variable. It is based on a simple mathematical model, which predicts one variable (*y*) with the help of other (*x*), with the help of simple regression equation, which presents the direct relation between *x* and *y*. This relationship is along the straight line which is a line of best fit for the data set on the scatter plot which reveals the relation between variable using equations of the variables (Hazra & Gogtay, 2016).

The general formula for the linear regression is:

$$y = a + bx \tag{1}$$

where a and b represent *y*-intercept and slope of the fitting line of the linear regression respectively. So, b is responsible for any change in y. The establishment of the value of a and b, makes the prediction of the y with respect to x. If the line of fit passes through the origin, the mathematical relation becomes:

$$y = bx \tag{2}$$

For the current problem, the linear regression model may be given as,

$$MRR = \beta_0 + \beta_1 V_g + \beta_2 N + \beta_3 T_{on} + \beta_4 I_d + \beta_5 D + \beta_6 P \tag{3}$$

Random Forest Regression

Random forest regression abbreviated as RF is a non-parametric, grouped regression which is based on CART i.e. classification and regression tree models. The RF is a Random number, who combines the performance to get dependent parameters. The independently created K trees $h_k(x)$, where $k = 1, \ldots,$ K. This technique models the input vector x for prediction simulating individual tree in the forest (Seo, Kim, Eo, Park, & Park, 2017).

RF regression is based on the relation given in the equation below:

$$\text{RF predictor} = \frac{1}{k}\sum_{k=1}^{k} h_k(x) \tag{4}$$

The out-of-bag data (OOB) combined with mean square error (MSE) is used to pick the unselected samples for exercise the kth tree in the prediction process, which is presented in equation (5)

$$OOB - MSE = \frac{1}{n}\sum_{i=i}^{n}(y_i - \overline{y_i}OOB)^2 \tag{5}$$

Where, y_i is the ith prediction and $\overline{y_i}OOB$ is the average i^{th} prediction of all the trees.

Since the OOB and MSE depend on the computational scales, OOB-R^2 considered using the relation given in equation (6).

$$OOB - R^2 = 1 - \left(OOB - MSE / V_{ar_y}\right) \tag{6}$$

Where, V_{ar_y} is the variance of the responses. RF regression gives measurements that can be used to rank the parameters based on their significance. Every parameter is subjected to permutation, and the regression trees are mature in the revised data sets. The difference between MSE and revised OOB of the data set is the vital measure.

Adaboost Regression

AdaBoost or adaptive boosting regression is a systematic grouping method which was developed to improve bifurcation and regression trees. It is based on the development of a set of weak learners with the help of training data obtained from multiple sampling of original data. It produces a set of separators using weak learners leading to the final classifier. In AdaBoost regression, the weightage obtained in the study of each hypothesis is used, which indicates the importance of each alternative and further used in calculating error in the hypothesis based on the data sets. Each stage revised this weightage for improvement, therefore at the end, it is focused on harder learners. This is a significant advantage of AdaBoost regression. In every regression model, the output is not necessarily exact, but it has some error that may be uniform. The predicted error compared with the terminal to call it as an error. This magnitude of error is used to define the probability of the data sets, saying that the higher error on predecessor learner has a large probability of being selected to the follower learner. The prediction of the various learners is them combined using the median or weighted average (Seo, Kim, Eo, Park, & Park, 2017).

Predictive Model Performance Measurement Metrics

If the actual values and predicted values are denoted as y_i and $\widehat{y_i}$, then the residues (εi) may be expressed as,

$$\varepsilon_i = y_i - \widehat{y_i} \tag{7}$$

If $\overline{y}$ is the mean of the response y, then the coefficient of determination R^2 can be calculated as

$$R^2 = 1 - \frac{\sum_{i=1}^{n}\left(y_i - \widehat{y_i}\right)^2}{\sum_{i=1}^{n}\left(y_i - \overline{y}\right)^2} \tag{8}$$

If n is the number of sample points, then the MAE (mean-absolute-error) calculated as

$$\text{MAE} = \frac{\sum_{i=1}^{n}\left|y_i - \hat{y}_i\right|}{n} \tag{9}$$

Similarly, MSE (mean-squared-error) is calculated as

$$\text{MSE} = \frac{\sum_{i=1}^{n}\left(y_i - \hat{y}_i\right)^2}{n} \tag{10}$$

The (RMSE) root-mean-squared-error is calculated as

$$\text{RMSE} = \sqrt{\frac{\sum_{i=1}^{n}\left(y_i - \hat{y}_i\right)^2}{n}} \tag{11}$$

The maximum error is calculated as

$$\text{Max.Error}\left(y, \hat{y}\right) = \max.\left(\left|y_i - \hat{y}_i\right|\right) \tag{12}$$

The median error is calculated as

$$\text{Med.Error}\left(y,\hat{y}\right) = \text{median}\left(\left|y_1 - \hat{y}_1\right|,\ldots,\left|y_n - \hat{y}_n\right|\right) \tag{13}$$

RESULTS AND DISCUSSION

Statistical Relation Between the Process Parameters and the Responses

Figure 1 shows the scatter plot of the MRR for the process parameters. Figure 2 shows the Pearson's correlation between the process parameters and the MRR. Both the figures indicate that the process parameters have a weak correlation between themselves, which is indicative that the selected process parameters for the study are appropriate. Further, all the process parameters have mild to moderately strong correlation with the MRR. From the analysis of the central composite design based experimental data, it is seen that the discharge current (I_d) has the highest correlation with MRR. Similarly, among the considered process parameters voltage gap (V_g) has the least correlation with the MRR.

Figure 1. Scatter plot of process parameters and output responses

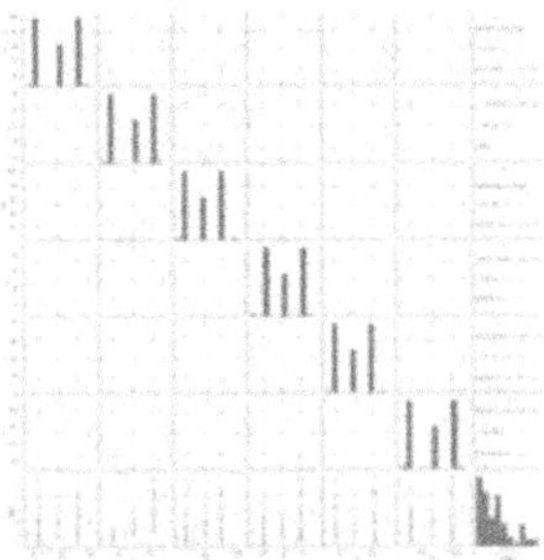

Figure 2. Correlation between process parameters and output responses

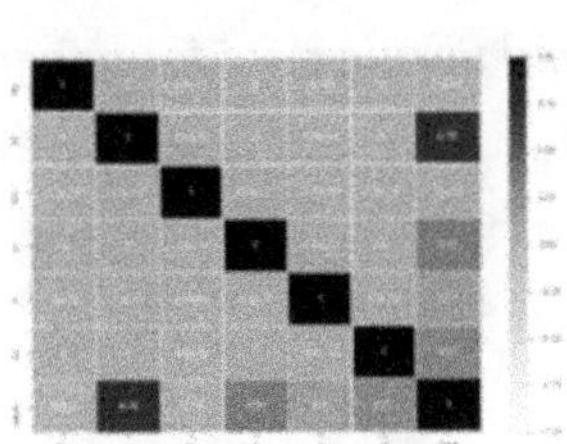

Comparison of Various Machine Learning Predictive Models

Using linear regression, a first-order polynomial regression-based predictive model is established between the six independent process parameters and the MRR. The coefficients of the linear regression model are listed in Table 1. A negative intercept of -3.65 is seen in the model. The performance of the predictive model is evaluated on both training and independent test data. Figure 3 shows the prediction error plot for linear regression. The scatter of the data points (both training and testing) across the identity line is seen to be random, with data points appearing on both sides of the identity line. This is indicative that the linear regression predictive model is not biased as the errors are random i.e. the model has perhaps the equal probability of underpredicting or overpredicting the MRR. The best fits lines are also drawn in Figure 3 and it is seen that the R^2 is 82.9% and 62.7% for training and testing dataset, respectively. The predicted versus the residuals for the linear regression-based predictive model of MRR is showed in Figure 4. It is seen that the residuals are fairly randomly scattered for various predicted values indicating that the assumption of a linear relationship is reasonable. It should be noted that any data point lying on the horizontal residue = 0 line in Figure 4 indicates that the prediction value and the actual value are the same. However, if a best line fit is tried to fit through the data points of residuals in Figure 4, an uphill inclined line will be perhaps encountered as a high cluster of residue datapoints is seen between predicted values of -0.5 to 2 in an uphill trend. This indicates that the current predictive model could be further improved, such that Figure 4 would more closely look like an 'ideal' residual versus predicted value plot. In an ideal residual versus predicted value plot, the best fit line drawn through the residues should be a line parallel to the residue = 0 line i.e. ideally the residues should form a horizontal band around the residue = 0 line.

Table 1. Coefficients of linear regression

Coefficient	Value
β_0	-3.65324673
β_1	0.00135882
β_2	0.08878135
β_3	0.00022332
β_4	0.02328535
β_5	0.00408296
β_6	0.00079991

Figure 3. Prediction error on training and testing data for linear regression

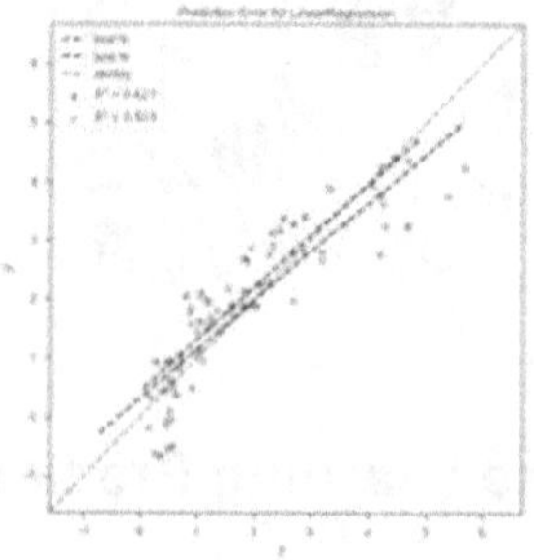

Figure 4. Residuals for linear regression on training and testing data

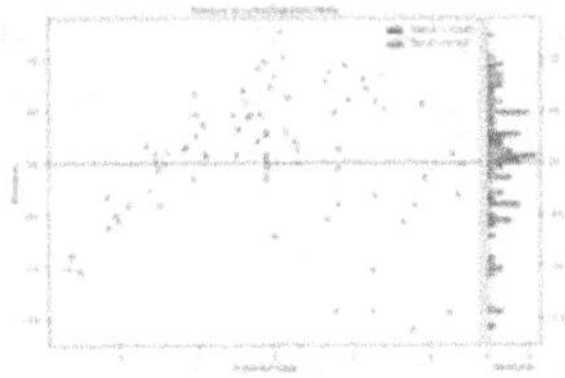

Table 2. Performance of machine learning algorithms on training data

Metric	Linear Regression	Random forest Regression	AdaBoost Regression
R^2	0.829306	0.964804	0.898046
MSE	0.324125	0.066832	0.193597
RMSE	0.569319	0.258519	0.439997
MAE	0.437118	0.180335	0.360374
Max. Error	1.607663	1.009393	0.965455
Median Error	0.378311	0.11985	0.3525

Table 3. Performance of machine learning algorithms on testing data

Metric	Linear Regression	Random forest Regression	AdaBoost Regression
R^2	0.627239	0.904966	0.841931
MSE	0.521629	0.132987	0.221196
RMSE	0.722239	0.364674	0.470315
MAE	0.591618	0.253175	0.347273
Max. Error	1.427398	0.9018	1.08
Median Error	0.587773	0.1639	0.21875

The performance of the linear regression predictive model for MRR is further evaluated by using various other metrics like MSE, RMSE, MAE, Max. Error and Median Error. Comparison of the metrics on training data (Table 2) and testing data (Table 3) show that the MSE increases for the testing case to 0.5216 from 0.3241 in training. Thus, it is clear that the performance of the linear regression predictive model for MRR deteriorates on independent test data. Similar fall in prediction capability is measured by all the other metrics except Max. Error. For testing data, the Max. Error of the linear regression predictive model is seen to be better than for the training data.

Figure 5 shows the prediction error for random forest regressor predictive model of MRR. It is seen that the random forest regressor predictive model has a R^2 of 96.5% and 90.5% for training and testing data, respectively. Figure 6 shows the residuals for the random forest regressor predictive model of MRR. The random forest regressor predictive model residuals are seen to smaller and more randomly distributed as compared to Figure 4. The 2D distribution shows that the residuals are more densely located near the residue = 0 line indicating much better prediction capability of random forest regressor predictive model of MRR. However, for a few datapoints, the residuals are seen to more spread as compared to the majority of the residuals. The prediction performance of the random forest regressor predictive model of MRR as per various accuracy metrics is reported in Table 2 and Table 3 for training and testing data, respectively.

Figure 5. Prediction error on training and testing data for random forest regression

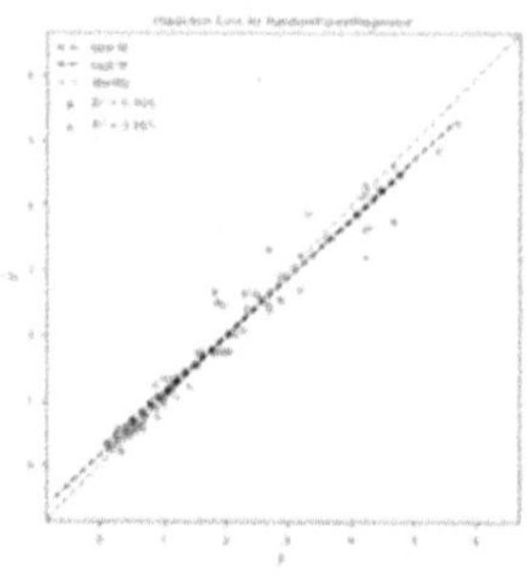

Figure 6. Residuals for random forest regression on training and testing data

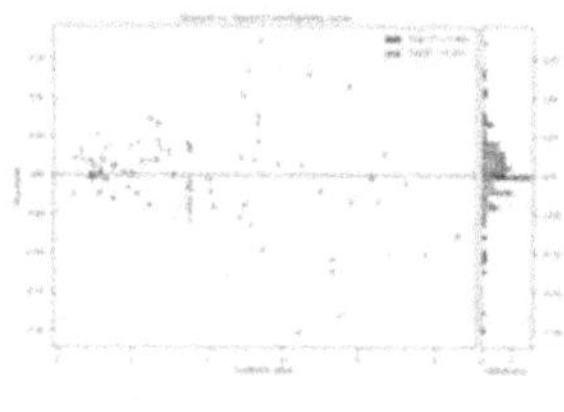

Figure 7 shows that the prediction error of the AdaBoost regressor predictive model of MRR. It is seen that the R^2 is 89.8% and 84.2% for training and testing data, respectively. Figure 8 shows the variation of the residues for the AdaBoost regressor predicted values of MRR. It is clear from the figure that if a best-fit line is drawn through the residuals, it would probably be a horizontal one. The performance of the AdaBoost regressor predictive model of MRR on various metrics is shown in Table 2 and Table 3 for training and testing data, respectively.

Figure 7. Prediction error on training and testing data for AdaBoost regression

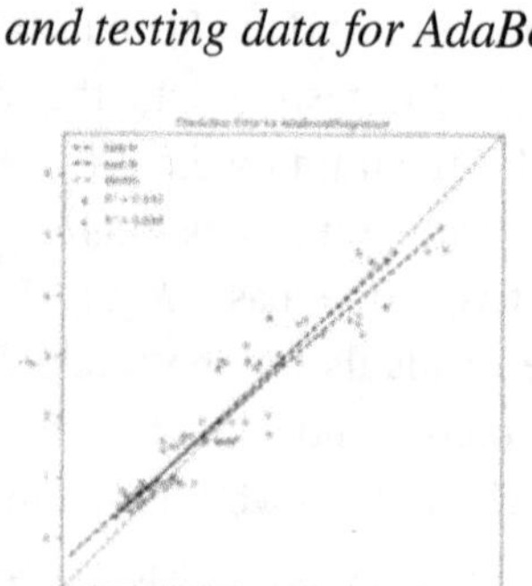

Figure 8. Residuals for AdaBoost regression on training and testing data

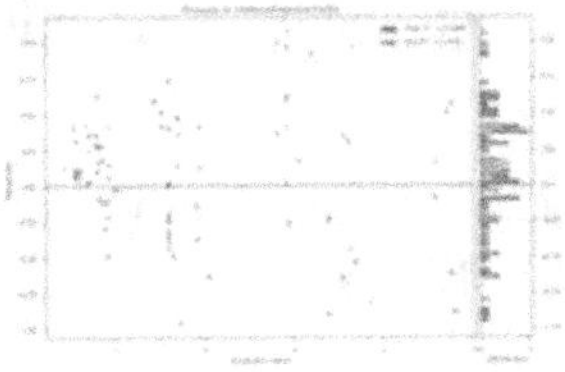

CONCLUSION

In this research, three different machine learning techniques namely, linear regression, AdaBoost regressor and random forest regressor are used to build predictive models for material removal rate in the dry electric discharge machining process. Six different process parameters are considered to express the MRR as a function of these parameters. It is seen that machine learning techniques like AdaBoost and random forest can be effectively used to build accurate predictive models for machining or manufacturing processes. The predictive performance of random forest regressor was found to the best among the three. Further, it is also realized that the reliance on multiple error metrics like MAE, MSE, Median Error etc., other than conventionally used R^2 is desirable.

REFERENCES

Dhakar, K., & Dvivedi, A. (2016). Parametric evaluation on near-dry electric discharge machining. *Materials and Manufacturing Processes*, *31*(4), 413–421. doi:10.1080/10426914.2015.1037905

Gholipoor, A., Baseri, H., & Shabgard, M. R. (2015). Investigation of near dry EDM compared with wet and dry EDM processes. *Journal of Mechanical Science and Technology*, *29*(5), 2213–2218. doi:10.100712206-015-0441-2

Hazra, A., & Gogtay, N. (2016). Biostatistics series module 6: Correlation and linear regression. *Indian Journal of Dermatology*, *61*(6), 593. doi:10.4103/0019-5154.193662

Islam, M. M., Li, C. P., & Ko, T. J. (2017). Dry electrical discharge machining for deburring drilled holes in CFRP composite. *International Journal of Precision Engineering and Manufacturing-Green Technology*, *4*(2), 149–154. doi:10.100740684-017-0018-x

Joshi, S., Govindan, P., Malshe, A., & Rajurkar, K. (2011). Experimental characterization of dry EDM performed in a pulsating magnetic field. *CIRP Annals*, *60*(1), 239–242. doi:10.1016/j.cirp.2011.03.114

Kant, G., & Sangwan, K. S. (2015). Predictive modeling for power consumption in machining using artificial intelligence techniques. *Procedia CIRP*, *26*, 403–407. doi:10.1016/j.procir.2014.07.072

Kao, C. C., Tao, J., & Shih, A. J. (2007). Near dry electrical discharge machining. *International Journal of Machine Tools & Manufacture*, *47*(15), 2273–2281. doi:10.1016/j.ijmachtools.2007.06.001

Kumar, J., Verma, R. K., & Mondal, A. K. (2020). Predictive modeling and machining performance optimization during drilling of polymer nanocomposites reinforced by graphene oxide/carbon fiber. *Archive of Mechanical Engineering, 67*.

Kunieda, M., Takaya, T., & Nakano, S. (2004). Improvement of dry EDM characteristics using piezo-electric actuator. *CIRP Annals*, *53*(1), 183–186. doi:10.1016/S0007-8506(07)60674-X

Kunleda, M., Miyoshi, Y., Takaya, T., Nakajima, N., ZhanBo, Y., & Yoshida, M. (2003). High speed 3D milling by dry EDM. *CIRP Annals*, *52*(1), 147–150. doi:10.1016/S0007-8506(07)60552-6

Kurniawan, R., Kumaran, S. T., Prabu, V. A., Zhen, Y., Park, K. M., Kwak, Y. I., ... Ko, T. J. (2017). Measurement of burr removal rate and analysis of machining parameters in ultrasonic assisted dry EDM (US-EDM) for deburring drilled holes in CFRP composite. *Measurement*, *110*, 98–115. doi:10.1016/j.measurement.2017.06.008

Martínez-Alvarado, R., Alba, G., Leyva-Bravo, J., Praga-Alejo, R., Chias-Sánchez, P., & Hernández-Rodríguez, A. (2018). Radial Basis Function Neural Network for modelling an Electrical Discharge Machining drilling process. *2018 IEEE International Autumn Meeting on Power, Electronics and Computing (ROPEC)*, 1–6. 10.1109/ROPEC.2018.8661380

Mirkoohi, E., Bocchini, P., & Liang, S. Y. (2019). Analytical temperature predictive modeling and non-linear optimization in machining. *International Journal of Advanced Manufacturing Technology*, *102*(5-8), 1557–1566. doi:10.100700170-019-03296-y

Pragadish, N., & Kumar, M. P. (2016). Optimization of dry EDM process parameters using grey relational analysis. *Arabian Journal for Science and Engineering*, *41*(11), 4383–4390. doi:10.100713369-016-2130-6

Puthumana, G., & Joshi, S. S. (2011). Investigations into performance of dry EDM using slotted electrodes. *International Journal of Precision Engineering and Manufacturing*, *12*(6), 957–963. doi:10.100712541-011-0128-2

Saha, S. K., & Choudhury, S. K. (2009). Experimental investigation and empirical modeling of the dry electric discharge machining process. *International Journal of Machine Tools & Manufacture*, *49*(3-4), 297–308. doi:10.1016/j.ijmachtools.2008.10.012

Seo, D. K., Kim, Y. H., Eo, Y. D., Park, W. Y., & Park, H. C. (2017). Generation of radiometric, phenological normalized image based on random forest regression for change detection. *Remote Sensing*, *9*(11), 1163. doi:10.3390/rs9111163

Singh, N. K., Singh, Y., Kumar, S., & Sharma, A. (2019). Comparative study of statistical and soft computing-based predictive models for material removal rate and surface roughness during helium-assisted EDM of D3 die steel. *SN Applied Sciences*, *1*(6), 529. doi:10.100742452-019-0545-x

Tao, J., Shih, A. J., & Ni, J. (2008). Experimental study of the dry and near-dry electrical discharge milling processes. *Journal of Manufacturing Science and Engineering*, 130.

Teimouri, R., & Baseri, H. (2012). Improvement of dry EDM process characteristics using artificial soft computing methodologies. *Production Engineering*, *6*(4-5), 493–504. doi:10.100711740-012-0398-2

Teimouri, R., & Baseri, H. (2013). Experimental study of rotary magnetic field-assisted dry EDM with ultrasonic vibration of workpiece. *International Journal of Advanced Manufacturing Technology, 67*(5-8), 1371–1384. doi:10.100700170-012-4573-6

Yadav, V. K., Kumar, P., & Dvivedi, A. (2019). Effect of tool rotation in near-dry EDM process on machining characteristics of HSS. *Materials and Manufacturing Processes, 34*(7), 779–790. doi:10.1080/10426914.2019.1605171

ADDITIONAL READING

Dixit, S. R., Das, S. R., & Dhupal, D. (2019). Parametric optimization of Nd: YAG laser microgrooving on aluminum oxide using integrated RSM-ANN-GA approach. *Journal of Industrial Engineering International, 15*(2), 333–349. doi:10.100740092-018-0295-1

Diyaley, S., Shilal, P., Shivakoti, I., Ghadai, R. K., & Kalita, K. (2017). PSI and TOPSIS based selection of process parameters in WEDM. *Periodica Polytechnica Mechanical Engineering, 61*(4), 255–260. doi:10.3311/PPme.10431

Ghadai, R. K., Kalita, K., Mondal, S. C., & Swain, B. P. (2018). PECVD process parameter optimization: Towards increased hardness of diamond-like carbon thin films. *Materials and Manufacturing Processes, 33*(16), 1905–1913. doi:10.1080/10426914.2018.1512114

Han, T., Jiang, D., Zhao, Q., Wang, L., & Yin, K. (2018). Comparison of random forest, artificial neural networks and support vector machine for intelligent diagnosis of rotating machinery. *Transactions of the Institute of Measurement and Control, 40*(8), 2681–2693. doi:10.1177/0142331217708242

Kalita, K., Nasre, P., Dey, P., & Haldar, S. (2018). Metamodel based multi-objective design optimization of laminated composite plates. *Structural Engineering and Mechanics, 67*(3), 301–310.

Kalita, K., Shivakoti, I., & Ghadai, R. K. (2017). Optimizing process parameters for laser beam micro-marking using genetic algorithm and particle swarm optimization. *Materials and Manufacturing Processes, 32*(10), 1101–1108. doi:10.1080/10426914.2017.1303156

Masoudi, S., Mirabdolahi, M., Dayyani, M., Jafarian, F., Vafadar, A., & Dorali, M. R. (2019). Development of an intelligent model to optimize heat-affected zone, kerf, and roughness in 309 stainless steel plasma cutting by using experimental results. *Materials and Manufacturing Processes, 34*(3), 345–356. doi:10.1080/10426914.2018.1532579

Ragavendran, U., Ghadai, R. K., Bhoi, A. K., Ramachandran, M., & Kalita, K. (2018). Sensitivity analysis and optimization of EDM process parameters. *Transactions of the Canadian Society for Mechanical Engineering, 43*(1), 13–25. doi:10.1139/tcsme-2018-0021

Shivakoti, I., Kibria, G., & Pradhan, B. B. (2019). Predictive model and parametric analysis of laser marking process on gallium nitride material using diode pumped Nd: YAG laser. *Optics & Laser Technology, 115*, 58–70. doi:10.1016/j.optlastec.2019.01.035

Wang, Z., Chegdani, F., Yalamarti, N., Takabi, B., Tai, B., El Mansori, M., & Bukkapatnam, S. (2020). Acoustic Emission Characterization of Natural Fiber Reinforced Plastic Composite Machining Using a Random Forest Machine Learning Model. *Journal of Manufacturing Science and Engineering*, *142*(3), 031003. doi:10.1115/1.4045945

KEY TERMS AND DEFINITIONS

Predictive Model: Predictive models (also referred to as metamodels) are approximate functions of the actual function. They are inexpensive and can be used as surrogates to the actual functions.

Process Parameter: The various independent variables involved in a physical process or phenomenon may be referred to as process parameters.

Process Response: The measured output (i.e., dependent variable) of a process is known as its response. Example—material removal rate, surface roughness, etc.

Regression: It is a statistical approach to express the process response as a function of the process parameters.

Chapter 11
Non-Traditional Machining Process Selection:
A Holistic Approach From a Customer Standpoint

Manish Kumar Roy
https://orcid.org/0000-0002-3323-8619
Sikkim Manipal Institute of Technology, Sikkim Manipal University, Majhitar, India

Partha Protim Das
Sikkim Manipal Institute of Technology, Sikkim Manipal University, Majhitar, India

Premchand Kumar Mahto
Sikkim Manipal Institute of Technology, Sikkim Manipal University, Majhitar, India

Ankit Kumar Singh
Indian Institute of Science, Bangalore, India

Manish Oraon
Birla Institute of Technology, Mesra, India

ABSTRACT

In the contemporary world, manufacturing by means of non-traditional machining (NTM) processes is gaining profound importance. However, there remains a plethora of choices for machining a particular product with desired accuracy and surface finish. This research delves into the need for selecting the best possible NTM machine which can produce a specific shape on a particular job material. In the work, initially analytic hierarchy process has been applied to find the relative importance of various NTM processes on the basis of combination of features pertaining to a product as well as process. Finally, using quality function deployment methodology a complete solution in terms of scores have been estimated for various NTM processes considering various shape features and work material combination. Also, the possible variance in the process capability features are taken to account during the analysis. The result obtained shows that electro chemical machining process overrules other NTM processes with reference to production time, radii at corner, surface finish, and tolerance.

DOI: 10.4018/978-1-7998-7206-1.ch011

INTRODUCTION

In order to cater the growing demand to machine various difficult-to-machine hard materials with a higher degree of precision and quality surface finish has led to the development of non-traditional machining (NTM) processes pool. When comparing to the conventional machining methods, NTM processes have boundless capabilities being extensively used in different application fields. The conventional methods of machining generally apply forces on the work material which in turn cause plastic deformation of the specimen that leads to the shear deformation in the shear plane as a result of which chip formation takes place. In these processes, the material removal is done in the form of chips because of which extreme preciseness along with accuracy cannot be jointly realized (Jain, 2002). Also continuous evolution in development of advanced materials coupled with pioneering and the demand to produce intricate shape feature put a lot of strain on the use of conventional machining processes. To overcome such problems, NTM processes because of its efficiency and enhanced capabilities find significant role in various manufacturing industries (Das & Chakraborty, 2020a; Das & Chakraborty, 2020b). However, the process characteristics of particular NTM process for a specific work material may or may not be in lieu with the product characteristics as required by a customer. Moreover a large hoard of NTM processes exists now-a-days but at the same time, there remains a criticality of selecting the best NTM set up when there exists a criticality in a shape feature to be developed on a specific work material. Also, it is worth mentioning that an NTM process which will be the most apt for a given set of conditions may or may not be as efficient under another condition set. Hence a need is perceived for the development of an approach which assists in selecting the best suited NTM machines so as to fulfill the demand and is feasible for the product.

Existing literature reflect that investigators harbored on various multi criteria decision making (MCDM) methods for selecting NTM process as computer programmed method of selection using hexadecimal characterization code. For the process of elimination, a computer program was coupled with a database (Cogun, 1993; Cogun, 1994). Analytic hierarchy process (AHP) framework is for the effective selection of suitable NTM process in restrained conditions of machining and material along with the evolution of a decision support system which automates the selection process (Chakraborty & Dey, 2006). A multi-attribute selection helps the manufacturing personnel in selecting the suitable NTM process for given requirements (Yurdakul & Cogun, 2003). The selection procedure uses a combination of TOPSIS (technique for order preference by similarity to ideal solution) and AHP. Chakladar and Chakraborty envisaged the problem of NTM selection that is suitable in respect of a combination for shape feature, specific work material and an expert system on the proposed methodology (Chakladar & Chakraborty, 2008). Chakraborty and Dey (Chakraborty & Dey, 2007) devised a method for selection of NTM utilizing quality function deployment (QFD) and for machining out a typical shape. They considered the unique characteristics of process and product and used a feature and work material combination followed by the development of an expert system. A web enabled expert system backed by database was also developed (Chandraseelan et al., 2008) which was three-layer architecture for NTM selection. Chakladhar et al. (Chakladar et al., 2009) did some work on a digraph-based approach for a definite NTM set up assessment and supplemented it with a decision system. Das and Chakraborty (Das & Chakraborty, 2011) applied analytic network process (ANP) for the selection of best NTM process. Also, Chakraborty achieved an excellent outcome based on multi-objective optimization on the basis of ratio analysis (MOORA) method which further aids in unraveling of six distinct difficulties in decision making including NTM selection (Chakraborty, 2011). By utilizing the amalgamated outlook of

PROMETHEE (preference ranking organization method for enrichment evaluation) and GAIA (geometrical analysis for interactive aid), a few problems of NTM selections have been solved (Karande & Chakraborty, 2012). Further by integrating AHP and TOPSIS methods, Choudhury et al. (Choudhury et al., 2013) selected the best NTM process. Chatterjee et al. (Chatterjee & Chakraborty, 2013) discussed the suitability, applicability and potentiality for evaluation of mixed data method using three illustrative examples for the selection of NTM processes. For appropriate selection of NTM processes, a study was done wherein carbon structural steel was cut to provide a distinct systematic approach in crisp and fuzzy environments (Temuçin et al., 2014). The NTM process selection can also be automated by software with suitable graphic user interfaces; this was incorporated in a decision-making model (Prasad & Chakraborty, 2014). The selection of best NTM process for drilling a hole in aluminum based on fuzzy AHP and QFD was determined by Roy et al. (Roy et al., 2014). For solving the non-conventional machining process selection, evaluation of suitability, applicability, and associated computations using OCRA, operational competitiveness ratings analysis was done (Madić et al., 2015). Various computer based decision guidance frameworks were also developed for ease in selection of best NTM process for a said machining application (Chakraborty & Dey, 2007; Prasad & Chakraborty, 2018). Recently, Chakraboty et al. (Chakraborty et al., 2020) applied rough multi-attributive border approximation area comparison (MABAC) approach for selection of best NTM process taking into consideration the machining capabilities and characteristics of each of the processes.

For manufacturing a product, the process engineer making the decision must choose the most appropriate NTM process and which is dependent on varied mutually supporting but contradictory attributes. This paper makes an attempt to deal with decision-making problems by the application of an integrated model of QFD and AHP to find the most suitable NTM process on the basis of an individuality set with respect to product and process. The present layout first explains AHP as well as QFD method and is trailed by a short description regarding product and process characteristics. Henceforth procedural steps involved in the proposed method has been put forward on the integration of AHP and QFD to estimate the solution for a specific NTM selection problem.

AHP METHOD

Saaty (Saaty, 1988) developed the AHP method, which is one of the basic techniques of decision making. As an analytical technique, it can be deemed to be one of the most popular MCDM technique being applied to solve unorganized and intricate decision making problems. The basic steps involved in AHP includes: structuring a complex issue of decision making in hierarchical form, making assessment pairwise among alternatives for approximating the comparative importance of the different elements on every line of the hierarchical structure, integration of significances for determination of a complete score of decision substitutes. AHP helps in comparing alternatives through pairwise comparison at a particular hierarchy level which allows to decision maker to assign priorities based on the preference.

The stated method is usually applicable for examining intricate alternatives with multiple parameters, including subjective criteria. To assign weights based on technical requirements to every single product characteristics as well as to the various NTM process alternatives, and for the generation of a decision matrix and pairwise comparison matrices, common phrases like 'much more important' are often used to determine preferences of the decision-makers. Saaty (Saaty, 1988) created a weight scale of importance for the pair-wise comparison as shown in Table 1.

Table 1. Weightage for relative pair-wise assessment

Weightage of relative importance	Description
1	Likewise preferable
3	Reasonably important
5	Highly preferable
7	Very highly preferable
9	Enormously Preferable
2, 4, 6, 8	Transitional ratings amid adjacent ratings

Generally, the following steps are involved for AHP models:

Step 1: A matrix is constructed for the comparison by the use of pair-wise assessment weightages on a relative basis. Considering there to be 'n' criteria, the comparative assessment of the ith criterion with the jth one provides a symmetric matrix AI. Here in this matrix the diagonal elements stand to be unity when i=j and also it is worth mentioning here that $a_{ji}=1/a_{ij}$.

$$AI = \begin{bmatrix} 1 & a_{12} & \cdots & a_{1n} \\ a_{21} & 1 & \cdots & a_{2n} \\ \cdots & \cdots & \cdots & \cdots \\ a_{n1} & a_{n2} & \cdots & 1 \end{bmatrix} \tag{1}$$

Step 2: By dividing each number of the pairwise comparison in a column matrix by its column sum, the relative normalized matrix is obtained.

Step 3: Priority vector (PV) which is the average of each row of the normalized matrix is then calculated for each alternative. All the row averages make up the priority vector of the alternative preferences in accordance to a criterion. The total sum of all the calculated PVs equals to1.

Step 4: For the obtained pairwise comparison matrix, for each alternative the weighted sum is then calculated by adding the multiples for the entries by the priority vector of its corresponding column alternative.

Step 5: Now for the rows, the weighted sum that has been obtained in the previous step is divided by its priority vector of corresponding row alternative.

Step 6: The maximum Eigen value, λ_{max} is found by averaging the result of the previous step.

Step 7: The consistency index (CI) is then determined for the *n* alternatives.

$$CI = \frac{(\lambda_{max} - n)}{n - 1} \tag{2}$$

Step 8: Consistency ratio (CR) is then computed. It is used to specify the degree of inconsistency for a pairwise comparison matrix. It is expected to be consistent if the calculated CR value is less than 0.10. Where the random index (RI) depends on number of alternatives as illustrated in Table 2.

$$CR = \frac{CI}{RI} \tag{3}$$

Table 2. Random index for the number of alternatives

n	3	4	5	6	7	8
RI	0.58	0.90	1.12	1.24	1.32	1.41

QFD PROCESS

Quality function Deployment (QFD) was explored by Akao (Akao, 1990); and Mitsubishi Industries first implemented his work in 1972. It is an efficient strategy for the planning and development of various products based on the technical and customer need. The basic function of QFD is translating subjective quality criteria quoted by the customer into objective criteria so that it is quantified and used suitably for designing and manufacturing the products. For product planning and development the development team can specify the customer's requirement with clarity why using it, and then the evaluation of each proposed product is done in a set pattern, in order to identify the degree up to which it meets the customer requirements (Hauser & Clausing, 1988; Wasserman, 1993).

The first phase for QFD implementation of is the formation of the HOQ, or the House of Quality matrix, which was named so by Hauser and Clausing (Hauser & Clausing, 1988). HOQ depicts the relation between the technical requirements i.e. quality characteristics (HOWs) and the customer requirements or the voice of customers (WHATs) (Chan & Wu, 2002; Chuang, 2001; Cohen, 1995; Govers, 2001). In this paper, HOQ is established in its most basic from, removing the planning matrices and the technical correlation. For decision-making, QFD can be taken as the representation of a market-oriented and customer-driven process when we combine the competitive analysis i.e. the WHYs with the WHATs and HOWs (Saaty, 1994). Also, the technical requirements can be considered as process characteristics whereas the voice of customers is considered as the product characteristics.

Product Characteristics

The characteristics of QFD provide a finer discernment amongst the different attributes or criteria more easily and help in examining the point of influence for the selection of suitable NTM processes. The decisive criteria can be linked to the variety of attainable independent product characteristics, in order to meet the customer's requirement in full capacity. The following product characteristics have been assessed in this paper are: product economy (PE), shape feature (ShFe), tolerance (T), production time (PT), surface damage depth (SDD), workpiece material (WPM), corner radii (CR), surface finish (SF). All product characteristics have been taken to be independent of one another, in order to avoid repetition in analysis.

Process Characteristics

The extent to which its process characteristics can meet desired product characteristics influences the optimal NTM process selection. Thus, the process characteristics given below are the ones responsible for achieving the product characteristics required:

1. **Shape Application:** The capability of an NTM process for the specific shape generation on a material is showed.
2. **Capital Investment:** The entirety of the initial cost and related investment required for NTM process installation.
3. **Tooling and fixtures:** Deals with is the replacement of any fixture and tooling and fixture of an NTM process.
4. **Power Requirement:** The power rating of an NTM process.
5. **Tool Consumption:** The tool changing requirement for machining a product and any cost involved with it is taken care of.
6. **Efficiency:** The ratio of the energy applied for the removal of material on NTM to the amount of input energy given to machine.
7. **Material Application:** The frequency at which an NTM process is going to be used for a given material is defined.
8. **Process Capability:** The capability of an NTM process to attain minimum surface damage depth, maximum possible material removal rate, high surface finish and precision.

The QFD Matrix

In the world today, a company's implementation of QFD is mainly focused on integrating the CR, or customer requirements, to the TR i.e. technical requirements in order to put great thrust to the customer's requirements. To begin, the QFD team first makes a list of the TRs which are likely to impact the CRs. Technical targets can be fixed with the help of the customer's views on competitor's product. And in the case of any remaining discrepancies between the QFD team's work on relating of CR and TR, and the customer's perspective then it can be understood by QFD matrix with ease. Looking at the vertical part of QFD matrix, we can understand the response of a company to corresponding customer requirements. In this work, the customer requirements are taken in standing with the product characteristics whereas the technical requirements are corresponding to the process characteristics.

PROPOSED METHODOLOGY

This methodology involves the amalgamation of AHP and QFD for the selection of NTM process problem for the combination of work material and shape feature. The steps which have been followed for the proposed are given below:-

Step 1: Customer requirements are identified and are recognized as product characteristics.
Step 2: Then the technical requirements are identified and considered to be process characteristics.

Step 3: A team of technical experts forms a QFD team such that a central relationship matrix is constructed based on their knowledge.
Step 4: The level of importance of various customer requirements i.e. product characteristics are computed using AHP.
Step 5: Level of importance of all the individual technical requirements i.e. process characteristics are calculated using equation:

$$w_j = \sum_{i=1}^{m} R_{ij} c_i \tag{4}$$

where w_j is level of importance of the jth technical requirement (j=1, 2, …, n), R_{ij} is the quantified relationship between the ith customer requirement and the jth technical criteria in the central relationship matrix; and ci is the importance weighing of the ith customer requirement.

Step 6: The normalization of level of importance for various technical requirement as calculated in previous step is carried out using the equation:

$$\bar{w}_j = \frac{w_j}{\sum_{j=1}^{n} w_j} \times 100 \tag{5}$$

Step 7: Using Saaty's (Govers, 2001; Prasad & Chakraborty, 2018) nine point scale as detailed in Table 1, the pairwise comparison between various NTM processes are done based on each technical requirement.
Step 8: The PV (e_{ij}) value is calculated for each NTM process based on various technical requirements.
Step 9: The overall score of various NTM processes are calculated using the equation:

$$S_j = \sum_{j=1}^{n} \bar{w}_j e_{ij} \tag{6}$$

where, S_j is the overall score of the jth NTM process alternative (j= 1, 2, …, n) which is again based on five different process capabilities viz. surface finish, corner radii, production time, tolerance and surface damage depth; w_j is the normalized level of importance of the jth technical criteria (j= 1, 2, …, n) and e_{ij} is the PV value of the jth alternative based on the ith technical criteria.

The ranking of the NTM processes is done in the descending order of the calculated overall scores. Ranking order is decided on the basis five different process capabilities parameters viz. surface damage depth, corner radii, surface finish, tolerance and production time.

CASE STUDY

The case of a medium size manufacturing company was taken for the validation of the proposed methodology; with the main objective of the present study being to select the best suited NTM process for a particular work material (aluminium in present case), and shape feature, drilling a hole (slenderness ratio £ 20) combination so as to achieve simplicity in manufacturing and thereby increasing productivity. An expert committee, consisting of a researcher, an academician and an industrial expert, considers following seven alternative NTM processes from which the best suited process is to be selected. The processes considered are: ultrasonic machining (USM), abrasive jet machining (AJM), electrochemical machining (ECM), electric discharge machining (EDM), electron beam machining (EBM), laser beam machining (LBM) and plasma arc machining (PAM).

The product characteristics which have been identified for specific manufacturing processes are shape feature (ShFe), workpiece material (WPM), surface damage depth (SDD), surface finish (SF), corner radii (CR), tolerance (T), product economy (PE), and production time (PT). In the same way, eight major process characteristics have been described which are: shape application, material application, capital investment, power requirement, efficiency, tooling and fixtures, tool consumption and process capability. This process capability has been further sub-divided in five different aspects viz., corner radii, surface finish, tolerance, surface damage depth and production time.

A decision matrix Table 3, of 8 × 8 elements, is made using AHP, to compute relative priority for each product characteristics. The calculated priority vector (PV) values for the decision matrix [D] are [0.388 0.240 0.109 0.060 0.068 0.059 0.051 0.021]$^{T.}$ And the calculated values of λ_{max} = 8.706; CI= 0.100 and CR= 7.15% respectively. Since CR < 10%, so the PV values are acceptable.

Table 3. Decision matrix

1	3	5	7	5	8	8	9
1/3	1	3	5	3	7	7	9
1/5	1/3	1	2	2	3	3	5
1/7	1/5	1/2	1	1/2	2	2	3
1/5	1/3	1/2	2	1	1/2	1/2	5
1/8	1/7	1/3	1/2	2	1	2	3
1/8	1/7	1/3	1/2	2	1/2	1	3
1/9	1/9	1/5	1/3	1/5	1/3	1/3	1

Table 4. Interrelationship scale

Scale value	Significance
0	No relationship
1	Very weak relationship
3	Slightly weak relationship
5	Moderately Weak
9	Strong relationship

Table 5. The QFD matrix for the selection of NTM process

		Technical Requirement								Scale
		Material Application	Shape Application	Capital Investment	Tooling and Fixture	Power Requirement	Efficiency	Process Capability	Tool Consumption	
Customer requirement	Workpiece material	9	1							0.388
	Shape feature	1	9		5					0.240
	Surface finish	1						9		0.109
	Surface damage depth							5		0.060
	Tolerance						1	9		0.068
	Corner radii		9		3			9		0.059
	Production time		5				3	5		0.051
	Production economy			9	5	3	3		3	0.021
Degree of Importance for selection criteria		3.846	3.344	0.197	1.489	0.065	0.288	2.693	0.065	
Normalized degree of importance for selection criteria		32.073	27.890	1.644	12.424	0.548	2.405	22.464	0.548	

Table 6. Overall NTM process scores taking surface finish as process capability

Technical requirement	Weight	Importance weight for NTM							C.I.	C.R.
		USM	AJM	ECM	EDM	EBM	LBM	PAM		
Material application	32.0734	0.0663	0.0175	0.1867	0.1555	0.1299	0.1083	0.3354	0.0877	0.0665
Shape application	27.8906	0.0967	0.0814	0.4859	0.0675	0.0549	0.0433	0.1699	0.0609	0.0461
Capital investment	1.6448	0.0586	0.0327	0.4151	0.1287	0.2291	0.1095	0.0259	0.0703	0.0532
Tooling and fixture	12.4243	0.1121	0.0942	0.2023	0.3980	0.0783	0.0639	0.0509	0.0565	0.0428
Power requirement	0.5482	0.1804	0.1501	0.3389	0.1252	0.1041	0.0560	0.0450	0.0547	0.0414
Efficiency	2.4052	0.0911	0.0774	0.2600	0.0649	0.0366	0.0288	0.4407	0.0696	0.0527
Process capability	22.4648	0.3414	0.1865	0.1554	0.0564	0.1299	0.1084	0.0217	0.0579	0.0438
Tool consumption	0.5482	0.2023	0.1121	0.0942	0.3980	0.0783	0.0639	0.0509	0.0565	0.0428
Overall score		14.4178	8.5800	27.0956	13.7421	10.1601	8.2290	17.7750		

Table 4 depicts the scale values for the interrelationships that exist between the customer requirement and technical requirement. Then the mentioned values of PV are used by QFD experts to calculate the weights of individual technical requirements, this has been detailed in Table 5. The choice of the scale of relationships was done based on the customer requirements versus technical requirement relationship.

The next thing that needs to be calculated is the ranking of various NTM processes, using AHP comparison matrix, with respect to the various technical requirements viz. material application, surface of revolution, surface finish, surface damage depth, tolerance, corner radii, production time, capital investment, tooling and fixture, power requirement, efficiency and tool consumption.

Table 7. Overall NTM process score considering different process capabilities

Process capability	USM	AJM	ECM	EDM	EBM	LBM	PAM
Surface	14.4	8.580	27.0	13.74	10.16	8.229	17.77
finish	178	0	956	21	01	0	50
Corner radii	14.0358	7.6537	26.3468	18.3961	8.7597	7.0559	17.7517
Production time	9.7730	5.3104	28.8224	15.0653	7.9944	6.3905	26.6438
Tolerance	16.1240	5.9052	24.8608	16.3396	10.4830	8.5168	17.7702
Surface damage depth	11.0019	13.1410	27.1963	14.5036	8.9322	7.1816	18.0431

Figure 1. Ranking of alternatives on the basis of sensitivity analysis

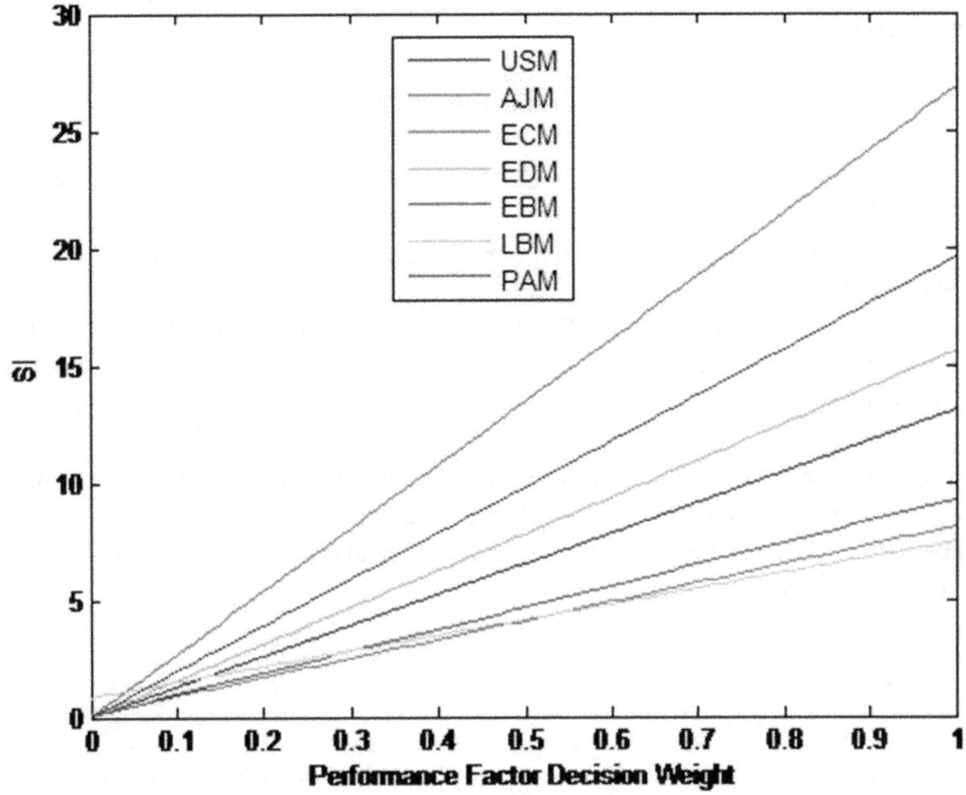

RESULTS AND DISCUSSIONS

The magnitudes of different technical requirements have been utilized to compute the total scores for the various non-traditional machining processes. These overall scores are found keeping in mind the characteristics of process capabilities, which are corner radii, tolerance, surface finish, surface damage depth and production time. Process capability parameter selection is mutually exclusive. While the Table 6 details the computation of overall score of various non-traditional machining processes on the basis of surface finish taking it as a process capability feature; the Table 6 shows the overall scores for various NTM processes based on a variety of process capabilities. Also, Table 7 highlights the results as acquired based on different process capabilities.

Here it is worth mentioning that the results thus obtained here are in good agreement with that obtained by past researchers. Also when the sensitivity analysis was carried out for the NTM processes the results were in good agreement with those obtained as has been highlighted in Figure 1.

CONCLUSION

The results obatined based on different process capabilities, are listed in the Table 7, from which the following can be concluded:

1. Considering surface finish as process capability ECM (27.0956%) is very much applicable succeeded by PAM (17.7750%).
2. Based on corner radii as process capability ECM (26.3468%) is much more dependable; followed by EDM (18.3961%).
3. Based on production time as process capability ECM (28.8224%) is more steadfast when followed by PAM (26.6438%).
4. Based on tolerance as process capability ECM (24.8608%) which is followed by PAM (17.7702%).
5. Taking the surface damage depth as the process capability ECM (27.1963%) is accompanied by PAM (18.0431%).

The main advantage of the present work has been to sort out the numerous NTM operations depending on the relative degree of importance with the five discreet process capabilities. Also need is felt to design an expert system that can provide solution for any particular product and process characteristic which is the future scope of the proposed work.

FUNDING

This research received no specific grant from any funding agency in the public, commercial, or not-for-profit sectors.

REFERENCES

Akao, Y. (1990). *Quality function deployment.* Productivity Press.

Chakladar, N. D., & Chakraborty, S. (2008). A combined TOPSIS- AHP method based approach for non-traditional machining pro- cesses selection. *Proceedings of the Institution of Mechanical Engineers. Part B, Journal of Engineering Manufacture*, *222*(12), 1613–1623. doi:10.1243/09544054JEM1238

Chakladar, N. D., Das, R., & Chakraborty, S. (2009). A digraph-based expert system for non-traditional machining processes selection. *International Journal of Advanced Manufacturing Technology*, *43*(3-4), 226–237. doi:10.100700170-008-1713-0

Chakraborty, S. (2011). Applications of the MOORA method for decision making in manufacturing environment. *International Journal of Advanced Manufacturing Technology*, *54*(9-12), 1155–1166. doi:10.100700170-010-2972-0

Chakraborty, S., Dandge, S. S., & Agarwal, S. (2020). Non-traditional machining processes selection and evaluation: A rough multi-attributive border approximation area comparison approach. *Computers & Industrial Engineering*, *139*, 106201. doi:10.1016/j.cie.2019.106201

Chakraborty, S., & Dey, S. (2006). Design of an analytic- hierarchy-process-based expert system for non-traditional machining process selection. *International Journal of Advanced Manufacturing Technology, 31*(5-6), 490–500. doi:10.100700170-005-0216-5

Chakraborty, S., & Dey, S. (2007). QFD-based expert system for non-traditional machining processes selection. *Expert Systems with Applications, 32*(4), 1208–1217. doi:10.1016/j.eswa.2006.02.010

Chan, L. K., & Wu, M. L. (2002). Quality function deployment: A literature review. *European Journal of Operational Research, 143*(3), 463–497. doi:10.1016/S0377-2217(02)00178-9

Chandraseelan, E. R., Jehadeesan, R., & Raajenthiren, M. (2008). Web-based knowledge base system for selection of non- traditional machining processes. *Malaysian Journal of Computer Science, 21*(1), 45–56. doi:10.22452/mjcs.vol21no1.5

Chatterjee, P., & Chakraborty, S. (2013). Non-traditional machining processes selection using evaluation of mixed data method. *International Journal of Advanced Manufacturing Technology, 68*(5-8), 1613–1626. doi:10.100700170-013-4958-1

Choudhury, T., Das, P. P., Roy, M. K., Shivakoti, I., Ray, A., & Pradhan, B. B. (2013). Selection of non-traditional machining process- a distance based approach. In *IEEE International Conference on Industrial Engineering and Engineering Management* (pp. 852-856). IEEE. 10.1109/IEEM.2013.6962532

Chuang, P. T. (2001). Combining the analytic hierarchy process and quality function deployment for a location decision from a requirement perspective. *International Journal of Advanced Manufacturing Technology, 18*(11), 842–849. doi:10.1007001700170010

Cogun, C. (1993). Computer-aided system for the selection of nontraditional machining operations. *Computers in Industry, 22*(2), 169–179. doi:10.1016/0166-3615(93)90063-7

Cogun, C. (1994). Computer-aided preliminary selection of non- traditional machining processes. *International Journal of Machine Tools & Manufacture, 34*(3), 315–326. doi:10.1016/0890-6955(94)90002-7

Cohen, L. (1995). *Quality Function Deployment- how to make QFD work for you.* Addison-Wesley.

Das, P. P., & Chakraborty, S. (2020a). Lexicographic method-based parametric optimization of non-traditional machining processes for ceramic materials. *OPSEARCH, 57*(3), 700–715. doi:10.100712597-020-00439-8

Das, P. P., & Chakraborty, S. (2020b). Application of grey correlation-based EDAS method for parametric optimization of non-traditional machining processes. *Scientia Iranica.* Advance online publication. doi:10.24200/SCI.2020.53943.3499

Das, S., & Chakraborty, S. (2011). Selection of non-traditional machining processes using analytic network process. *Journal of Manufacturing Systems, 30*(1), 41–53. doi:10.1016/j.jmsy.2011.03.003

Govers, C. P. M. (2001). QFD not just a tool but a way of Quality management. *International Journal of Production Economics, 69*(2), 151–159. doi:10.1016/S0925-5273(00)00057-8

Hauser, J. R., & Clausing, D. (1988). The house of quality. *Harvard Business Review*, 63–73.

Jain, V. K. (2002). *Advanced machining processes.* New Delhi: Allied Publishers Private Limited.

Karande, P., & Chakraborty, S. (2012). Application of PROMETHEE-GAIA method for non-traditional machining process selection. *Management Science Letters*, *2*(6), 2049–2060. doi:10.5267/j.msl.2012.06.015

Madić, M., Petković, D., & Radovanović, M. (2015). Selection of non-conventional machining processes using the OCRA method. *Serbian Journal of Management*, *10*(1), 61–73. doi:10.5937jm10-6802

Prasad, K., & Chakraborty, S. (2014). A decision-making model for non-traditional machining processes selection. *Decision Science Letters*, *3*(4), 467–478. doi:10.5267/j.dsl.2014.7.002

Prasad, K., & Chakraborty, S. (2018). A decision guidance framework for non-traditional machining processes selection. *Ain Shams Engineering Journal*, *9*(2), 203–214. doi:10.1016/j.asej.2015.10.013

Roy, M. K., Ray, A., & Pradhan, B. B. (2014). Non-traditional machining process selection using integrated fuzzy AHP and QFD techniques: A customer perspective. *Production & Manufacturing Research*, *2*(1), 530–549. doi:10.1080/21693277.2014.938276

Saaty, T. L. (1988). *The Analytic Hierarchy Process: Planning, Priority Setting, Resource Allocation.* RWS Publication.

Saaty, T. L. (1994). How to make decision: The analytic hierarchy process. *Interfaces*, *24*(6), 19–43. doi:10.1287/inte.24.6.19

Temuçin, T., Tozan, H., Vayvay, Ö., Harničárová, M., & Valíček, J. (2014). A fuzzy based decision model for non-traditional machining process selection. *International Journal of Advanced Manufacturing Technology*, *70*(9-12), 2275–2282. doi:10.100700170-013-5474-z

Wasserman, G. S. (1993). On how to prioritize design requirements during the QFD planning process. *IIE Transactions*, *25*(3), 59–65. doi:10.1080/07408179308964291

Yurdakul, M., & Cogun, C. (2003). Development of a multi-attribute selection procedure for non-traditional machining processes. *Proceedings of the Institution of Mechanical Engineers. Part B, Journal of Engineering Manufacture*, *217*(7), 993–1009. doi:10.1243/09544050360686851

Chapter 12
Parametric Optimization of Electrochemical Discharge Machining Using Particle Swarm Optimization Algorithm

Arindam Debroy
Indian Institute of Technology, Kharagpur, India

ABSTRACT

It is very important to select the optimal parametric values for various non-traditional machining processes (NTM) for improving their performance. The performance measures of NTM processes include material removal rate (MRR), radial overcut (ROC), heat affected zone (HAZ), etc. In this chapter, particle swarm optimization has been used to find out the optimal parametric settings for electrochemical discharge machining (ECDM) to improve its performance measure. Both single-objective as well as multi-objective optimization has been performed and the results have been compared with those obtained by other researchers.

INTRODUCTION

The advancement of technology during the last few decades, in industries, such as aeronautics, nuclear reactors, automobiles have been demanding materials, like high strength temperature resistant (HSTR) alloys having high strength-to-weight ratio to be used by them [Ghosh and Mallik (2008)]. In conventional machining processes, the increase in hardness of the work material causes a significant decrease in economic cutting speed. In order to machine such hard and difficult-to-machine materials the hardness and strength of the cutting tool has to be increased, which increase in the cost of machining. The increasing level of strength of the work materials has had a dangerous effect on the total cost of machining. Due to these reasons advancement in machining processes were made. The advanced machining processes also known as non-traditional machining processes (NTM) are different from conventional processes, as they

DOI: 10.4018/978-1-7998-7206-1.ch012

do not employ traditional tools, which are used by traditional machining processes. These machining processes employ unusual methods for removing metals, such as abrasives, water, electrons etc.

Electrochemical discharge machining (ECDM) process is a hybrid machining technology, which combines the machining processes of electrochemical machining and electrical discharge machining. It is a reproductive machining process in which the shape of the tool electrode is produced on the workpiece in exactly the same manner. It uses two electrodes, one connected to the positive terminal, i.e. cathode, where the tool is connected and the other connected to the negative terminal, i.e. anode where the auxiliary electrode is connected. The workpiece is placed just below the tool and along with the auxiliary electrode, is submerged in an electrolytic solution in the machining chamber. In ECDM process, the thermal erosive effect of the electric discharge action is followed by the electrochemical reaction. The electrochemical reaction helps in generating positively charged ions and gas bubbles, i.e. hydrogen. These gas bubbles accumulate across the interface of the tool and the workpiece. The electrode discharge action takes place between the tool and the electrolyte across the gas bubble layers. DC power supply is used. If the DC power supply is greater than the breakdown voltage of the insulating layer, then a spark is initiated. The intensity and the energy of the spark increases with increase in the applied voltage between the two electrodes. A non-conducting ceramic workpiece is placed in the closed vicinity of the electrical discharge, and the material of the workpiece gets melted, vaporized and eroded, due to the transmission of a fraction of the spark energy to the workpiece. This raises the temperature of the region dramatically, and a part of the molten portion of the workpiece is removed due to the mechanical shock resulting from the sudden phase change and the electrical spark discharge. Some additional material removal also takes places due to thermal spalling. The thermal spalling of ceramics is usually known as a mechanical failure of the material, which takes place without melting due to a localized, thermally-induced internal stress caused by a rapid change in temperature that exceeds the bond strength of the materials. In ECDM process, the material undergoes thermal cycling under a spark discharge condition when the pulsed DC voltage is applied and a complex temperature gradient is established. These results in internal thermal stress and leads to thermal spall.

The variety of the NTM processes has increased over the last few decades. All these NTM processes have their own optimal parametric settings in which their performance measures, such as material removal rate, surface roughness, etc. have their optimal values. Many optimization algorithms have been used by past researchers in order to find out the optimal parametric settings for these non-traditional machining processes. Sumanta and Chakraborty (2011) used artificial bee colony optimization algorithm for finding the optimal parametric conditions for three NTM processes. They found the optimal process parameters of electrochemical machining, electrochemical discharge machining and electrochemical micro machining. Rao. et. al. (2008) applied particle swarm optimization algorithm for finding optimal process parameters for electrochemical machining and compared their results with the results obtained by using other optimization processes.

In this chapter, we have shown the application of particle swarm optimization to find the optimal parameters for electrochemical discharge machining. We have used the experimental outputs found by Sarkar et al. (2006) for finding the process parameters and optimizing them. Both single and multi-objective optimization of the process has been conducted in this work and results have been shown.

The rest of the paper is organized as follows. In section 2, brief discussion about particle swarm optimization has been showed followed by the illustrative example in section 3. Section 4 states the conclusion and future scope of the work.

PARTICLE SWARM OPTIMIZATION

Particle swarm optimization algorithm (PSO) was first developed by Kennedy and Eberhart (1999) in the year 1995. PSO is based on the social behavior of a flock of birds, school of fishes and a group of people. In PSO, the group is also known as swarm. Swarm has been defined as a population of interactive elements or agents who are able to optimize some global objective through collaborative search in space. The swarm consists of number of individuals called the particles. These particles fly in n-dimensional space. In PSO algorithm, the flight trajectory of a particle is influenced by the trajectory of neighborhood particles as well as the flight experience of the particle itself. Each particle is treated as a point in the n-dimensional space. Each particle keeps information of its best coordinates in the problem space. These coordinates are the best solutions that the particle has achieved till that time. It is termed as pbest (personal best). The particle also keeps track of other particles in the neighborhood. So, it is important to check whether there is any other particle in the neighborhood that has better coordinates than this particle. If there is another particle with better coordinates than its coordinates, then that particle's coordinates are termed as local best, (lbest), otherwise, the coordinates of the considered particle are taken as the local best (lbest). Now, if this method is continued and all the particles individually are considered, the option of checking the coordinates of all the particles and changing the lbest and pbest values accordingly can be adopted. The particle with the best coordinate values after all the particles have been compared is termed as the global best value (gbest). This global best value gives the best solution of the optimization problem. Every particle has a velocity component and a position component. Say, for i^{th} particle the position component is represented by x_i and the velocity component is represented by v_i. Now, this particle will keep on changing its velocity, along with which the position of the particle also changes. The velocity component is represented by $v_i = v_{i1}, v_{i2}, v_{i3}, \ldots\ldots, v_{in}$ and the position component is represented by $x_i = x_{i1}, x_{i2}, \ldots\ldots, x_{in}$. The particles change their velocity and position according to the following equation:

$$v_i^{k+1} = wv_i^k + c_1 r_1 \left(p_i - x_i^k\right) + c_2 r_2 \left(p_g - x_i^k\right) \quad (1)$$

$$x_i^{k+1} = x_i^k + v_i^{k+1} \quad (2)$$

where i = 1, 2, . . ., N, N is the size of the population, w is the inertia weight, c_1 and c_2 are two positive constants, called the cognitive and social parameter respectively, and r_1 and r_2 are random numbers uniformly distributed within the range (0,1). The above equations are used to determine the new velocity and position of ith particle. The movement of a particle in a two dimensional space is shown in Figure 1. This type of optimization process is used to search the entire problem space and thus helpful in solving complex problems. This optimization method has its own advantages, i.e.

1. It is insensitive to scaling of the design variables.
2. Its implementation is simple.

Figure 1. Description of velocity and position updates in PSO

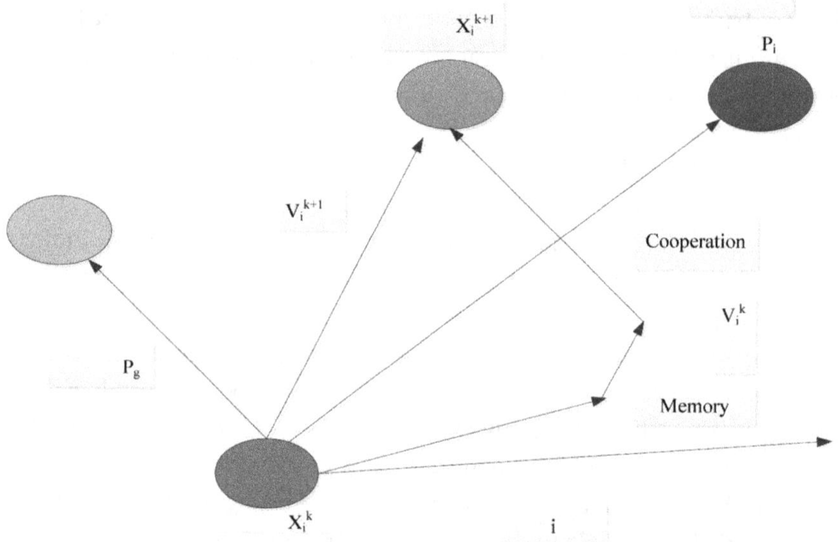

The pseudo codes for PSO algorithm has been shown in Table 1 [Singh, C. and Wang, L (2007)]:

Table 1. Pseudo codes for PSO algorithm

```
For each particle:
Initialize particle
END
Do
For each particle:
Calculate fitness value
If the fitness value is better than the best fitness value (pbest) in history
Set the new value as the pbest
END
Choose the particle with the best fitness value of all the particles as the gbest value
For each particle:
Calculate the particle velocity according to Eqn. (1)
Update the particle position according to Eqn. (2)
END
```

Review of Past Researches on Application of Particle Swarm Optimization Technique

Abd-El-Wahed et al. [2012] proposed a hybrid multi-objective evolutionary algorithm combining two heuristic optimization techniques, i.e. GA and PSO. That algorithm had two features, i.e. firstly, the algorithm was nitialized by a set of random particles which were flown through the search space, and secondly, the local search scheme was implemented as a neighbourhood search engine to improve the solution quality. The authors used various kinds of multi-objective benchmark problems to show the importance of hybridization algorithm in generating Pareto-optimality sets for multi-objective optimization problems.

Afzulpurkar and Navalertporn [2011] developed an integrated optimization approach using an ANN and a bidirectional particle swarm. The authors used ANN to obtain the relationship between the decision variables and performance measures of interest, and the bidirectional particle swarm was then adopted to perform the optimization with multiple objectives. The approach was used to solve a process parameter design problem in cement roof tile manufacturing.

Singh and Wang [2007] proposed a fuzzified multi-objective PSO algorithm and applied that algorithm to dispatch the electric power, considering both economic as well as environmental issues. The environmental issues arising from the use of fossil fuels in the power generation in conventional economic power dispatch could not meet the environmental protection requirements. It had considered economic goal more important as compared to the environmental problems. The algorithm developed by the authors considered both the objectives and provided satisfactory results.

Malviya and Pratihar [2011] applied PSO for tuning of the neural networks that were utilized for carrying out both forward and reverse mapping of different metal inert gas welding processes. Four different approaches were developed and their performances were compared while solving the problems considered. The first and second approaches dealt with PSO-tuned multilayer fed forward neural networks (MLFFNN) and radial basis function neural networks (RBFNN) respectively, whereas, some hybrid schemes were adopted in the last two approaches. Both the third and fourth approaches showed better results as compared to first and second approaches.

Zhao et al. [2011] used two lbests multi-objective particle swarm optimization (2LB-MOPSO) and applied it to design multi-objective robust proportional integral derivative (PID) controller for distillation column plant and longitudinal control system of the super manoeuvrable F18/HARV aircraft. The authors formulated a multi-objective robust PID controller design problem by minimizing integral squared error and balanced robust performance criteria.

Tripathi et al. [2007] described a novel PSO approach for multi-objective optimization called time-variant multi-objective particle swarm optimization (TV-MOPSO) for solving multi-objective optimization problems. The developed approach was made adaptive by allowing the vital parameters of the algorithm, i.e. inertial weight and acceleration coefficient to change along with the iterations. A mutation operator was used to resolve the problem of premature convergence to the local Pareto-optimal front. The performance measure of the new approach was compared with other multi-objective PSO techniques and it was found that the new approach gave better optimization results.

Jiang et al. [2007] presented a stochastic convergence analysis of the standard PSO algorithm. The authors considered each particle's position as a stochastic vector and analysed the standard PSO algorithm using stochastic process theory. The results of the analysis had led to a convergent condition for the standard particle swarm system. The results obtained by the authors proved helpful in understanding the mechanism of standard PSO algorithm and selecting appropriate parameters to make PSO algorithm more powerful.

Zhang et al. [2012] proposed a new bare-bones multi-objective PSO technique to solve the economic dispatch problems. The authors compared the developed algorithm with three other multi-objective particle swarm techniques and found that it was capable of generating excellent approximation of the true Pareto front. The authors considered three objectives, i.e. the minimal fuel cost, the minimal emission and the compromise solution, and found that the proposed algorithm gave better solutions than the other known multi-objective PSO algorithms.

Coelho and Lee [2008] proposed an improved PSO approach for solving economic dispatch problems, taking into account non-linear generator features, such as ramp-rate limits and prohibited operating zones in the power system operations. The authors combined PSO algorithm with Gaussian probability distribution functions and chaotic sequences to develop a new algorithm. Two test systems were considered and solved using the proposed method. The proposed method outperformed the other modern meta heuristic-based optimization techniques.

Safari and Shayeghi [2011] used iterative PSO to determine the feasible optimal solution of the economic load dispatch problem considering various generator constraints. The authors considered many realistic constraints, such as ramp rate limit, generation limitation, prohibited operating zone and transmission loss. Two test power systems were considered and the algorithm was applied along with some other algorithms. The proposed algorithm gave better results as compared to other algorithms.

El-Zonkoly [2011] proposed a multi-objective index approach that could be used for optimally determining the size and location of multi-distributed generation units in distribution systems with different load models. An optimization technique based on PSO algorithm was also used. The proposed optimization algorithm was applied to two test systems. The results showed the efficiency of the algorithm in improving voltage profile, reduction of power losses and reduction of maga-volt-ampere flow intake from the grids.

Deepa and Sugumaran [2011] proposed an algorithm for model order formulation of an absolutely stable higher order linear time invariant multivariable discrete system, using a new version of evolutionary computing technique, which was named as modified particle swarm optimization (MPSO) algorithm. A nth order linear time invariant dynamic multivariable system with q inputs and r outputs was considered. The desired results were compared with those obtained by the earlier techniques in order to validate its ease of computation.

Kuo and Han [2011] applied bi-level linear programming to supply chain distribution problems and developed an efficient method based on hybrid of GA and PSO. Three hybrid methods were developed integrating GA and PSO. The methods were applied to solve four problems. The hybrid algorithms were found to give better results as compared to individual algorithms from which they were developed.

Zhang et al. [2008] proposed an improved PSO algorithm considering different constraints to solve the flow shop scheduling problem with the goal of minimizing the make span. The PSO algorithm was combined with genetic operators. The improved PSO was tested on different scale problems and was also compared with the proposed algorithm. The authors observed from the results that the improved PSO gave better performance as compared to other algorithms.

Galvez and Iglesias [2012] investigated the use of PSO to recover the shape of a surface from clouds of noisy 3-D data points. The authors applied PSO approach in order to reconstruct a non-uniform rational B-spline surface of a certain order from a given set of 3-D data points. Seven examples were considered including open, semi closed, zero-genus, high-genus surface and real world scanned objects. The results obtained proved that the method was superior to the other methods in terms of accuracy and generality.

ILLUSTRATIVE EXAMPLE

To demonstrate the application and effectiveness of PSO algorithm for parametric optimization of ECDM process is established by considering the observations of Sarkar et al. (2006).

Sarkar et al. (2006) developed a setup to study the influence of various process parameters in an electrochemical discharge machining operation. A silicon nitride ceramics workpiece of dimension, 20 x 20 mm^2 and 5 mm thickness was considered for experimental purpose. A stainless steel tool of 400 µm diameter was chosen for the experiments. Aqueous sodium hydroxide solution was used as the electrolyte, as it has high electrical conductivity. Figure 2 shows the schematic diagram of various components of the experimental setup. Sarkar et al. (2006) considered three process parameters, i.e. applied voltage, electrolyte concentration and inter-electrode gap, and three process performances, i.e. material removal rate (MRR), radial overcut (ROC) and heat affected zone (HAZ). These process parameters were set at five different levels, as shown in Table 2. A central composite rotatable second order-experimental plan was adopted, and 20 experiments were conducted based on that plan. Based on the observed data, three RSM-based second order equations were developed using the coded values of the process parameters, which are shown in Eqns. (3)-(5). PSO algorithm is now used to solve and optimize these RSM-based

Figure 2. Schematic diagram of an ECDM setup [Sarkar et al. (2006)]

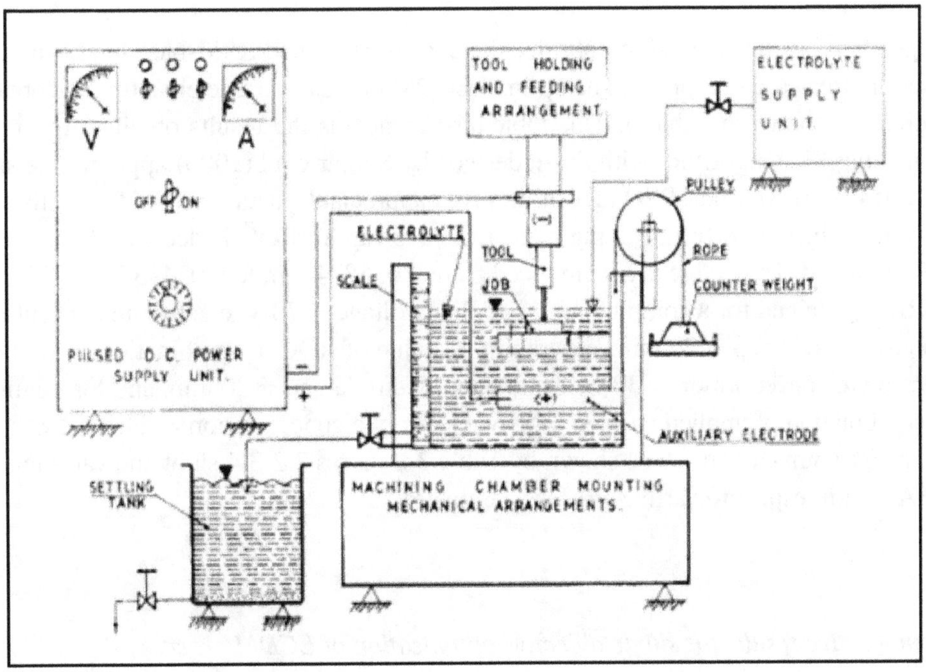

Table 2. Process parameters with their levels for ECDM process

Process parameters	Levels				
	-1.682	-1	0	+1	+1.682
Applied voltage (V)	50	54	60	66	70
Electrolyte concentration (wt %)	10	14	20	24	30
Inter-electrode gap (mm)	20	24	30	36	40

$$Y_u(MRR) = 0.60266 + 0.16049 * x_1 - 0.04044 * x_2 - 0.03481 * x_3 + 0.087813 * x_1^2 - 0.03060 * x_2^2 + 0.01358 * x_3^2 - 0.065 * x_1 * x_2 - 0.0375 * x_1 * x_3 + 0.045 * x_2 * x_3 \quad (3)$$

$$Y_u(ROC) = 0.16114 + 0.05333 * x_1 - 0.01017 * x_2 - 0.00716 * x_3 + 0.02454 * x_1^2 + 0.01727 * x_2^2 + 0.00598 * x_3^2 + 0.02603 * x_1 * x_2 - 0.00940 * x_1 * x_3 + 0.01493 * x_2 * x_3 \quad (4)$$

$$Y_u(HAZ) = 0.07835 + 0.01583 * x_1 - 0.00418 * x_2 - 0.00599 * x_3 + 0.00523 * x_1^2 + 0.00857 * x_2^2 + 0.00905 * x_1 * x_2 - 0.00060 * x_1 * x_3 + 0.00382 * x_2 * x_3 \quad (5)$$

where x_1 = applied voltage, x_2 = electrolyte concentration and x_3 = inter-electrode gap.

Single Response Optimization

The PSO algorithm is now adopted to optimize the above-mentioned RSM-based equations, where all the responses are treated individually. The optimal ECDM process parameter settings as obtained from PSO algorithm are shown in Table 3. This table also compares the results obtained from parametric optimization using PSO algorithm with those derived by Sarkar et al.(2006) applying the steepest ascent method. It is observed that PSO algorithm gives appreciably better results for all the responses. The MRR value is increased from 1.2 mg/hr to 1.602273 mg/hr, ROC is decreased from 0.1086 mm to 0.0591503 mm and HAZ value is also reduced from 0.055874 mm to 0.054955 mm. The maximum value of MRR is obtained for a combination of applied voltage = 70 V, electrolyte concentration = 10 wt% and inter-electrode gap = 20 mm. For minimum value of ROC, a combination of applied voltage = 50 V, electrolyte concentration = 30 wt% and inter-electrode gap = 20 mm, and for minimum value of HAZ, a combination of applied voltage = 50 V, electrolyte concentration = 24.5012 wt% and inter-electrode gap = 40 mm can be set, as shown by Table 3. Figures 3.2-3.4 show the variations of MRR, ROC and HAZ with respect to various process parameters.

Table 3. Comparative results for single objective optimization of ECDM process

Process parameters	Results obtained by Sarkar et al. [2006]			PSO algorithm		
	MRR (mg/hr)	**ROC (mm)**	**HAZ (mm)**	**MRR (mg/hr)**	**ROC (mm)**	**HAZ (mm)**
Applied voltage (V)	70	50	30	70	50	50
Electrolyte concentration (wt %)	18	22	22	10	30	24.5012
Inter-electrode gap (mm)	27	39	39	20	20	40
Optimal value	1.20	0.1086	0.055874	1.62273	0.0591503	0.054955

Figure 3 exhibits the variations of MRR with applied voltage, electrolyte concentration and inter-electrode gap. The figure clearly reveals that an increase in applied voltage is responsible for increase in MRR. A decrease in electrolyte concentration and inter-electrode gap would increase the value of MRR, as observed from this figure.

Figure 3. Variations of MRR with respect to different ECDM process parameters

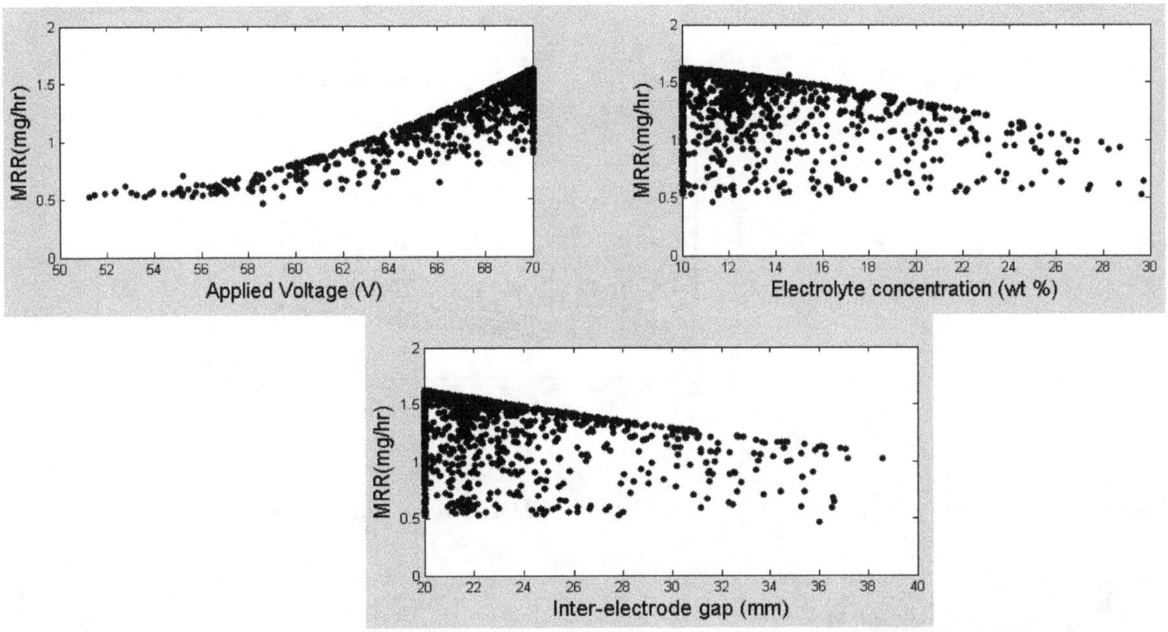

Figure 4 shows the influences of various process parameters on ROC. It is observed from the figure that ROC would decrease with decrease in applied voltage, increase in electrolyte concentration and decrease in inter-electrode gap.

Figure 5 clearly exhibits the variations of HAZ with various ECDM process parameters. It is observed from the figure that HAZ would decrease with decrease in applied voltage. An increase in electrolyte concentration and inter-electrode gap would cause a decrease in HAZ, as shown in this figure.

Figures 6 – 8 show the convergence of PSO algorithm for MRR, ROC and HAZ responses respectively.

Multi-response Optimization

For multi-response optimization of the responses for ECDM process, the following equation is developed.

$$Min(Z_1) = w_1 * \frac{Y_u(ROC)}{ROC_{min}} + w_2 * \frac{Y_u(HAZ)}{HAZ_{min}} - w_3 * \frac{Y_u(MRR)}{MRR_{max}} \tag{6}$$

where Y_u(ROC), Y_u(HAZ) and Y_u(MRR) are the three RSM-based equations of ROC, HAZ MRR respectively as given in Eqns. (3)-(5). ROC_{min}, HAZ_{min} and MRR_{max} are the minimum values of ROC,

Figure 4. Variations of ROC with respect to different ECDM process parameters

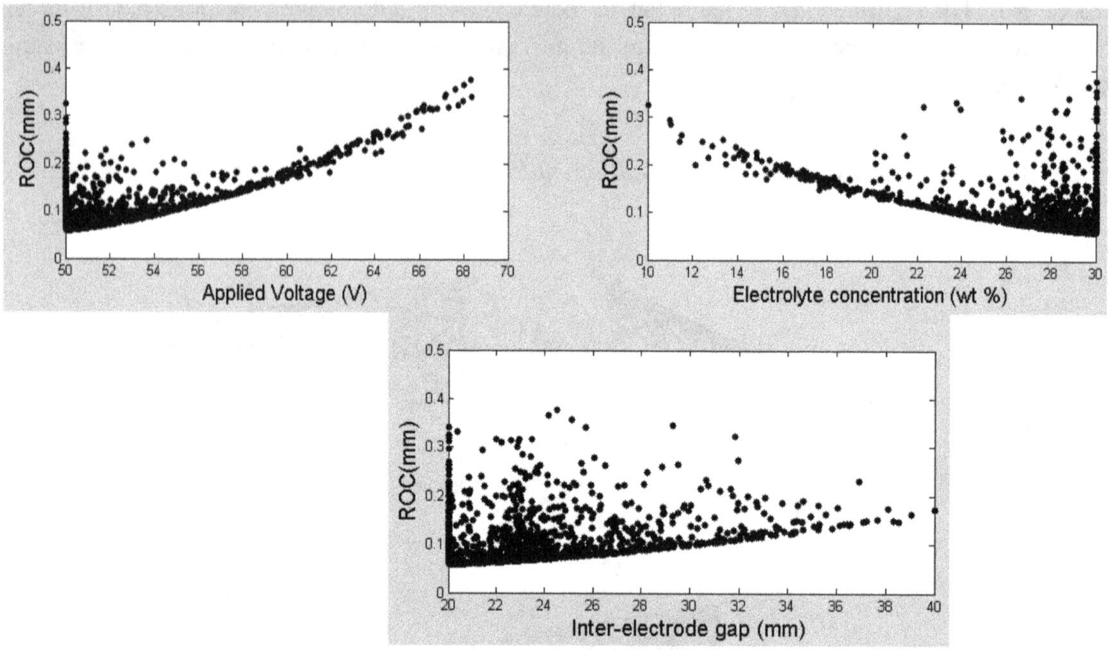

Figure 5. Variations of HAZ with respect to different ECDM process parameters

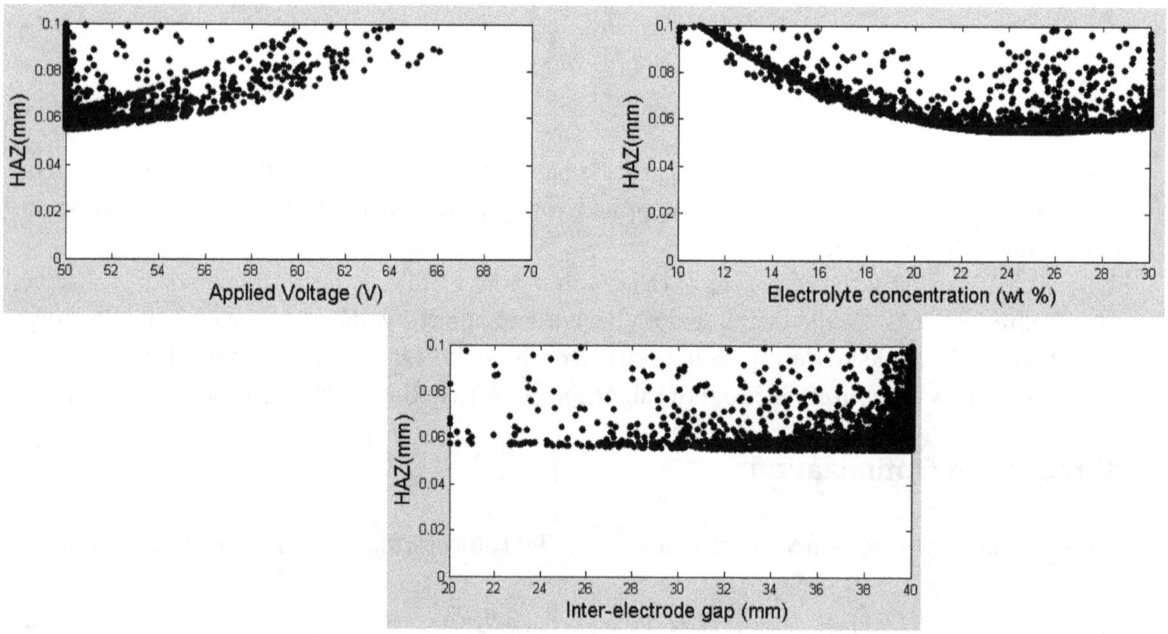

HAZ, and maximum value of MRR respectively, and w_1, w_2 and w_3 are the weights assigned to ROC, HAZ and MRR respectively. These weights can have any value, provided $w_1 + w_2 + w_3 = 1$. The values assigned to these weights depend on the priorities given to the responses by the process engineers. In this case, equal priority is allotted to all the responses, i.e. $w_1 = w_2 = w_3 = 0.3333$. The results obtained

Figure 6. Convergence of PSO for MRR

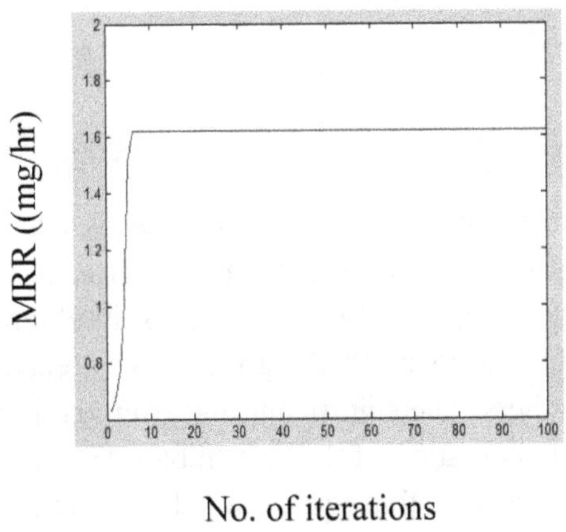

Figure 7. Convergence of PSO for ROC

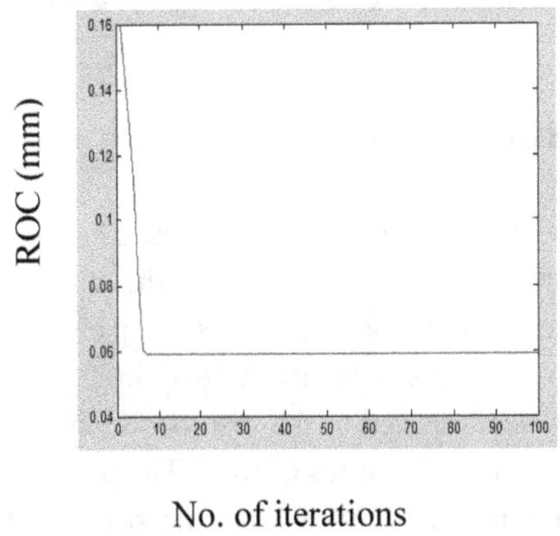

Figure 8. Convergence of PSO for HAZ

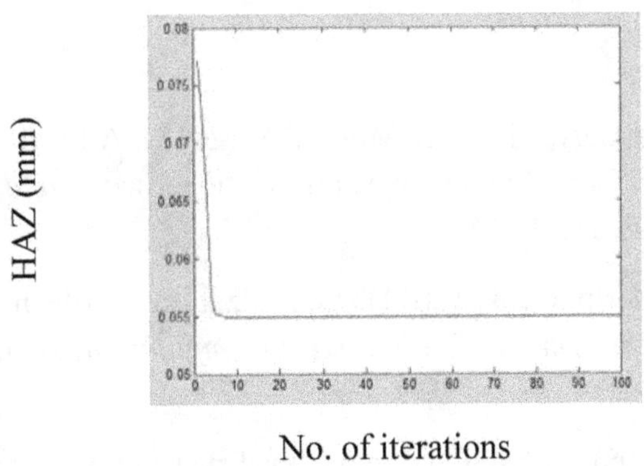

Table 4. Multi objective optimization of ECDM process

Process parameters and objective function	PSO algorithm
Applied voltage (V)	50
Electrolyte concentration (wt%)	30
Inter-electrode gap (mm)	20
MRR (mg/hr)	0.47401
ROC (mm)	0.05910
HAZ (mm)	0.05742
Z_1	0.5796

from multi-response optimization of ECDM process by applying PSO algorithm is shown in Table 4. It is observed that at a combination of applied voltage = 50 V, electrolyte concentration = 30 wt % and inter-electrode gap = 20 mm, the optimal values of all the responses are obtained. The values of MRR, ROC and HAZ are obtained as 0.47401 mg/hr, 0.05910 mm and 0.05742 mm respectively. The minimum value of the objective function (Z_1) is 0.5796. As Sarkar et al. [2006] did not perform any multi-response optimization of the ECDM process, thus the results cannot be compared here.

CONCLUSION

The selection of optimal parametric conditions for the non-traditional machining processes is vital for improving their performance. In this paper particle swarm optimization algorithm has been used to find out the optimal process parameters for electrochemical discharge machining (ECDM). The results obtained by performing single objective optimization of this NTM process by using PSO algorithm is a great deal better as compared to the results obtained by previous researchers. The significant improvement in the performance measures of the NTM process is due to the implementation of the swarm based optimization algorithm PSO. PSO has also been used for solving the multi-objective equation of the process. The future scope of this research work is, the algorithm can be used to find the optimal parametric settings for other non-traditional machining processes.

REFERENCES

Abd-El-Wahed, W. F., El-Shorbagy, M. A., & Mousa, A. (2012). A local search based hybrid particle swarm optimization algorithm for multi objective optimization. *Swarm and Evolutionary Computation*, *3*, 1–14. doi:10.1016/j.swevo.2011.11.005

Afzulpurkar, N. V., & Navalertporn, T. (2011). Optimization of tile manufacturing process using particle swarm optimization. *Swarm and Evolutionary Computation*, *1*(2), 97–109. doi:10.1016/j.swevo.2011.05.003

Coelho, L. S., & Lee, C. (2008). Solving economic load dispatch problems in power systems using chaotic and Gaussian particle swarm optimization approaches. *Electrical Power and Energy Systems*, *30*(5), 297–307. doi:10.1016/j.ijepes.2007.08.001

Deepa, S. N., & Sugumaran, G. (2011). Model order formulation of a multivariable discrete system using a modified particle swarm optimization approach. *Swarm and Evolutionary Computation*, *1*(4), 204–212. doi:10.1016/j.swevo.2011.06.005

El-Zonkoly, A. M. (2011). Optimal placement of multi-distributed generation units including different load models using particle swarm optimization. *Swarm and Evolutionary Computation*, *1*(1), 50–59. doi:10.1016/j.swevo.2011.02.003

Galvez, A., & Iglesias, A. (2012). Particle swarm optimization for non-uniform rational B-spline surface reconstruction from clouds of 3D data points. *Information Sciences*, *192*, 174–192. doi:10.1016/j.ins.2010.11.007

Ghosh, A., & Mallik, A. K. (2008). *Manufacturing Science*. East West Press Private Limited.

Jiang, M., Luo, Y. P., & Yang, S. Y. (2007). Stochastic convergence analysis and parameter selection of the standard particle swarm optimization algorithm. *Information Processing Letters*, *102*(1), 8–16. doi:10.1016/j.ipl.2006.10.005

Kuo, R. J., & Han, Y. S. (2011). A hybrid of genetic algorithm and particle swarm optimization for solving bi-level linear programming problem – A case study on supply chain model. *Applied Mathematical Modelling*, *35*(8), 3905–3917. doi:10.1016/j.apm.2011.02.008

Malviya, R., & Pratihar, D. K. (2011). Tuning of neural networks using particle swarm optimization to model MIG welding process. *Swarm and Evolutionary Computation*, *1*(4), 223–235. doi:10.1016/j.swevo.2011.07.001

Rao, R. V., Pawar, P. J., & Shankar, R. (2008). Multi-objective optimization of electrochemical machining process parameters using a particle swarm optimization algorithm, Proceedings of the Institution of Mechanical Engineers. *Proceedings of the Institution of Mechanical Engineers. Part B, Journal of Engineering Manufacture*, *222*(8), 949–958. doi:10.1243/09544054JEM1158

Safari, A., & Shayeghi, H. (2011). Iteration particle swarm optimization procedure for economic load dispatch with generator constraints. *Expert Systems with Applications*, *38*(5), 6043–6048. doi:10.1016/j.eswa.2010.11.015

Samanta, S., & Chakraborty, S. (2011). Parametric optimization of some non-traditional machining processes using artificial bee colony algorithm. *Engineering Applications of Artificial Intelligence*, *24*(6), 946–957. doi:10.1016/j.engappai.2011.03.009

Sarkar, B. R., Doloi, B., & Bhattacharyya, B. (2006). Parametric analysis on electrochemical discharge machining of silicon nitride ceramics. *International Journal of Advanced Manufacturing Technology*, *28*(9-10), 873–881. doi:10.100700170-004-2448-1

Shi, Y., & Eberhart, R. C. (1999). *Emperical study of particle swarm optimization*. Congress on Evolutionary Computation.

Singh, C., & Wang, L. (2007). Environmental/economic power dispatch using a fuzzified multi-objective particle swarm optimization algorithm. *Electric Power Systems Research*, *77*(12), 1654–1664. doi:10.1016/j.epsr.2006.11.012

Singh, C., & Wang, L. (2007). Environmental/economic power dispatch using a fuzzified multi-objective particle swarm optimization algorithm. *Electric Power Systems Research*, *77*(12), 1654–1664. doi:10.1016/j.epsr.2006.11.012

Tripathi, P. K., Bandyopadhyay, S., & Pal, S. K. (2007). Multi-objective particle swarm optimization with time variant inertia and acceleration coefficients. *Information Sciences*, *177*(22), 5033–5049. doi:10.1016/j.ins.2007.06.018

Zhang, C., Sun, J., Zhu, X., & Yang, Q. (2008). An improved particle swarm optimization algorithm for flow shop scheduling problem. *Information Processing Letters*, *108*(4), 204–209. doi:10.1016/j.ipl.2008.05.010

Zhang, Y., Gong, D., & Ding, Z. (2012). A bare-bones multi-objective particle swarm optimization algorithm for environmental/economic dispatch. *Information Sciences*, *192*, 213–227. doi:10.1016/j.ins.2011.06.004

Zhao, S.-Z., Iruthayarajan, M. W., Baskar, S., & Suganthan, P. N. (2011). Multi-objective robust PID controller tuning using two lbests multi-objective particle swarm optimization. *Information Sciences*, *181*(16), 3323–3335. doi:10.1016/j.ins.2011.04.003

ADDITIONAL READING

Ghadai, R. K., Kalita, K., Mondal, S. C., & Swain, B. P. (2018). PECVD process parameter optimization: Towards increased hardness of diamond-like carbon thin films. *Materials and Manufacturing Processes*, *33*(16), 1905–1913. doi:10.1080/10426914.2018.1512114

Ghadai, R. K., Kalita, K., Mondal, S. C., & Swain, B. P. (2019). Genetically optimized diamond-like carbon thin film coatings. *Materials and Manufacturing Processes*, *34*(13), 1476–1487. doi:10.1080/10426914.2019.1594273

Goswami, D., & Chakraborty, S. (2014). Differential search algorithm-based parametric optimization of electrochemical micromachining processes. *International journal of industrial engineering computations*, *5*, 41–54.

Kalita, K., Ragavendran, U., Ramachandran, M., & Bhoi, A. K. (2019). Weighted sum multi-objective optimization of skew composite laminates. *Structural Engineering and Mechanics*, *69*, 21–31.

Kalita, K., Shivakoti, I., & Ghadai, R. K. (2017). Optimizing process parameters for laser beam micro-marking using genetic algorithm and particle swarm optimization. *Materials and Manufacturing Processes*, *32*(10), 1101–1108. doi:10.1080/10426914.2017.1303156

Prakash, C., Singh, S., Singh, M., Antil, P., Aliyu, A. A., Abdul-Rani, A. M., & Sidhu, S. S. (2018). Multi-objective optimization of MWCNT mixed electric discharge machining of Al–30SiC p MMC using particle swarm optimization. In *Futuristic Composites* (pp. 145–164). Springer. doi:10.1007/978-981-13-2417-8_7

Ragavendran, U., Ghadai, R. K., Bhoi, A. K., Ramachandran, M., & Kalita, K. (2018). Sensitivity analysis and optimization of EDM process parameters. *Transactions of the Canadian Society for Mechanical Engineering*, *43*(1), 13–25. doi:10.1139/tcsme-2018-0021

Rahman, M. A., Anwar, S., & Izadian, A. (2016). Electrochemical model parameter identification of a lithium-ion battery using particle swarm optimization method. *Journal of Power Sources*, *307*, 86–97. doi:10.1016/j.jpowsour.2015.12.083

Rao, R. V., & Kalyankar, V. D. (2011). Parameters optimization of advanced machining processes using TLBO algorithm. *EPPM, Singapore*, *20*, 21–31.

Rao, R. V., Pawar, P. J., & Shankar, R. (2008). Multi-objective optimization of electrochemical machining process parameters using a particle swarm optimization algorithm. *Proceedings of the Institution of Mechanical Engineers. Part B, Journal of Engineering Manufacture*, *222*(8), 949–958. doi:10.1243/09544054JEM1158

KEY TERMS AND DEFINITIONS

Process Parameter: The various independent variables involved in a physical process or phenomenon may be referred to as process parameters.

Process Parameter Optimization: The search for suitable process parameter combination such that the desired process response is obtained.

Process Response: The measured output (i.e., dependent variable) of a process is known as its response. Example—material removal rate, surface roughness, etc.

Chapter 13
Optimization of Drilling Parameters for Composite Laminate Using Genetic Algorithm

Subham Pal
Indian Institute of Engineering Science and Technology, Shibpur, India

Salil Haldar
Indian Institute of Engineering Science and Technology, Howrah, India

ABSTRACT

Composite materials are preferred mostly in recent times due to their durability and ample space of applicability. Drilling is also an essential process in manufacturing, and it is frequently done to assemble products. Delamination due to drilling of the CFRP composite is considered as the primary concern in the manufacturing and assembly process. In this chapter, the empirical model for thrust force, torque, entry-delamination factor, exit-delamination factor, and eccentricity of drilling of CFRP composite is developed based on the extensive experiment. Response surface methodology is accounted to formulate a mathematical modal considering thrust force, torque, entry and exit-delamination factor, and eccentricity as response parameter and spindle speed, feed rate, and point angle as a process parameter. ANOVA is performed to check the statistical significance of the mathematical model. GA is employed to trace the optimum values of three process parameters to minimize the five response parameters. The Pareto front curve for various combinations of process parameters is also examined.

INTRODUCTION

Composite materials are mostly used in the aerospace engineering and marine engineering field. In recent time composite material are widely used in many engineering fields due to its high strength to weight ratio, high tensile modulus to weight ratio, and high fatigue strength. Among the different types

DOI: 10.4018/978-1-7998-7206-1.ch013

of composite material, fiber-reinforced composite materials are mostly applied to design aircraft components because of their high stiffness, low maintenance cost, lightweight, and it shows high internal damping. Due to advancements in manufacturing, the composite material can be designed as required. Among the manufacturing process, drilling is the most crucial process in mechanical assembly. Drilling in the fiber-reinforced composite material is challenging and also takes time and costly. Drilling in the fiber-reinforced composite leads to defects in a composite material such as delamination. Delamination can lead to failure of the component in use. Failure of an element due to drilling is depending on spindle speed, feed rate, and point angle. So, it is very obvious to trace the optimum drilling parameters to minimize the thrust force, torque, entry-delamination factor, exit-delamination factor, and eccentricity.

The literature review of drilling for the composite plate is vast. Some literature regarding the drilling of composite plate and optimization of drilling parameters are reported in this paper. Lazar and Xirouchakis have done an experimental analysis on drilling of composite material. They have used three types of drill bits to drill carbon-fiber-reinforced plastics (CFRP) and glass-fiber reinforced composite plastics (GFRP) (Lazar & Xirouchakis, 2011). They have focused on the delamination of the composite plate due to thrust force, torque, and different tool geometry. They have presented a variation of thrust force and axial force for different tool geometry at a different speed and feed rate. Quan and Zhong have examined the drilling and grinding of carbon fiber reinforced plastics (Quan & Zhong, 2009). They have investigated effects in different machining parameters, tool life, and material removing rate. Piquet et al. have used a special drill to experiment with drilling damage in the composite plate (Piquet, Ferret, Lachaud, & Swider, 2000). An experimental and statistical study of drilling in CFRS has been investigated by Davim and Reis. They have investigated the delamination tendency of a laminate due to the action of cutting force. They have examined the variation of power, specific cutting pressure, and delamination for various cutting parameters (speed and feed rate) (Davim & Reis, 2003). Delamination in the drilling of Armour-GFRP sandwich composite due to different cutting parameters has been examined by Bosco et al. They have presented the variation of delamination factor with respect to feed rate and speed through the graph (Bosco, Palanikumar, Prasad, & Velayudham, 2013). They have concluded that the feed rate is a highly influencing parameter in the drilling of the composite laminate. Hocheng and Tsao have studied to find a path towards delamination free drilling of composite material (Hocheng & Tsao, 2005). They have selected different tool geometry, various methods (traditional and non-traditional), and operating conditions. Krishnamoorthy et al. (Krishnamoorthy, Boopathy, & Palanikumar, 2009) have studied the delamination of CFRP composite due to drilling. They have used response surface methodology to develop a second-order mathematical equation for predicting delamination in drilling. Ghabezi and Khoran have investigated the effect in composite material due to drilling parameters such as cutting speed, feed rate, and tool geometry (GHABEZI1a & Khoran, 2014). Experimental results presented by them shows that the feed rate and drill diameter are the most influencing parameters for the delamination of composite material. They have found that the delamination factor is decreased with the increase of tool diameter. Davim and Reis have examined delamination due to drilling in CFRP. They have presented a new approach to select the cutting parameters to produce a damage-free drilling operation in CFRP. Taguchi's technique and ANOVA have been introduced in a new approach (Davim & Reis, 2003). They have presented a correlation between cutting velocity and feed rate with the delamination factor of CFRP. Study of damage-free drilling of carbon fiber reinforced thermoset using Taguchi's technique (Gowda, Ravindra, Prakash, Nishanth, & Ugrasen, 2015) (Sasikumar, 2015), and multi-objective optimization has been conducted by Enemuoh et al. (Enemuoh, El-Gizawy, & Okafor, 2001). Hocheng and Tsao (Hocheng & Tsao, 2006) have examined delamination due to drilling of composite laminate using a special drill bit. They have presented a variation of delamination with respect to thrust force for different drill bits. Delamination-free drilling of composite laminate has been examined by Upadhyay and Lyons to estimate the critical thrust force in drilling (Upadhyay & Lyons, 1999). Taguchi analysis for drilling of carbon fiber-reinforced plastics (CFRP) has been analysed by Tsao and Hocheng (Tsao & Hocheng, 2004). They have made a correlation between spinel speed, feed rate, and drill bit diameter with delamination factor using multi-variable linear regression. Krishnamoorthy et al. (Krishnamoorthy, Boopathy, Palanikumar, & Davim, 2012) have examined an optimization analysis to find the optimum drilling parameters for CFRP composite. They have used grey fuzzy logic (Vinayagamoorthy, et al., 2018) for the optimization of drilling parameters. Multi-objective optimization in the drilling of GFRP composite has been examined by Sonkar et al. (Sonkar, Abhishek, Datta, & Mahapatra, 2014) using the Degree of Similarity Measure approach. Saravanan et al. (Saravanan, Ramalingam, Manikandan, & Kaarthikeyen, 2012) have studied multi-objective optimization for drilling of the composite laminate. They have formulated the Genetic Algorithm (GA) (Sardinas, Reis, & Davim, 2006) (Krishnaraj, et al., 2012) to find the optimum value of hole-eccentricity and material removal rate. Optimization of drilling parameters for composite a using a response surface methodology has been performed by Chaudhary et al. (Chaudhary, Kumar, Verma, & Srivastav, 2014). They have considered MRR, oversize, and surface roughness as response parameter and spindle speed, feed rate, and point angle as process parameters. Abhishek et al. (Abhishek, Datta, & Mahapatra, 2016) have performed a multi-objective optimization for a CFRP composite using a fuzzy embedded Harmony Search (HS) algorithm. They have examined the efficiency of the Harmony Search Algorithm by comparing it with the Genetic Algorithm (GA) and Taguchi's robust optimization technique.

Although there have been many papers regarding the experimental drilling of composite and process optimization of drilling parameters, there is still ample scope to search for improved process combinations. In this paper, an attempt is made to find the optimum drilling parameters to minimize the thrust force, feed rate, entry-delamination factor, exit-delamination factor, and eccentricity. Genetic Algorithm (GA) is used to find the optimum values of the process parameter and the response parameter. A mathematical model is developed from experimental data using response surface methodology (RSM). Single and multi-objective optimization is carried out for different response parameters. The Pareto front curve for different response parameters is examined.

MATERIALS AND METHOD

Table 1. Material properties of the laminate (Krishnamoorthy, Boopathy, Palanikumar, & Davim, 2012)

CFRP	
Materials	Standard grade of carbon fiber
Thickness of carbon fiber filaments	0.05 mm
Tensile strength (GPa)	3.5
Tensile modulus (GPa)	230
Density (gm/ccm)	1.75
Specific strength (GPa)	2.00
Epoxy	
Material	EPON resin 8132
Viscosity (Poise)	5-7
Weight per epoxide	192-215
Density (lb/gal)	9.2

Table 2. Three sets of process parameters (Krishnamoorthy, Boopathy, Palanikumar, & Davim, 2012)

Process parameters	Level 1	Level 2	Level 3
Point angle, θ (°)	100	118	135
Feed rate, f (mm/min)	100	300	500
Drilling speed, v (rpm)	1000	2000	3000

Experimental Details

In the present study, the carbon fiber reinforced plastic is considered for drilling. Material properties of the carbon fiber and epoxy are described in Table 1. The CFRP plate is manufactured through a hand layup process. The CFRP plate has a thickness of 3 mm, and drilling is done using a 6 mm diameter high-speed steel (HSS) drill bit (Krishnamoorthy, Boopathy, Palanikumar, & Davim, 2012). Three drill bits of different point angles are used in this experiment. A combination of three sets of cutting parameters reported in Table 2 is used for the experiment. Piezo-electric Kistler dynamometer is used to measure the thrust force and drilling. Nikon D-200 camera is used to take photographs of the drilled hole. Images are feed to coreldraw software to measure the maximum drill diameter. A universal measuring machine (UMM) is used to calculate the eccentricity.

The delamination factor (F_d) of entry and exit is calculated (Krishnamoorthy, Boopathy, Palanikumar, & Davim, 2012) using Eqn. 1.

$$F_d = \frac{D_{max}}{d} \quad (1)$$

Where D_{max} = Maximum diameter at the drilling zone, d= diameter of the drill bit.

Table 3. Experimental data for drilling of CFRP (Krishnamoorthy, Boopathy, Palanikumar, & Davim, 2012)

Spindle speed (rpm)	Point angle (°)	Feed rate (mm/min)	Thrust force (N)	Torque (Nm)	Entry-delamination factor	Exit-delamination factor	Eccentricity (mm)
1000	100	100	99.6900	0.73	1.3418	1.4378	0.0728
1000	100	300	165.2033	0.84	1.3759	1.6373	0.0619
1000	100	500	198.3633	1.12	1.4368	1.5410	0.0609
1000	118	100	156.2500	0.99	1.3921	1.2628	0.0517
1000	118	300	253.2933	1.34	1.4400	1.4658	0.0431
1000	118	500	310.4667	1.37	1.5211	1.4137	0.0619
1000	135	100	155.4333	1.37	1.3398	1.1851	0.0437
1000	135	300	261.2300	1.52	1.3587	1.3692	0.0302
1000	135	500	310.0600	1.87	1.4756	1.2739	0.0251
2000	100	100	92.3667	0.48	1.3900	1.4455	0.0623
2000	100	300	154.0100	0.68	1.3439	1.5100	0.0815
2000	100	500	192.8733	0.87	1.3817	1.3607	0.1113
2000	118	100	140.1767	0.57	1.4287	1.4000	0.0652
2000	118	300	231.5233	0.92	1.4300	1.4562	0.0821
2000	118	500	271.8100	0.93	1.4474	1.3794	0.0799
2000	135	100	150.7533	0.64	1.4021	1.3296	0.0671
2000	135	300	234.7800	0.94	1.3798	1.3585	0.0655
2000	135	500	299.1833	0.95	1.4110	1.4500	0.0671
3000	100	100	84.2300	0.39	1.4287	1.4100	0.0156
3000	100	300	152.3867	0.47	1.3974	1.3807	0.032
3000	100	500	165.8133	0.60	1.3600	1.1688	0.0588
3000	118	100	130.6167	0.40	1.4347	1.3534	0.0308
3000	118	300	191.8567	0.54	1.4098	1.5100	0.0342
3000	118	500	270.3867	0.70	1.4224	1.4000	0.0411
3000	135	100	143.6367	0.48	1.4601	1.4400	0.0448
3000	135	300	226.0333	0.55	1.4264	1.5100	0.0601
3000	135	500	283.0000	0.78	1.4018	1.4774	0.0770

The experimental results of drilling of CFRP are presented in Table 3. All three process parameters and five response parameters are reported in Table 3.

Design of Experiment

Three process parameters (drilling speed, feed rate, and point angle) and five response parameters (thirst force, torque, entry-delamination, exit-delamination, and eccentricity) are used to build a mathematical model using a Response surface methodology (RSM). The process parameters are considered as the independent variables (input), and response parameters are regarded as dependent variables (output). The basic of the regression model can be written as following from

$$y_i = C_0 + C_i x_i + \epsilon_i \tag{2}$$

Where y_i is the regressand and x_i is the regressor and ϵi represent normally distributed statistical error, and $C0$ represents intercept and Ci represents slopes. Here i=1,2,3,...,n.

$$\epsilon_i = y_i - C_0 - C_i x_i \tag{3}$$

$$SSE = \sum_{i=1}^{n} \epsilon_i^2 = \sum_{i=1}^{n} \left(y_i - C_0 - C_i x_i \right)^2 \tag{4}$$

Minimization of the sum of the square of error (SSE) gives the estimates of C_0, C_i values. The second-order mathematical model can be written as follows

$$y = C_0 + \sum_{i=1}^{n} C_i x_i + \sum_{i=1}^{n} \sum_{i>j}^{n} C_{ij} x_i x_j + \sum_{i=1}^{n} C_{ii} x_i^2 + \epsilon_i \tag{5}$$

A second-order mathematical model is formulated in this article to establish a relationship between the three process parameters and five response parameters. Further, an analysis of variance (ANOVA) test is carried out to select and exclude the statically non-significant terms from the RSM model.

Genetic Algorithm

The quadric equation formulated by the RSM method is optimized using a genetic algorithm. GA is a nature-inspired search heuristic. The algorithm works on Charles Darwin's theory of natural evolution. It imitates the process of genetic and natural selection. The five primary operations of the genetic algorithm are the initial population, fitness function, selection, crossover, and mutation. GA select fitness individual from a set of data called the population. One individual is called a chromosome. Each chromosome is the solution to the problem. Set of variable called Genes to characterize Each chromosome. The algorithm selects the fittest individuals from the population. They produced offspring which have a similar character to the parents and send them to the next generation. The parents having better fitness value produce better offspring then them and have a high chance of surviving. This iterative process goes on, and in the end, a generation with a better chromosome will be found. GA can solve constrained and unconstrained both types of problems. GA can search for optimal parameters and predict the solution from a vast search space. It does not get trapped in local optima as it search parallel from the population of points. Pseudocode of the meta-modelling and optimization technique used in this study is given below.

1. Identify input and output parameters.
2. Build the RSM model.
3. Perform ANOVA

{

Check $R^2, R_{adj}{}^2$ *and Significance F*

```
If it is desirable, then use the model
Else model reduction and repeat ANOVA
}
```

4. Use metamodel as an objective function in GA.

    4.1 Set GA parameters
    4.2 Initialize random population
    4.3 Evaluate fitness and perform non-dominated sorting

```
{
Selection, Crossover, Mutation
Update fitness of each individual
Record best individual so far
If generation limit exhausted
Then predict optimal parameters
Else repeat from selection, crossover, and mutation
}
End
```

RESULT AND DISCUSSION

Formulation of the RSM model

In this article, a second-order regression model is formulated by using the RSM model to establish the relationships among input and output variables from the experimental data. The quadratic model for different objective functions (Thrust force, Torque, Entry-delamination, Exit-delamination, and Eccentricity) can be written as follows.

$$\text{Objective function} = C_0 + C_1 * v + C_2 * \theta + C_3 * f + C_4 * v * \theta + C_5 * v * f + C_6 * \theta * f + C_7 * v^2 + C_8 * \theta^2 + C_9 * f^2 \quad (6)$$

Here C_0 represents intercept and $C_1, C_2, C_3, C_4, C_5, C_6, C_7, C_8, C_9$ represents the slopes, v represent drilling speed, θ represent point angle of the drill bit, f represent feed rate of the tool.

Table 4. ANOVA table for second order RSM

Coefficient of intercept	*Objective function*				
	Thrust Force	**Torque**	**Entry-delamination factor**	**Exit-delamination factor**	**Eccentricity**
C_0	-1353.89464	-1.77584	-0.41773	3.04807	0.28913
C_1	-0.00577	0.00021	- 0.00002	- 0.00061	2.94831E-06
C_2	23.65598	0.03147	0.03074	- 0.01646	- 0.00377
C_3	0.13986	0.00122	- 6.82064E-06	0.00054	0.00009
C_4	- 0.00007	- 8.18553E-06	3.78790E-07	5.99196E-06	8.14037E-07
C_5	-0.00002	- 1.91667E-07	- 2.07958E-07	-2.08375E-07	4.41667E-08
C_6	0.00388	2.85183E-06	1.66585E-06	0.00001	-1.59299E-06
C_7	1.27760E-06	+ 1.16667E-07	1.07722E-08	-7.93889E-09	- 2.8783E-08
C_8	-0.09478	- 0.00003	- 0.00013	-0.000002	0.00001
C_9	-0.00040	- 4.99999E-07	4.888889E-07	-2.31431E-06	7.70833E-08
R^2	0.98597	0.96414	0.85626	0.85597	0.86516
$R^2_{adj.}$	0.97854	0.94515	0.78016	0.77972	0.79378
Significance F	5.88308E-14	1.59621E-10	0.00001	0.000015	8.84110E-06

ANOVA test results are given in Table 4. Values of intercept and slopes foe all objective functions are presented in Table 4. The applicability and correlation between input and output variables of the present mathematical models, correlation coefficient values are checked. Correlation co-efficient such as $R^2, R_{adj}^{\ 2}$ value close to 1 gives a good prediction of output variables. $R^2, R_{adj}^{\ 2}$ and *Significance F* values for all five response parameters are also reported in Table 4. From Table 4, it is clear that for every response parameters, the values of R^2 *and* $R_{adj}^{\ 2}$ are much closed to 1 and values of *Significance F* is less than 0.001. So, formulated mathematical model is accepted. The correlation between actual and predicted values of Thrust force, Torque, Entry-delamination factor, Exit–delamination factor, and eccentricity are shown in Figure 1, Figure 2, Figure 3, Figure, 4, Figure 5 respectively.

Figure 1. Actual with respect to predicted plot for thrust force

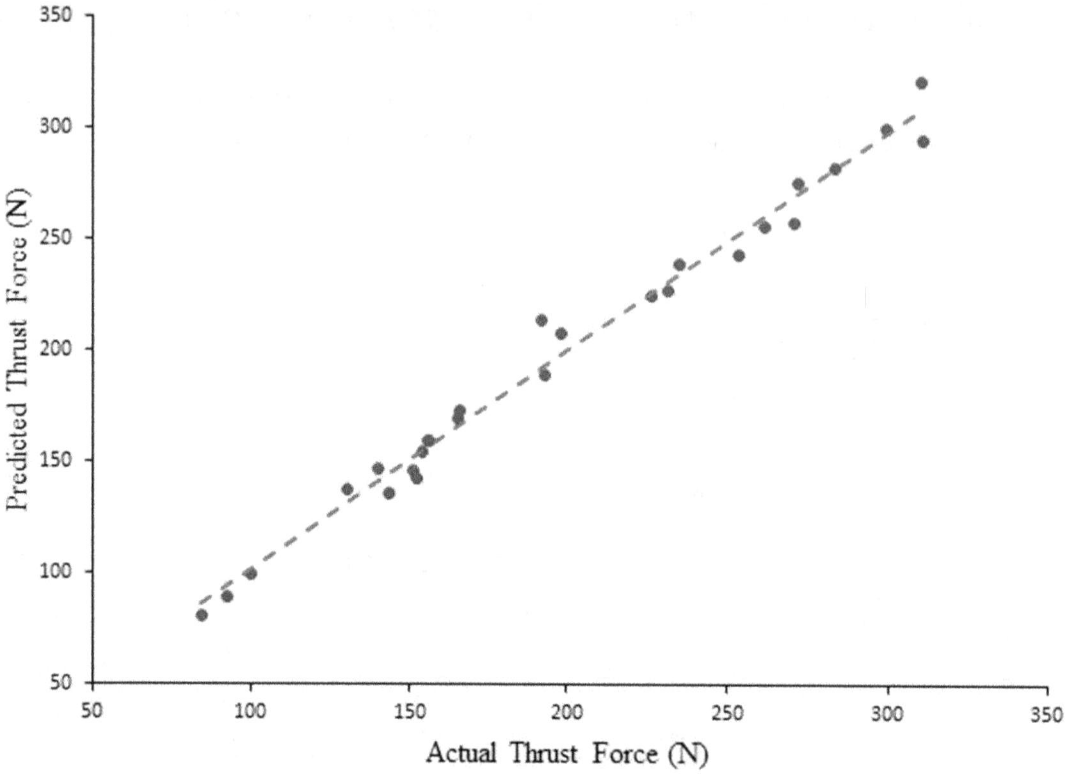

OPTIMIZATION USING GA

Single Objective Optimization

Single objective optimization is studied using the objective function (Eqn. 6) formulated for thrust force, torque, entry, and exit-delamination factor, and eccentricity of a CFRP composite laminate. The genetic algorithm is used to optimize the response parameters considering drilling speed (v), point angle (θ), and feed rate (f) as variables. The genetic algorithm optimization technique is performed using MATLAB R2018a software. For GA analysis, a population size of 50, generation of 500, and selection function as a stochastic uniform are considered. Again, a cross over probability of 0.8 and a mutation probability of 0.05 is considered.

The objective of the present work is to minimize the thrust force, torque, entry and exit-delamination, and eccentricity.

Minimize objective function
Subjected to the limits, $1000 \leq v \leq 3000;\ 100 \leq f \leq 500;\ 100 \leq \theta \leq 135;$

Figure 2. Actual with respect to predicted plot for torque

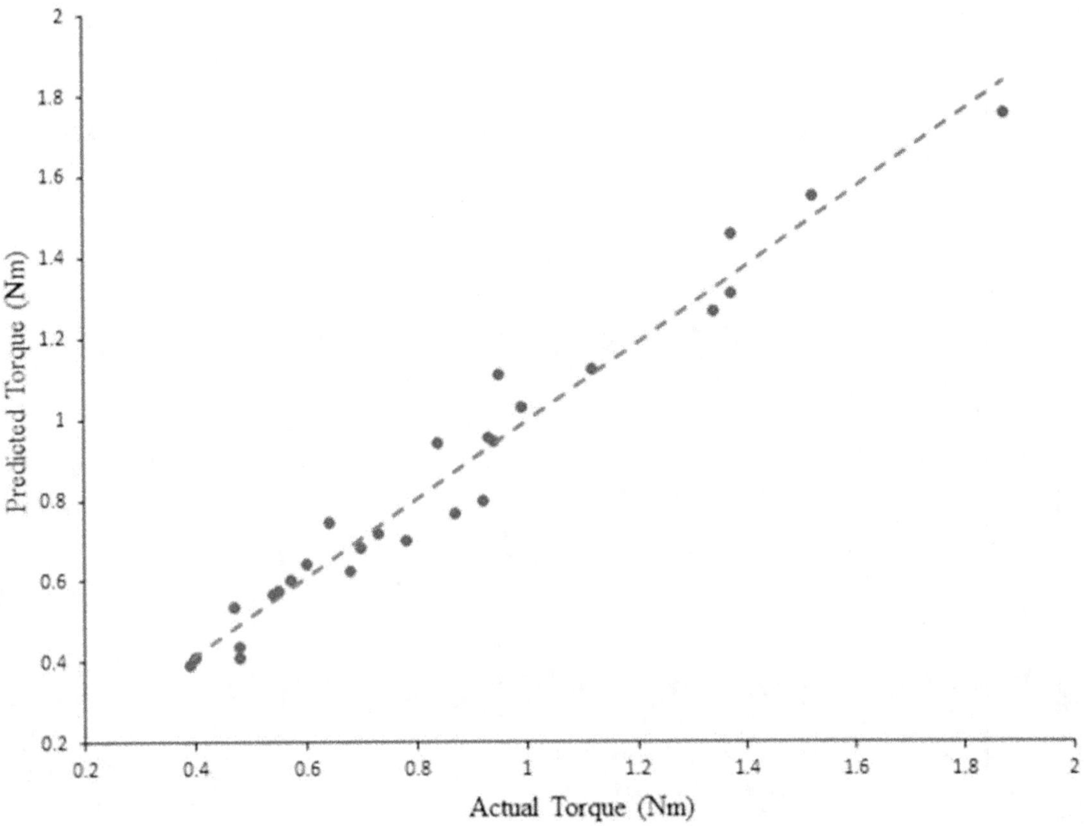

The optimum process parameters and minimum values of objective functions are reported in Table 5. From Table 5, it is clear that for the minimization of thrust force and torque, a maximum amount of cutting speed and minimum values of point angle and feed rate is required. At the end of 55th generation, it gives a minimum value for thrust force, and at the end of 164th generation, it provides a minimum amount for torque. For the minimization of the entry-delamination factor, minimum values of cutting speed, feed rate, and point angle are required, and at the end of 51th generation, it gives a minimum value. For minimization of the Exit-delamination factor, the same process parameter like entry-delamination is required except point angle. The minimum value of exit-delamination is obtained at the end of 237th generation. For minimization of the eccentricity, a high value of cutting speed and low value of feed rate and point angle are required, and at the end of 390th generation, it gives optimal values. The minimum value of the feed rate is required for the minimization problem.

Multi-Objective Optimization

In this example, the multi-objective optimization genetic algorithm (MOGA) is performed to find the optimal combination of multiple objective functions.

Figure 3. Actual with respect to predicted plot for entry-delamination factor

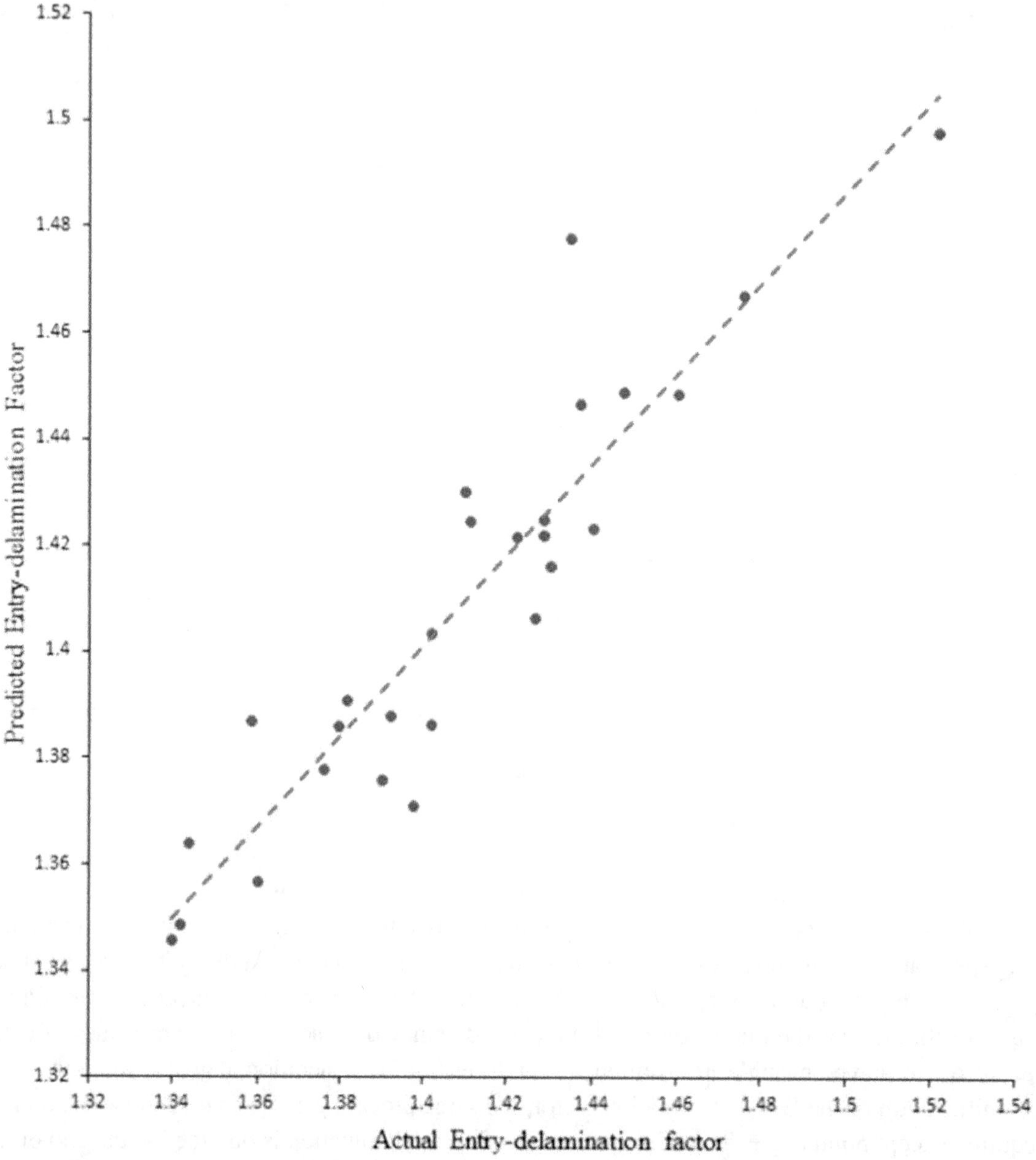

Example of multi-objective optimization can be given as follows,

Minimize Thrust Force
Minimize Torque
Subjected to the limits, $1000 \leq v \leq 3000$; $100 \leq f \leq 500$; $100 \leq \theta$ £135;

Figure 4. Actual with respect to. predicted plot for exit-delamination factor

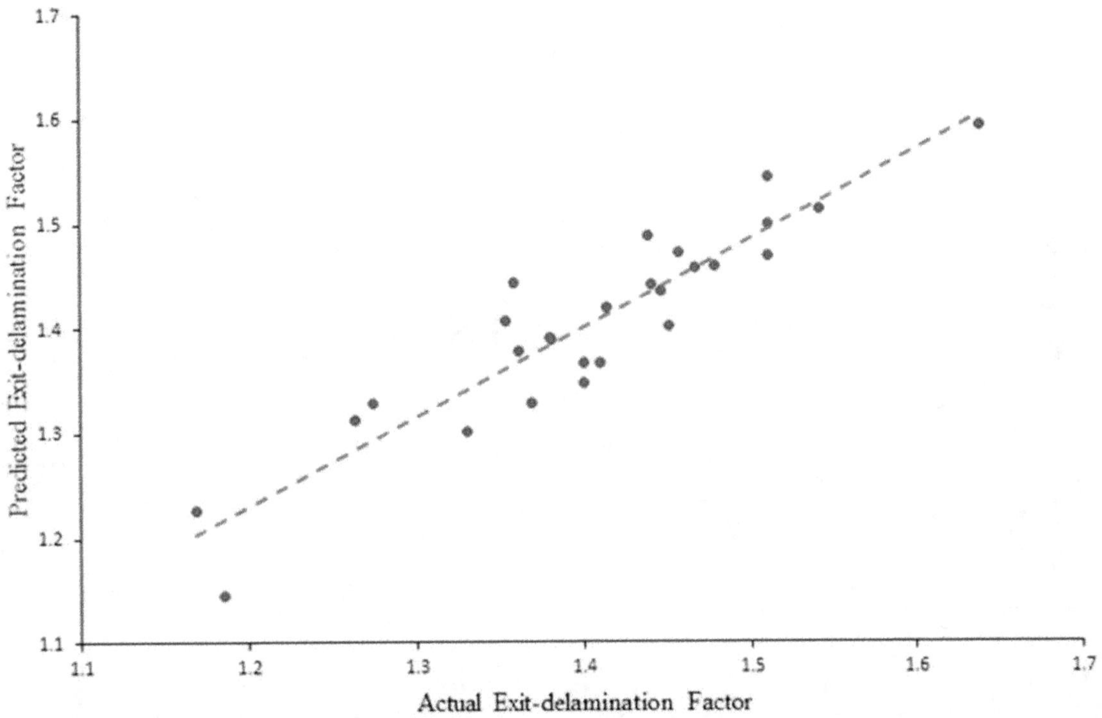

The Pareto front for various combinations of response parameters is performed and analysed. Pareto front for objective function minimization, such as thrust force and torque, is shown in Figure 6, and it shows one optimal solution. Again, Pareto front for minimization of thrust force versus entry-delamination is reported in Figure 7. Figure 7 gives a wide range of optimal values to the researchers to select according to application. So, the optimal value for maximum thrust force and minimum entry-delamination or maximum entry-delamination and minimum thrust force can be selected from the set of the optimum solution.

Farther, Pareto front curve for minimization of entry-delamination factor versus minimization of eccentricity is performed and presented in Figure 8. It gives the optimal sets of the objective function values, which are non-conflicting. So, the minimization of one objective function does not affect the other objective function.

The minimization of exit-delamination factor versus minimization of eccentricity is analysed and reported in Figure 9, and it gives a discreet type plot. Few other combinations of the objective function for multi-objective optimization are performed and reported such as minimization of force versus minimization of eccentricity (Figure 10), minimization of torque versus minimization of eccentricity (Figure 11) and minimization of torque versus minimization of entry-delamination factor (Figure 12). Among the Pareto front solution presented in this paper, torque with respect to force and eccentricity with respect to force gives only one Pareto optimal point.

Figure 5. Actual with respect to predicted plot for eccentricity

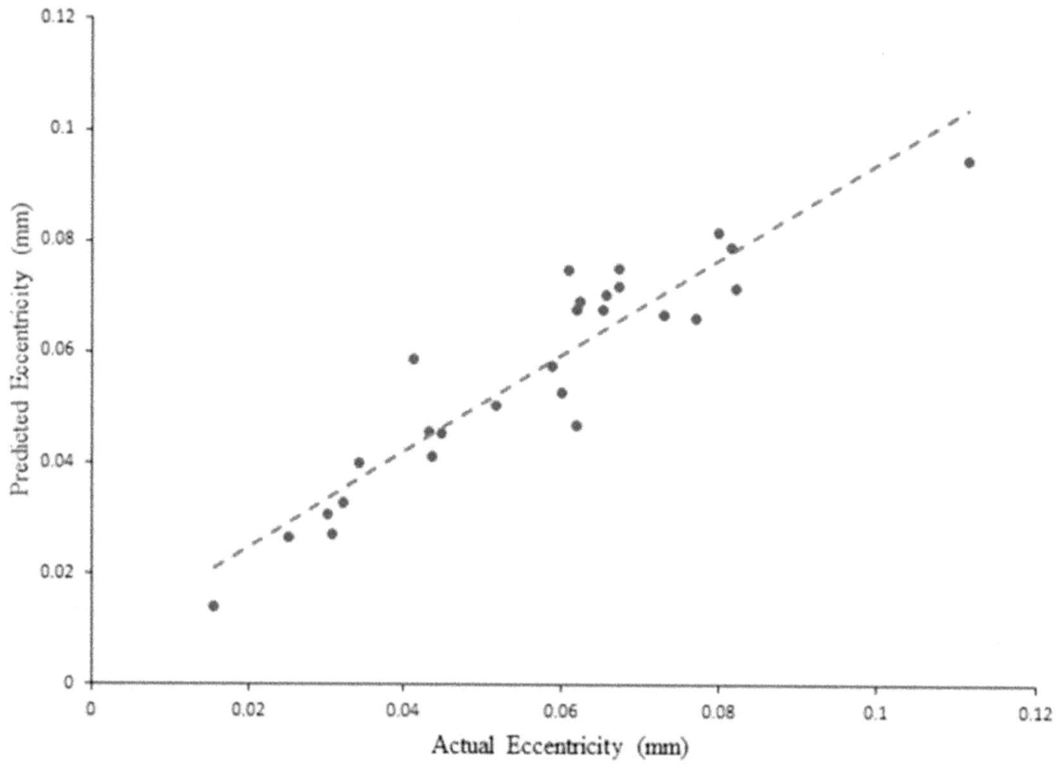

CONCLUSION

In the present study, the optimization of drilling parameters for a CFRP composite is performed using the genetic algorithm. The objective of the present study is to minimize the five response parameters. The results are summarized as follows:

Table 5. Optimum values of Drilling Parameters for single objective optimization

	Spindle speed (v) (RPM)	Feed rate (mm/min)	Point Angle (θ) (°)	Optimum objective function
Thrust Force (N)	2890.54	100	100	81.34
Torque (Nm)	2745.07	100	100	0.3806
Entry-delamination factor	1000	100	100	1.348
Exit-delamination factor	1000	100	135	1.146
Eccentricity (mm)	300	100	100	0.014

- The experimental value and the predicted value from the RSM model are very close, which means variability in output variables entirely explained by input variables.
- Genetic algorithm is advantageous to find the global optimal solution.
- The feed rate required for the minimization of response parameters is low. So, the feed rate has a great influence on the drilling of CFRP composite.
- Multi-objective optimization results give a set of data for the researchers to select according to their application.

Figure 6. Pareto front for force with respect to torque

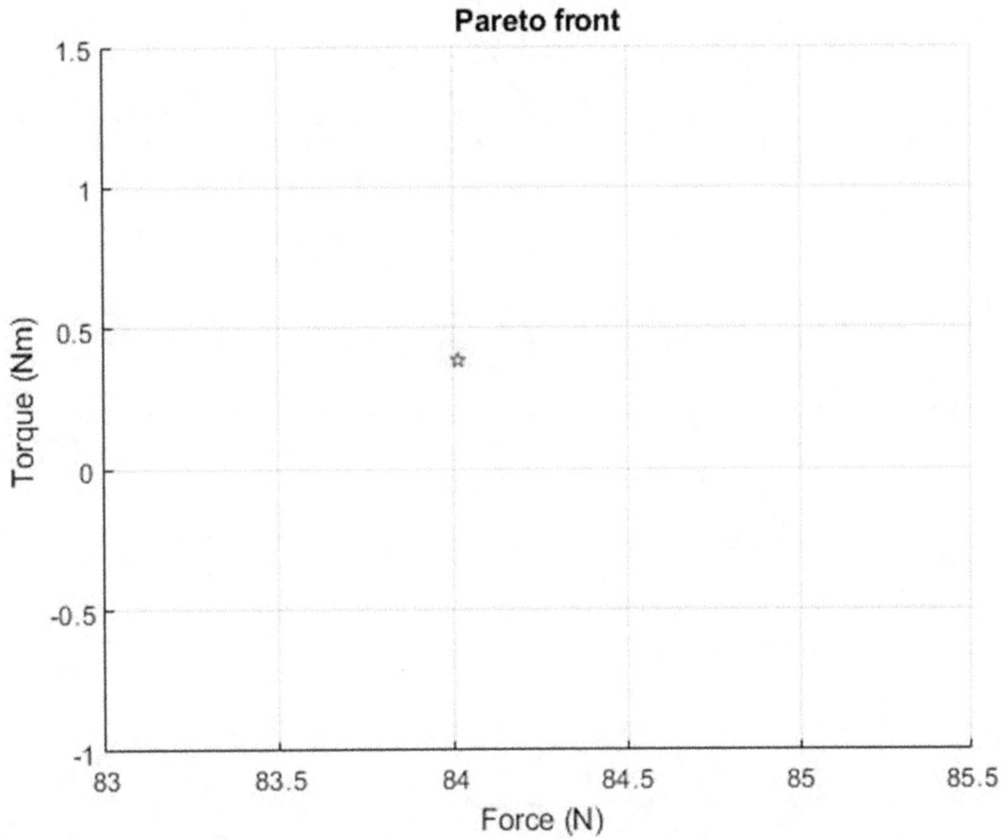

Figure 7. Pareto front for force with respect to entry-delamination

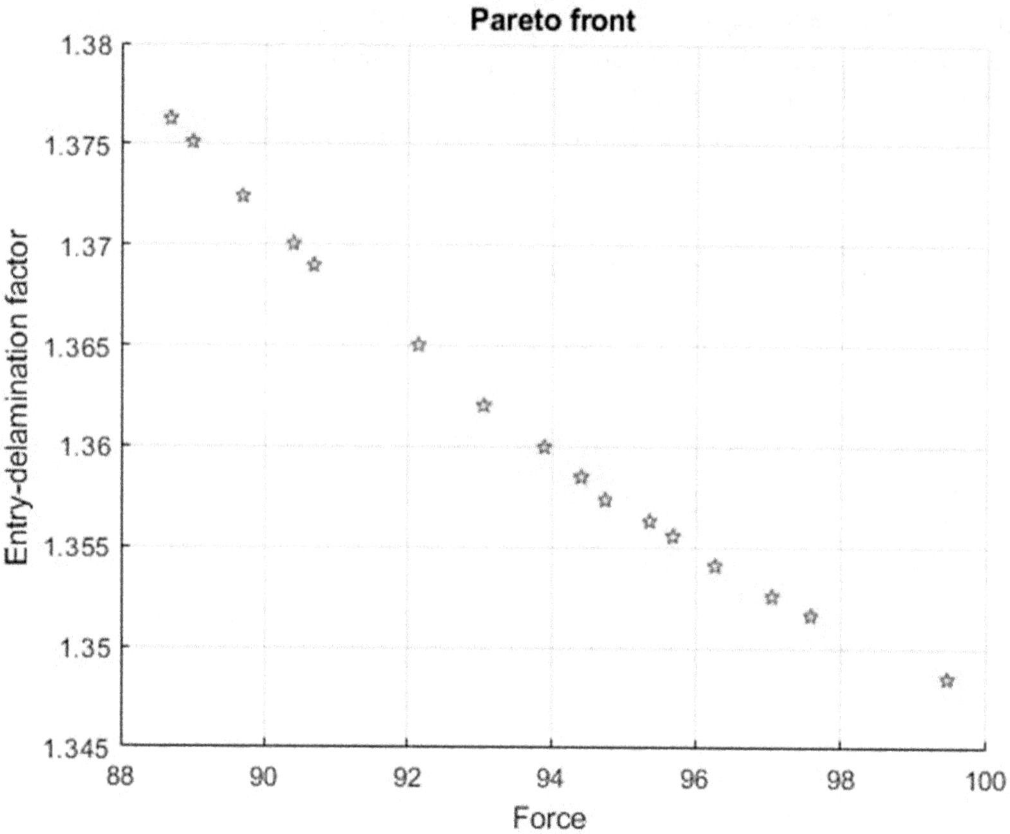

Figure 8. Pareto front for entry-delamination with respect to eccentricity

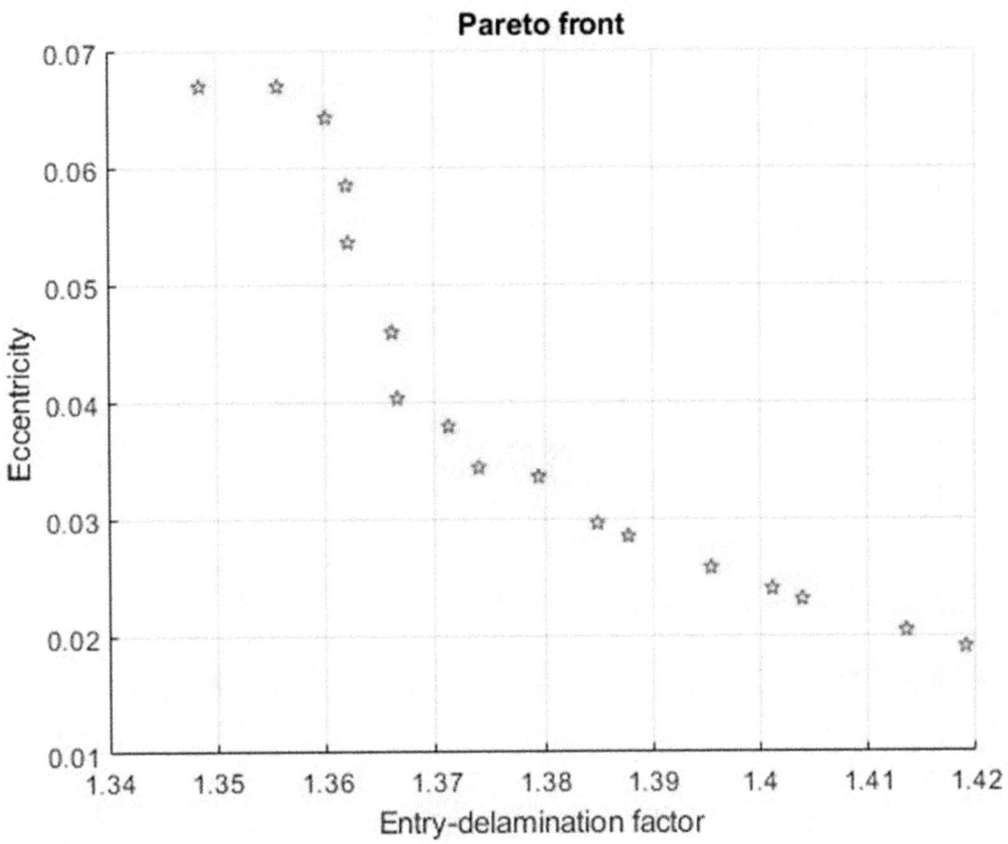

Figure 9. Pareto front for exit-delamination with respect to eccentricity

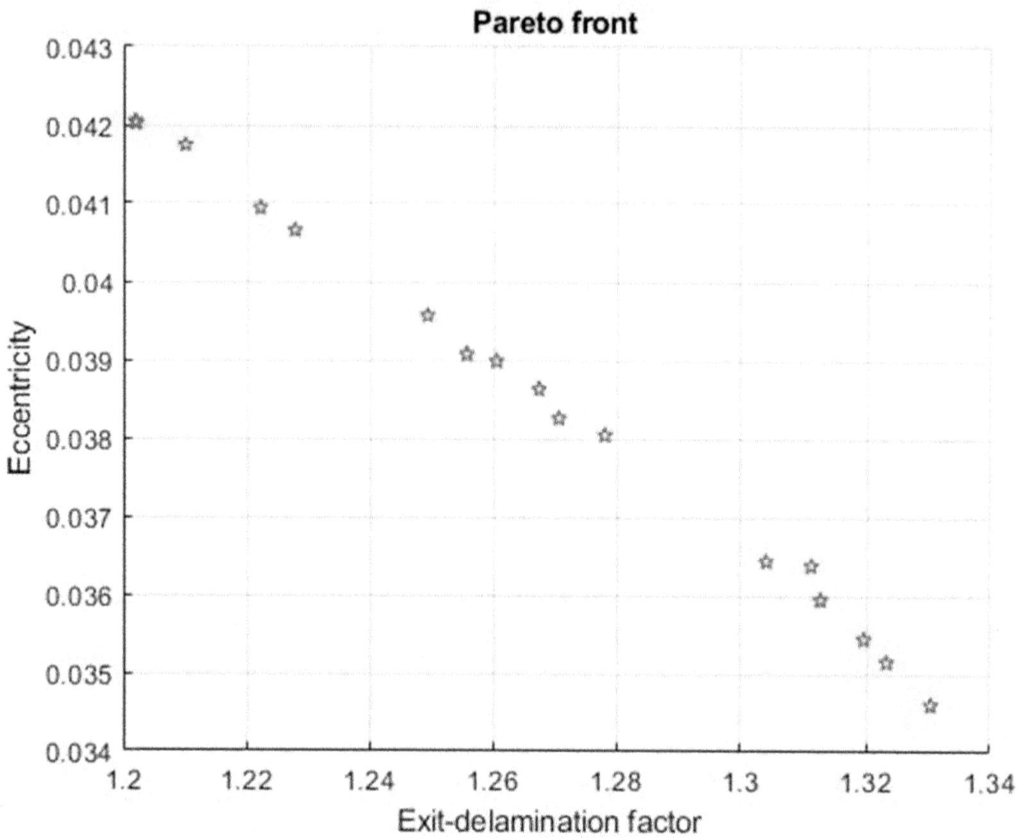

Figure 10. Pareto front for force with respect to eccentricity

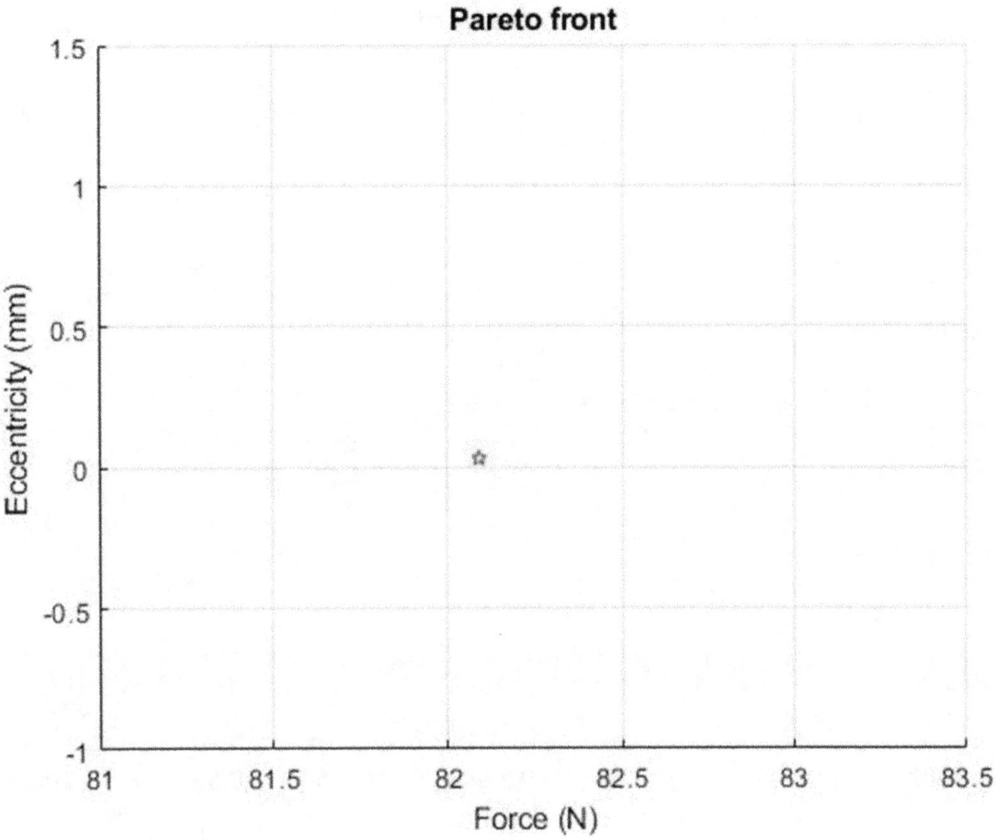

Figure 11. Pareto front for torque with respect to eccentricity

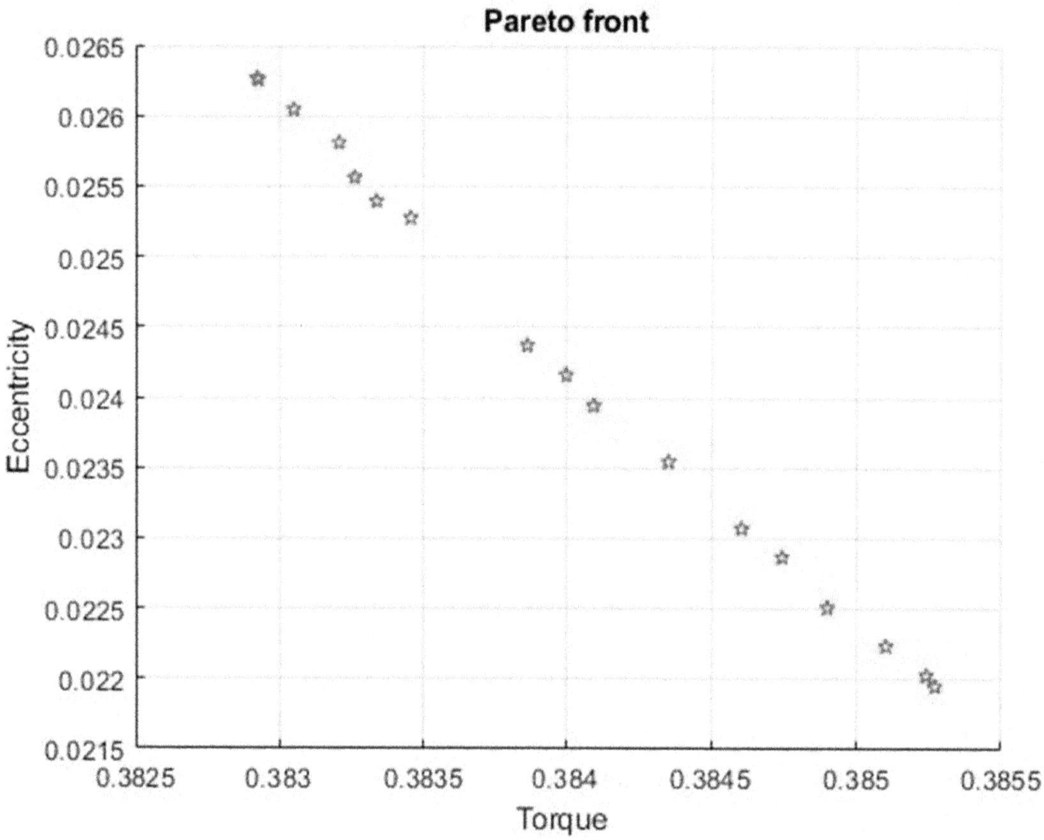

Figure 12. Pareto front for torque with respect to entry-delamination factor

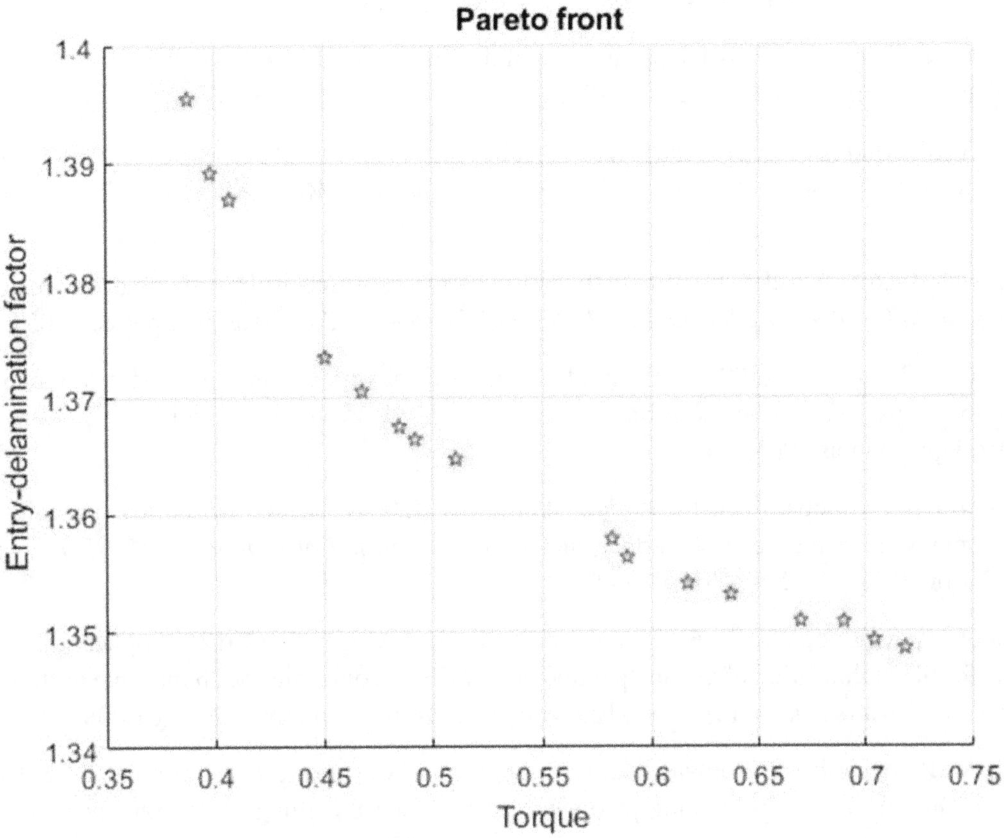

REFERENCES

Abhishek, K., Datta, S., & Mahapatra, S. S. (2016). Multi-objective optimization in drilling of CFRP (polyester) composites: Application of a fuzzy embedded harmony search (HS) algorithm. *Measurement, 77*, 222–239. doi:10.1016/j.measurement.2015.09.015

Bosco, M. A., Palanikumar, K., Prasad, B. D., & Velayudham, A. (2013). Influence of machining parameters on delamination in drilling of GFRP-armour steel sandwich composites. *Procedia Engineering, 51*, 758–763. doi:10.1016/j.proeng.2013.01.108

Chaudhary, G., Kumar, M., Verma, S., & Srivastav, A. (2014). Optimization of drilling parameters of hybrid metal matrix composites using response surface methodology. *Procedia Materials Science, 6*, 229-237.

Davim, J. P., & Reis, P. (2003). Drilling carbon fiber reinforced plastics manufactured by autoclave_experimental and statistical study. *Materials & Design, 24*(5), 315–324. doi:10.1016/S0261-3069(03)00062-1

Davim, J. P., & Reis, P. (2003). Study of delamination in drilling carbon fiber reinforced plastics (CFRP) using design experiments. *Composite Structures, 59*(4), 481–487. doi:10.1016/S0263-8223(02)00257-X

Enemuoh, E. U., El-Gizawy, A. S., & Okafor, A. C. (2001). An approach for development of damage-free drilling of carbon fiber reinforced thermosets. *International Journal of Machine Tools & Manufacture*, *41*(12), 1795–1814. doi:10.1016/S0890-6955(01)00035-9

Ghabezila, P., & Khoran, M. (2014). Optimization of drilling parameters in composite sandwich structures (PVC core). *Indian J. Sci. Res, 2*, 173-179.

Gowda, B. M., Ravindra, H. V., Prakash, G. V., Nishanth, P., & Ugrasen, G. (2015). Optimization of process parameters in drilling of epoxy Si3N4 composite material. *Materials Today: Proceedings*, *2*, 2852–2861.

Hocheng, H., & Tsao, C. C. (2005). The path towards delamination-free drilling of composite materials. *Journal of Materials Processing Technology*, *167*(2-3), 251–264. doi:10.1016/j.jmatprotec.2005.06.039

Hocheng, H., & Tsao, C. C. (2006). Effects of special drill bits on drilling-induced delamination of composite materials. *International Journal of Machine Tools & Manufacture*, *46*(12-13), 1403–1416. doi:10.1016/j.ijmachtools.2005.10.004

Krishnamoorthy, A., Boopathy, S. R., & Palanikumar, K. (2009). Delamination analysis in drilling of CFRP composites using response surface methodology. *Journal of Composite Materials*, *43*(24), 2885–2902. doi:10.1177/0021998309345309

Krishnamoorthy, A., Boopathy, S. R., Palanikumar, K., & Davim, J. P. (2012). Application of grey fuzzy logic for the optimization of drilling parameters for CFRP composites with multiple performance characteristics. *Measurement*, *45*(5), 1286–1296. doi:10.1016/j.measurement.2012.01.008

Krishnaraj, V., Prabukarthi, A., Ramanathan, A., Elanghovan, N., Kumar, M. S., Zitoune, R., & Davim, J. P. (2012). Optimization of machining parameters at high speed drilling of carbon fiber reinforced plastic (CFRP) laminates. *Composites. Part B, Engineering*, *43*(4), 1791–1799. doi:10.1016/j.compositesb.2012.01.007

Lazar, M.-B., & Xirouchakis, P. (2011). Experimental analysis of drilling fiber reinforced composites. *International Journal of Machine Tools & Manufacture*, *51*(12), 937–946. doi:10.1016/j.ijmachtools.2011.08.009

Piquet, R., Ferret, B., Lachaud, F., & Swider, P. (2000). Experimental analysis of drilling damage in thin carbon/epoxy plate using special drills. *Composites. Part A, Applied Science and Manufacturing*, *31*(10), 1107–1115. doi:10.1016/S1359-835X(00)00069-5

Quan, Y., & Zhong, W. (2009). Investigation on drilling-grinding of CFRP. *Frontiers of Mechanical Engineering in China*, *4*(1), 60–63. doi:10.100711465-009-0008-y

Saravanan, M., Ramalingam, D., Manikandan, G., & Kaarthikeyen, R. R. (2012). Multi objective optimization of drilling parameters using genetic algorithm. *Procedia Engineering*, *38*, 197–207. doi:10.1016/j.proeng.2012.06.027

Sardinas, R. Q., Reis, P., & Davim, J. P. (2006). Multi-objective optimization of cutting parameters for drilling laminate composite materials by using genetic algorithms. *Composites Science and Technology*, *66*(15), 3083–3088. doi:10.1016/j.compscitech.2006.05.003

Sasikumar, K. S. (2015). Optimization of drilling parameter on delamination based on taguchi method in drilling of natural fiber reinforced (Agave) composite. *International Journal of Recent Technology and Engineering*, *4*, 53–55.

Sonkar, V., Abhishek, K., Datta, S., & Mahapatra, S. S. (2014). Multi-objective optimization in drilling of GFRP composites: a degree of similarity approach. *Procedia Materials Science*, *6*, 538-543.

Tsao, C. C., & Hocheng, H. (2004). Taguchi analysis of delamination associated with various drill bits in drilling of composite material. *International Journal of Machine Tools & Manufacture*, *44*(10), 1085–1090. doi:10.1016/j.ijmachtools.2004.02.019

Upadhyay, P. C., & Lyons, J. S. (1999). On the evaluation of critical thrust for delamination-free drilling of composite laminates. *Journal of Reinforced Plastics and Composites*, *18*(14), 1287–1303. doi:10.1177/073168449901801402

Vinayagamoorthy, R., Manoj, I. V., Kumar, G. N., Chand, I. S., Kumar, G. V., & Kumar, K. S. (2018). A central composite design based fuzzy logic for optimization of drilling parameters on natural fiber reinforced composite. *Journal of Mechanical Science and Technology*, *32*(5), 2011–2020. doi:10.100712206-018-0409-0

ADDITIONAL READING

Baş, D., & Boyacı, I. H. (2007). Modeling and optimization I: Usability of response surface methodology. *Journal of Food Engineering*, *78*(3), 836–845. doi:10.1016/j.jfoodeng.2005.11.024

Gu, C. (2013). *Smoothing spline ANOVA models* (Vol. 297). Springer Science & Business Media. doi:10.1007/978-1-4614-5369-7

Harik, G. R., Lobo, F. G., & Goldberg, D. E. (1999). The compact genetic algorithm. *IEEE Transactions on Evolutionary Computation*, *3*(4), 287–297. doi:10.1109/4235.797971

Houck, C. R., Joines, J., & Kay, M. G. (1995). A genetic algorithm for function optimization: a Matlab implementation. *Ncsu-ie tr, 95*, 1-10.

Kalita, K., Mallick, P. K., Bhoi, A. K., & Ghadai, K. R. (2018). Optimizing Drilling Induced Delamination in GFRP Composites using Genetic Algorithm& Particle Swarm Optimisation. *Advanced Composites Letters*, *27*(1), 096369351802700101. doi:10.1177/096369351802700101

Khuri, A. I., & Mukhopadhyay, S. (2010). Response surface methodology. *Wiley Interdisciplinary Reviews: Computational Statistics*, *2*(2), 128–149. doi:10.1002/wics.73

Miller, R. G. Jr. (1997). *Beyond ANOVA: basics of applied statistics*. CRC press. doi:10.1201/b15236

Myers, R. H., Montgomery, D. C., & Anderson-Cook, C. M. (2016). *Response surface methodology: process and product optimization using designed experiments*. John Wiley & Sons.

Tabachnick, B. G., & Fidell, L. S. (2007). *Experimental designs using ANOVA*. Thomson/Brooks/Cole Belmont.

Upputuri, H. B., Nimmagadda, V. S., & Duraisamy, E. (2020). Optimization of drilling parameters on carbon fiber reinforced polymer composites using fuzzy logic. *Materials Today: Proceedings*, *23*, 528–535.

Whitley, D. (1994). A genetic algorithm tutorial. *Statistics and Computing*, *4*(2), 65–85. doi:10.1007/BF00175354

KEY TERMS AND DEFINITIONS

Analysis of Variance (ANOVA): ANOVA is a group of statistical models used to find the regression model or used to check the accuracy of the proposed regression model.

Carbon Fibre Reinforced Plastic (CFRP): CFRP is a type of composite material where carbon fibers are arranged in a polymer matrix. CFRP material has a high strength-to-weight ratio and stiffness.

Delamination: Delamination is a type of defect in composite material. In loading condition, laminated composite materials split apart into layer that is called the delamination of a composite laminate.

Drilling: Drilling is a cutting process that produces a circular hole in solid components by removing material from the surface.

Genetic Algorithm (GA): Genetic algorithm is a nature-inspired metaheuristic technique. GA mimic Charles Darwin's theory of natural evolution.

Optimization: Optimization is a mathematical procedure to select the best decision variables from a set of alternatives to get the optimum objective value.

Pareto Front: Pareto front gives the solution for multi-objective optimization. The Pareto front gives a set of optimal solutions that are non-dominating in nature.

Response Surface Methodology (RSM): RSM is a mathematical and statistical method that helps to build the regression model, which relates the input variables to output variables.

Chapter 14
Solution for Multi-Objective Single Row Facility Layout Problem Using PSO Algorithm

Lenin Nagarajan
Vel Tech Rangarajan Dr. Sagunthala R&D Institute of Science and Technology, India

Siva Kumar Mahalingam
Vel tech Rangarajan Dr. Sagunthala R&D Institute of Science and Technology, India

Gurusamy Selvakumar
Sri Sivasubramaniya Nadar College of Engineering, India

Jayakrishna Kandasamy
VIT University, India

ABSTRACT

An optimized facility layout design helps to ensure a high level of machine usability along with minimum cost and good performance. However, facility layout is a multi-objective design optimization problem, and as such, it is difficult to solve. This original research work aims to design an optimal linear machine sequence using particle swarm optimization algorithm that minimizes the following: the total investment cost of machines, the total number of machines in the final sequence, the total flow time of the products, and the total flow distance of the products. The effectiveness of the proposed algorithm is demonstrated with a reasonable number of problems. Maximum of 19.1% reduction in total flow distance of products, 12.8% reduction in total investment cost of machines, 28.4% reduction in total flow time of products, and reduction of two numbers of machines in the layout are achieved by proposed method compared with the previous approaches.

DOI: 10.4018/978-1-7998-7206-1.ch014

INTRODUCTION

Facility layout planning (FLP) plays a vital role in manufacturing processes and profoundly impacts a company's profitability. In this context, a facility is defined as an entity that facilitates the execution of a job. It may be a machine tool, a work center, a manufacturing cell, a machine shop, a department, or a warehouse (Heragu SS, 1997). The facility layout is the physical arrangement of "process elements" that are required for the production of goods or delivery of services. Depending on the specific application, the "process elements" that are to be arranged can be a collection of machines, work centers, entire departments, storage areas, and so on (Ali Taghavi and Alper Murat, 2011).

Facility layout design decisions require considerable financial investment and planning efforts. Furthermore, these decisions have a significant impact on the cost and efficiency of operations. Machinery that is arranged well, for example, contributes to the overall efficiency of operations (Tompkins JA et al., 1996). In most manufacturing systems, process configuration decisions determine the type and number of machines required and their capacities. It might be reduced the flow distance & time of products and investment cost of machines. Given this information, the facility layout planner's goal is to determine the optimal sequence of machines. Linear sequencing of machines is a popular choice because it results in a shorter flow distance and flow time and easier control of the production process. Hence, a linear machine sequencing method is applied in this work.

RELATED RESEARCH

The theoretical attractiveness and practical applications of the single-row facility layout problem (SRFLP) have created a rich and growing literature in this field. This literature has proven the SRFLP to be a non-polynomial (NP) complete problem (Suresh G and Sahu S, 1993). As such, the application of precise methods to large instances of the problem is cumbersome and time-consuming; therefore, heuristic methods have been developed to obtain a near-optimal solution to the problem. Several researchers have applied these heuristic methods to formulate the linear sequencing of machines for the SRFLP.

Heuristic Approaches to the SRFLP

Houshyar and McGinnis (1990) introduced a heuristic for assigning facilities to locations to minimize work-in-process travel distance along a straight track. They established that the performance of their heuristic was better than the modified and classical lower bound methods for the test problems.

Kouvelis and Chiang (1992) simulated an annealing procedure to determine a flow line (or single-row layout) under the assumption that the number of machines is fixed and backtrack movements are allowed. The authors aimed to determine a machine sequence with a minimum total backtrack distance. Braglia (1997) regarded the linear machine sequencing problem as an NP-hard combinatorial problem. The number of possible sequences grows exponentially because of duplicate machines. Braglia determined a linear machine sequence with the minimum expected movement of the machine handling devices that are located between machines in a machine cell.

Wang et al. (1998) formulated a model for minimizing the total material handling distance on a shop floor, in facilities with both inter- and intra-cell layouts, for cellular manufacturing systems. The authors used an improved simulated annealing algorithm to solve this problem. Ho and Moodie (1998) investigated the effect of flow line characteristics on machine layouts. They provided vital information on selecting appropriate flow line analysis methods and on determining appropriate evaluation criteria for different layout problems.

Chen et al. (2001) aimed to determine a common multi-product linear machine sequence for multiple products that have different operation sequences and facilities with a limited number of duplicate machine types. The authors used a modified simulated annealing algorithm to model and minimize the total flow distance traveled by the products on this linear flow line. Djellab and Gourgand (2001) focused their work on a single-row facility layout problem occurring in Flexible Manufacturing Systems (FMS) and proposed a new heuristic to minimize the total time required by material handling systems to transport the parts between machines.

Chrysostomos and Vlachos (2005) used the linear programming model for minimal backward flow to determine the optimal linear machine sequence for a manufacturing cell. They applied a modified Ant Colony System (ACS) algorithm to the conditions and parameters of the linear machine layout problem. Pillai et al. (2005) identified a linear sequence that minimizes the total distance traveled by multiple items with different operation sequences. The authors considered a limited availability of each type of machine and used an algorithm that simulates the annealing process to determine the best solution. Solimanpur (2005) et al. formulated the single-row machine layout problem as a non-linear, 0-1 programming model in which the distance between the machines is sequence-dependent. They developed an ACS algorithm to solve this problem.

Singh and Sharma (2006) discussed the current and future trends of research on facility layout problems. Noticing that facility layout software uses meta-heuristics such as simulated annealing processes, genetic algorithms, and concurrent engineering, the authors concluded that there is a trend toward multi-objective approaches. For apparel manufacturing, Lin (2009) proposed a hierarchical order-based genetic algorithm to minimize the moving distance of cut pieces in a U-shaped single-row machine layout.

Ramazan and Orhan (2009) proposed a simulated annealing algorithm that combines two objectives—minimization of the total material handling cost and maximization of total closeness rating scores—to address the facility layout problem. Samarghandi and Eshghi (2010) proposed a new algorithm based on the Tabu Search optimization technique to find optimal linear placement for rectangular facilities with varying dimensions on a straight line. Satheesh Kumar et al. (2010) employed an artificial immune system algorithm to minimize material handling costs for both single-row and loop layout problems in FMSs. Siva Kumar et al. (2011) developed a simple heuristic to determine the optimal linear sequence that minimizes the flow distance traveled by the products. Feng and Che (2018) maximized the material flow between the adjacent facilities in fixed-size rectangular departments by using two novel linear integer programming models.

Particle Swarm Optimization (PSO) Based SRFLP

Teo and Ponnambalam (2008) proposed a hybrid Ant Colony Optimization (ACO)/PSO heuristic to solve the SRFLP. ACO was used as a constructive heuristic and a new pheromone update was developed to improve the performance of the proposed algorithm. PSO, on the other hand, was used as an improvement heuristic to guide the ants to reach the best solution.

Semih Onut et al. (2008) developed a PSO algorithm for determining the optimal layout to minimize the annual carrying cost of products in a warehouse. Samarghandi et al. (2010) employed a PSO algorithm to solve the SRFLP. They arranged some rectangular facilities with varying lengths on one side of a straight line and minimized the weighted sum of the distance between all facility pairs. Lien and Cheng (2012) proposed an optimization hybrid swarm algorithm, namely, a particle-bee algorithm (PBA). The PBA is based on the particular intelligent behavior of honey bees and bird swarms and integrates this intelligence to solve hypothetical construction site layout problems with high dimensionality. Liu et al. (2018) developed a particle swarm optimization algorithm along with heuristic configured mutation operation to solve unequal-area facility layout problems. The constraint like non-overlapping of facilities was considered in this approach. The objectives undertaken in this work were material handling cost, adjacency value and the utilization ratio of the shop floor.

Other Algorithms to the SRFLP

Lenin et al. (2013) developed a genetic algorithm to identify the linear machine sequence (*lms*) for the simultaneous minimization of total flow distance of products, the total number of machines in the *lms* and total investment cost on machines. Lenin et al. (2014) proposed a heuristic-based Tabu Search algorithm to formulate *lms* that simultaneously minimizes the total flow distance of products, the total number of machines in the *lms*, total investment cost on machines and total material handling cost. Nagarajan et al. (2018) considered the bi-objectives namely flow distance of products and length of the flow line while arranging the facilities in a row. They developed a modified heuristic to identify the layout and further the optimal layout was identified using the Artificial Bee Colony algorithm.

Even though significant attempts have been made to solve the SRFLP, most of the studies focused on the minimization of a single parameter – flow distance. To the best of our knowledge, there is no published work on the SRFLP that especially uses the PSO algorithm to simultaneously minimize other equally important parameters such as the total number of machines in a layout, the investment cost of machines, and flow time of products between machines. This paper proposes a PSO algorithm to solve the SRFLP.

PROBLEM STATEMENT

The location and number of machines in a linear machine sequence for a single-row facility layout design are the keys to determine the flow distance of multiple products, the total investment cost of machines, and total flow time. In facilities with duplicate machines and multiple products, the single-row layout design is considered as an NP-hard problem (Braglia M, 1997).

The following assumptions are considered in the proposed method:

1. The number of products, flow distance of products, flow time of products between machines, machine type sequences of individual products, individual cost of the machine types, and availability of duplicate machine types are known.
2. The first machine that each product enters is exactly as specified in the final linear machine sequence.
3. The products' flow distances are considered up to the end machine type of the respective products without affecting their precedence.
4. A single machine type will not be adjacent to the same machine type in the final linear machine sequence.
5. The backtracking of products is not permitted.

PROBLEM MODEL

Model Notations and Definitions

D_{tot} total flow distance in units
M_{tot} total number of machines available in the final linear sequence
C_{tot} total investment cost of machines in the final linear sequence
T_{tot} total flow time of all products in the final linear sequence
d_i ith product flow distance
P_{il} ith product's last machine position in the final machine sequence
P_{if} ith product's first machine position in the final machine sequence
n number of products
P_{ij+1} ith product's j+1th machine position in the final machine sequence
P_{ij} ith product's jth machine position in the final machine sequence
m_k number of kth machine available in the final linear machine sequence
dm_k number of duplicate kth machine types available for use
b[…] final linear machine sequence
t_m total number of machines available for use
m_t number of machine types
k index to represent the machine type; $k = 1,2,3,\ldots mt$
c_k cost of the kth machine type
m_i number of machines for the ith product
$ft_{ij-ij+1}$ flow time between jth machine to j+1th machine for the ith product in the final linear sequence
$(ND_{tot})_l$ normalized value of the total flow distance of multi-products for the lth sequence of products
$(NM_{tot})_l$ normalized value of the total number of machines in the final linear machine sequence of lth sequence of products
$(NC_{tot})_l$ normalized value of the total investment cost of machines for the lth sequence of products
$(NT_{tot})_l$ normalized value of the total flow time of multi-products for the lth sequence of products
$(D_{tot})_{min}$ & $(D_{tot})_{max}$ minimum & maximum value of total flow distance for 1,2,3, …l number of sequences of products
$(M_{tot})_{min}$ & $(M_{tot})_{max}$ minimum & maximum number of machines in the final linear sequence for 1,2,3, …l number of sequences of products
$(C_{tot})_{min}$ & $(C_{tot})_{max}$ minimum & maximum value of total investment cost of machines for 1,2,3,…l number of sequences of products
$(T_{tot})_{min}$ & $(T_{tot})_{max}$ minimum & maximum value of total flow time for 1,2,3, …l number of sequences of products
$(D_{tot})_l$ total flow distance of multi-products for the lth sequence of products
$(M_{tot})_l$ total number of machines in the final sequence of the lth sequence of products
$(C_{tot})_l$ total investment cost of machines for the lth sequence of products
$(T_{tot})_l$ total flow time of multi-products for the lth sequence of products
AFF average fitness factor
D_i flow distance of ith product (in units)
Rem() function which will give a remainder or quotient or remainder of the quotient
Seq() function which will produce the sequence from the given non-sequence based on generating a random number within the product number that does not exist in the given non-sequence

Total Flow Distance in Units

The total flow distance of a product in units (D_{tot}) is determined using equation 1. The constraints are presented by equations 2 to 4:

$$D_{tot} = \sum_{i=1}^{n} d_i (P_{il} - P_{if}) \tag{1}$$

$$P_{ij+1} > P_{ij} \tag{2}$$

$$P_{ij} > P_{i1} \tag{3}$$

$$m_k \leq dm_k \tag{4}$$

Equation 2 shows that the position of the $j+1^{th}$ machine should always be larger than the position of the jth machine in the linear machine sequence. Equation 3 indicates that the position of the $j+1^{th}$ machine in the individual product machine sequence should always be larger than the position of the first machine in the linear machine sequence. According to equation 4, the number of kth machine types available in the final linear machine sequence should be less than or equal to the number of duplicate k^{th} machine types available for use.

Total Number of Machines in the Final Linear Sequence

The minimum number of machines in the final linear sequence (M_{tot}) of the single-row layout design reduces both flow distance and initial investment. This is expressed using,

$$M_{tot} = count(b[......]) \tag{5}$$

$$t_m = \sum_{k=1}^{m_t} dm_k \tag{6}$$

$$M_{tot} \leq t_m \tag{7}$$

The total number of machines is calculated using equation6 and is equal to the sum of the duplicates of individual machine types. Equation7 shows that the total number of machines in the final linear sequence must be less than or equal to the total number of machines available for use, including the duplicate machines.

Total Investment Cost of Machines

Companies want to reduce their operating/manufacturing costs as well as their initial investment. In the single-row layout design, the investment cost of machines can be expressed by,

$$C_{tot} = \sum_{k=1}^{m_t} c_k m_k \tag{8}$$

Total Flow Time of Products

The flow time of products between machines may reduce the manufacturing lead time. Therefore, the minimization of flow time is an important factor. Total flow time is estimated by equation 9.

$$T_{tot} = \sum_{i=1}^{n} \sum_{j=1}^{m} ft_{ij-ij+1} \tag{9}$$

Average Fitness Factor

The total flow distance, the total number of machines in the final linear sequence, the total investment cost of machines, and total flow time are expressed in different units and their values by different ranges. Summing up these values will not produce a meaningful result. Therefore, the average fitness factor method is applied to derive the objective values within the range of 0 to 1 (normalization). Details of this method can be found in Sivakumar K et al. (2011). The normalized values of total flow distance, the total number of machines, the total investment cost of machines, and total flow time are determined using equations 10 through 13.

$$(ND_{tot})_l = \frac{(D_{tot})_{\max} - (D_{tot})_l}{(D_{tot})_{\max} - (D_{tot})_{\min}} \tag{10}$$

$$(NM_{tot})_l = \frac{(M_{tot})_{\max} - (M_{tot})_l}{(M_{tot})_{\max} - (M_{tot})_{\min}} \tag{11}$$

$$(NC_{tot})_l = \frac{(C_{tot})_{\max} - (C_{tot})_l}{(C_{tot})_{\max} - (C_{tot})_{\min}} \tag{12}$$

$$(NT_{tot})_l = \frac{(T_{tot})_{\max} - (T_{tot})_l}{(T_{tot})_{\max} - (T_{tot})_{\min}} \tag{13}$$

The average fitness factor value is determined by equation 14. The maximum value of the average fitness factor is considered in the minimization problem.

$$AFF = \frac{(ND_{tot})_l + (NM_{tot})_l + (NC_{tot})_l + (NT_{tot})_l}{4} \tag{14}$$

The corresponding linear machine sequence which has the maximum average fitness factor value is the optimum one among the l number of sequences of products.

PROPOSED HEURISTIC AND ALGORITHM

Our research solves the SRFLP by using a simple heuristic and the PSO algorithm. PSO helps to determine the optimal product sequence, and the final linear machine sequence for the product's sequence is obtained using a simple heuristic.

Simple Heuristic to Evaluate the Linear Sequence of Machines

This research has developed a simple heuristic to evaluate the linear sequence of machines for a given product's sequence. This heuristic reduces computing time and improves the consistency of the solutions. The detailed algorithm is given below.

Step (a): Read the number of machine types (*m*), number of duplicate machines in each type (*mtn[]*), number of products (*n*), number of machine types for each product (*nm[]*), machine type sequence for each product (*pseq[][]*), product sequence *(ps[])*, flow distance of each product (*fd[]*), cost of each machine type *(c[])*, and flow time between machine to machine *(ft[])*.
Step (b): Assign the machine type for the first product's machine sequence and store in *b[]*. Update the availability of the machine type in *mtn[]*.
Step (c): For each of the remaining products in *pno[]*, do steps (d) through (k).
Step (d): For each machine type (*mno*) in the product sequence *pseq[][]*, do Step (e).
Step (e): If the machine *mno* is unassigned, add the machine to the front of the existing machine sequence *b[]* and update its availability. Go to Step (d).
Step (f): If the machine type is assigned, check the machine type *mno* in the existing sequence *b[]*.
Step (g): If available, check the availability of the remaining machine types of *pseq[][]* in *b[]*.
Step (h): If all the machine types are available then go to Step (c).

Step (i): If not, and if the machine type *mno* is unassigned, insert the *mno* in the appropriate position (the position that does not affect the existing product machine sequence in *b[]*) after the position of the previous machine type in the existing machine sequence *b[]* and update the availability. Go to Step (d).
Step (j): If the machine type *mno* is unassigned but its insertion in *b[]* affects the existing product sequence, add the *mno* at the end of *b[]* and update the availability. Go to Step (d).
Step (k): If the machine type *mno* is assigned, then the existing sequence is not feasible: stop the program.
Step (l): Display the linear sequence of machine type *b[]*.

The PSO Algorithm

The PSO algorithm simulates the behaviors of birds flocking. Consider the following scenario. A group of birds is searching for food in an area. There is only one piece of food in the area in which they are searching. None of the birds know where the food is; however, they know how far away the food is in each iteration. So, what is the best strategy for finding food? The effective strategy is to follow the bird which is nearest to the food. PSO learned from this scenario and used it to solve optimization problems. A single solution is one "bird" in the search space: in PSO terms, the "bird" is represented by a "particle". All of the particles have fitness values, which are evaluated by the fitness function to be optimized, and have velocities directing the flight of the particles. The particles "fly" through the problem space, following the best particles.

The PSO algorithm is initialized with a group of random particles (solutions) and then searches for optima by updating each generation. In every iteration, each particle is updated with two "best" values; both values are tracked and stored. The first value is the best solution (fitness) that the particle has achieved so far. This value is called *p-best*. The second value is the best value that has been obtained so far by any other particle in the population. This best value is the global best and is called the *g-best*. When a particle takes part in the population and its topological neighbors, the best value is the local best and is called the *i-best* (Kannan SM et al., 2009). The general view of the PSO is shown in Figure 1. The different stages of problem-solving by the PSO approach are detailed with numerical illustration in the following section.

NUMERICAL ILLUSTRATION

The following problem example is used to illustrate the effectiveness of the proposed method. For each machine type (M.No.), Table 1 shows the number of machines for that type, their availability, and the investment cost.

The product number (P.No.), the product's machine type sequences, and the demand for the product in units (flow distance) are represented in Table 2. The details of material flow time between machines are given in Table 3.

Table 1. Details of the machine, its availability, and cost

M.No.	1	2	3	4	5	6	7	8	9	10	11	12	13
Availability of duplicate machines	1	2	2	2	2	1	2	2	1	2	2	1	2
Machine cost (Rs.)	9570	17415	19323	1180	4951	16465	6091	15386	11243	2676	14342	5793	17246

Figure 1. General view of particle swarm optimization

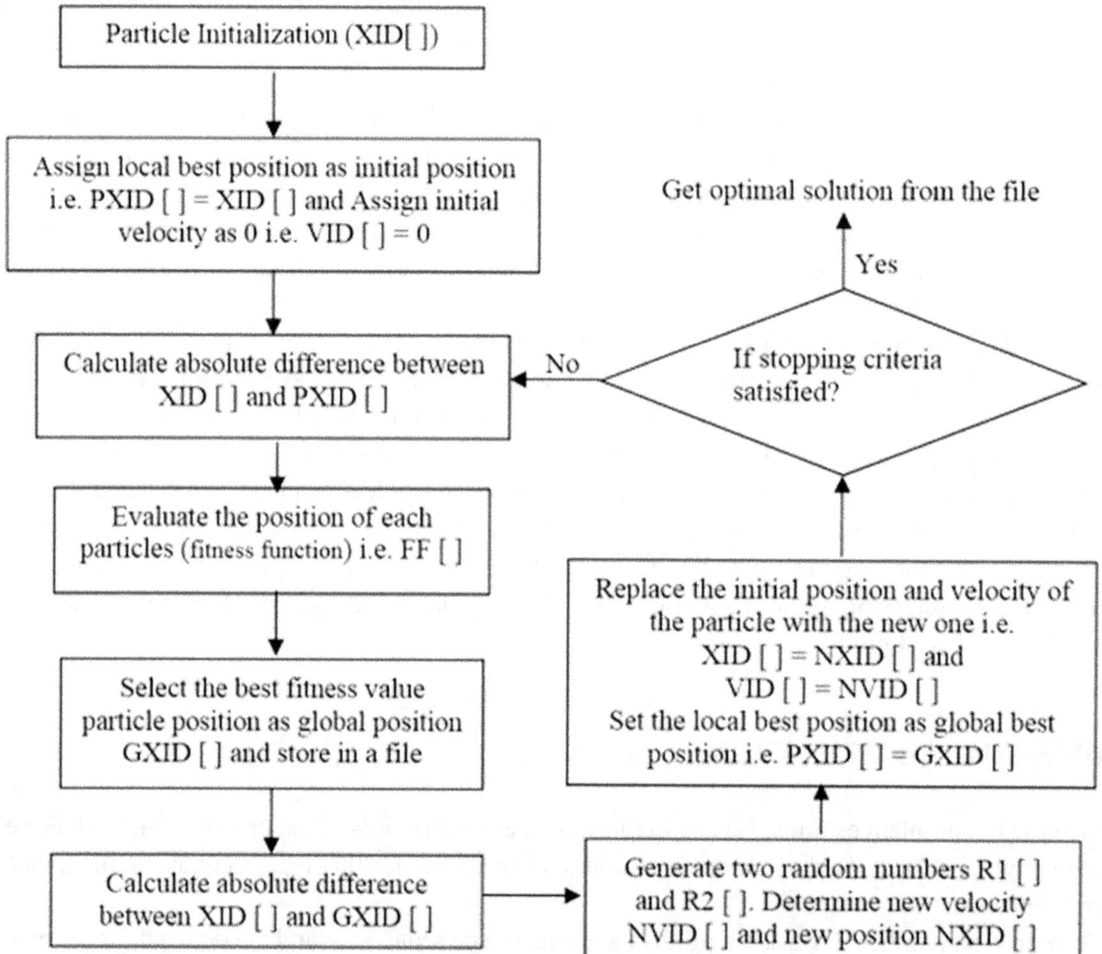

Table 2. Details of machine sequence and demand for individual products

P.No.	Machine sequence	Demand in units
1	8-3-4-12-5	48
2	9-10-7-11	18
3	13-10-7-5	32
4	4-12-2-10-8	41
5	9-10-7-3-2	1
6	11-1-10-2-3	3

PSO Representation

STAGE I – Particle Initialization

Step 1: The product sequences are generated for six products (number of products, *n=6*); the sequences are repeated for a particle size of six *(PS=6)* and stored in the same in XID[ ]. This is shown in Table 4.

STAGE II – Determination of local and global best position

Step 2: The individual particles are evaluated and the average fitness factor value (position) for each particle is determined as per Eq. 14.

Table 3. Details of material handling cost between machines

Material flow time in minutes		Machines												
		1	**2**	**3**	**4**	**5**	**6**	**7**	**8**	**9**	**10**	**11**	**12**	**13**
Machines	1	0	9	9	8	8	8	7	8	9	9	7	8	9
	2	8	0	8	7	7	8	7	8	9	8	8	7	9
	3	8	8	0	7	9	9	9	9	8	9	9	9	8
	4	7	9	7	0	7	9	7	7	9	9	7	8	7
	5	8	8	8	7	0	7	9	9	9	7	8	8	7
	6	9	7	8	7	7	0	9	9	9	7	8	7	8
	7	7	8	9	8	9	8	0	7	9	8	8	8	7
	8	8	8	7	7	7	9	8	0	8	8	7	9	8
	9	7	7	7	8	7	7	7	7	0	8	9	7	7
	10	8	7	8	9	7	7	8	8	9	0	9	8	8
	11	9	7	9	9	9	7	8	9	7	9	0	7	9
	12	8	7	8	9	8	8	8	9	9	9	7	0	9
	13	9	8	7	9	9	8	8	7	7	9	9	9	0

The initial particle product sequence 1-3-4-2-5-6 has been selected for demonstration purposes. The final linear machine sequence and the average fitness factor value for the selected product sequence are evaluated by the following simple heuristic procedure as explained in section 5.1.

Step (a): Select the first product's machine type sequence 8-3-4-12-5. Assign machine types 8,3,4,12 and 5. Store in b[]. Update the availability of machine types.

Table 4. Details of particle initialization XID[PS]

Particle number	Product sequence XID[ ]
1	1-3-4-2-5-6
2	2-1-4-3-5-6
3	3-1-4-6-5-2
4	4-1-2-3-5-6

Table 5. Updated unassigned machine types

M.No.	1	2	3	4	5	6	7	8	9	10	11	12	13
Availability of duplicate machines	1	2	**1**	**1**	**1**	1	2	**1**	1	2	2	**0**	2

The linear machine type sequence (*b[]*) after the first product is **8-3-4-12-5**.

Step (b): The next (3rd product) product's machine type sequence is 13-10-7-5. Machine type number (*mno*) 13 is unassigned, hence, add the machine type in front of *b[]* and update the availability of the machine type. Similarly, the other machine types (*mno*) 10,7, and 5 are unassigned; therefore add these machine types in front of the existing machine sequence *b[]* and update the availability of each machine type. The linear machine type sequence (*b[]*) after the third product is **13-10-7-5**-8-3-4-12-5.

Step (c): The next (4thproduct) product machine type sequence is 4-12-2-10-8. Machine type number (*mno*) 4 is unassigned. Therefore, add the machine type in front of the existing sequence. The next machine type number (*mno*) 12 is assigned and is therefore unavailable in unassigned machines and available in the existing sequence. The next machine type numbers (*mno*) 2,10, and 8 are unassigned and unavailable in the existing machine sequence *b[]*; therefore, add these to the end of theexisting machine sequence *b[]*and update the availability of each machine type. The linear machine type sequence (*b[]*) after the 4th product is **4**-13-10-7-5-8-3-4-**12**-5-**2**-**10**-**8**.

Step (d): The next (2ndproduct) product machine type sequence is 9-10-7-11. Machine type number (*mno*) 9 is unassigned. Therefore, add machine type number (*mno*) 9 to the front of the existing sequence. Machine type numbers *(mno)* 10 and 7 are assigned and available in the existing sequence. The next machine type number (*mno*) 11 is unassigned and unavailable in the existing machine sequence *b[]*; hence, add machine type number (*mno*) 11 to the end of the existing machine sequence *b[]* and update the availability of each machine type. The linear machine type sequence (*b[]*) after the 2nd product is **9**-4-13-**10-7**-5-8-3-4-12-5-2-10-8-**11**.

Step (e): The next (5thproduct) product machine type sequence is 9-10-7-3-2. Machine type numbers *(mno)* 9 and 10 are assigned and available in the existing sequence. Machine type numbers (*mno*) 7,3 and 2 are unassigned but available in the existing sequence. The linear machine type sequence (*b[]*) after the 5th product is **9**-4-13-**10-7**-5-8-**3**-4-12-5-**2**-10-8-11.

Step (f): The next (6thproduct) product machine type sequence is 11-1-10-2-3. Machine type numbers *(mno)* 11 and 1 are unassigned; therefore, add these machine types to the front of *b[]* and update the availability of each machine type. Machine type number *(mno)* 10 is assigned and available in the existing sequence. Machine type number *(mno)* 2 is unassigned but available in the existing sequence. The next machine type number *(mno)* 3 is unassigned and unavailable after machine type number *(mno)*2. Hence, add machine type number 3 at the end of the existing sequence. The linear machine type sequence (*b[]*) after the last product is **11-1**-9-4-13-**10**-7-5-8-3-4-12-5-**2**-10-8-11-**3**.

The total number of machines (M_{tot}) in the linear machine sequence is 18. The total flow distance (D_{tot}), the total investment cost of machines (C_{tot}), and the total flow time (T_{tot}) for the above linear machine sequence are listed in Tables 6, 7and 8. The normalized values of the individual objectives as per Eqs.10 through 13 and the average fitness factor as per equation14 are listed in Table 9.

Step 3: Initially, the local best position is assumed as the initial particle position. Hence, the local best particles (PXID[ ]) are the same as the initial particles, that is, PXID[ ] = XID[ ]. The initial velocity of each particle (VID[ ]) is 0. The best initial position is assumed as the global best position and hence, the global best particle (GXID[ ])is assumed as the best initial position of the particle. The largest average fitness factor value identifies the best initial position. This is the result of the first iteration and is stored in a separate file. The initial particle position (average fitness factor), local best particles, initial velocity, and global best particles are presented in Table 10.

STAGE III – Determination of new Position and Velocity of Particles

Step 4: The absolute difference between XID[ ] and PXID[ ] and also between XID[ ] and GXID[ ] are determined as per the following expressions.

Abs(XID[1]-PXID[1]) = Abs([1 3 4 2 5 6] - [1 3 4 2 5 6])
= [0 0 0 0 0 0]
Abs(XID[1]-GXID[1]) = Abs([1 3 4 2 5 6] - [4 1 2 3 5 6])
= Abs([1-4 3-1 4-2 2-3 5-5 6-6])
= [3 2 2 1 0 0]

Step 5: Two sets of random numbers (R1[ ] and R2[ ]) are generated within the number of products (*n*). The size of R1[ ] and R2[ ] are the same as the particle size (*PS*). This is represented in Table 11.

Table 6. Determination of flow distance for the final machine sequence of product sequence 1-3-4-2-5-6 (Initial particle)

P.No.	Product's machine sequence					Final machine sequence																		P_{i1}	P_{if}	d_i	D_i
						11	1	9	4	13	10	7	5	8	3	4	12	5	2	10	8	11	3				
						Position of final machine sequence																					
						1	2	3	4	5	6	7	8	9	10	11	12	13	14	15	16	17	18				
						List of machines involved in final machine sequence for the individual product's machine sequence																					
1	8	3	4	12	5									8	3	4	12	5						9	13	48	192
2	9	10	7	11				9	4	13	10	7	5	8	3	4	12	5	2	10	8	11		3	17	18	252
3	13	10	7	5						13	10	7	5											5	8	32	96
4	4	12	2	10	8											4	12	5	2	10	8			11	16	41	205
5	9	10	7	3	2			9	4	13	10	7	5	8	3	4	12	5	2					3	14	1	11
6	11	1	10	2	3	11	1	9	4	13	10	7	5	8	3	4	12	5	2	10	8	11	3	1	18	3	51
Total flow distance in units (D_{tot})																											807

Table 7. Determination of total investment cost of machines for the final machine sequence of product sequence 1-3-4-2-5-6 (Initial particle)

Machine type	1	2	3	4	5	6	7	8	9	10	11	12	13	C_{tot} in Rs.
No. of machine type available in final sequence	1	1	2	2	2	0	1	2	1	2	2	1	1	
Cost of machine type	9570	17415	19323	1180	4951	16465	6091	15386	11243	2676	14342	5793	17246	
Investment cost of each machine types	9570	17415	38646	2360	9902	0	6091	30772	11243	5352	28684	5793	17246	1,83,074

Table 8. Determination of total flow time of products for the final machine sequence of product sequence 1-3-4-2-5-6 (Initial particle)

P.No.	Machine sequence	Flow time in Minutes
1	8-3-4-12-5	30
2	9-10-7-11	111
3	13-10-7-5	26
4	4-12-2-10-8	40
5	9-10-7-3-2	88
6	11-1-10-2-3	138
Total flow time (T_{tot})		433

Step 6: The values of T1[ ] and T2[ ] are determined for the individual particles as per equations 15 and 16.

Table 9. Determination of average fitness factor for the final machine sequence of product sequence 1-3-4-2-5-6 (Initial particle)

C_{tot}	T_{tot}	M_{tot}	D_{tot}	$(NC_{tot})_l$	$(NT_{tot})_l$	$(NM_{tot})_l$	$(ND_{tot})_l$	AFF
183074	433	18	807	0.739616	0.607735	0.333333	1	0.670171

Table 10. Local and global best particles

P.No.	Initial position (i.e., AFF)	Local best particles PXID[ ]	Initial velocity VID[ ]	Global best particles GXID[ ]
1	0.670171	[1 3 4 2 5 6]	[0 0 0 0 0 0]	[4 1 2 3 5 6]
2	0.298552	[2 1 4 3 5 6]	[0 0 0 0 0 0]	[4 1 2 3 5 6]
3	0.204874	[3 1 4 6 5 2]	[0 0 0 0 0 0]	[4 1 2 3 5 6]
4	**0.921952**	[4 1 2 3 5 6]	[0 0 0 0 0 0]	[4 1 2 3 5 6]

T1[ ]= C1 * R1[ ] * Abs(XID[ ]-GXID[ ]) (15)

T2[ ]= C2 * R2[ ] * Abs(XID[ ]-PXID[ ]) (16)

C1 and C2 values are assumed as 2 based on sensitivity analysis and previous literature. Table 12 represents the values of T1[ ] and T2[ ].

T1[1] = 2 * R1[1] * Abs(XID[1]-GXID[1])
= 2 * [5 3 2 5 6 3] * [3 2 2 1 0 0]

Table 11 Details of random numbers

P.No.	Abs(XID[ ]-PXID[ ])	Abs(XID[ ]-GXID[ ])	R1[ ]	R2[ ]
1	[0 0 0 0 0 0]	[3 2 2 1 0 0]	[5 3 2 5 6 3]	[1 2 2 5 6 1]
2	[0 0 0 0 0 0]	[2 0 2 0 0 0]	[4 6 2 5 3 1]	[6 3 2 3 4 5]
3	[0 0 0 0 0 0]	[1 0 2 3 0 4]	[3 4 3 2 5 6]	[4 3 2 5 6 4]
4	[0 0 0 0 0 0]	[0 0 0 0 0 0]	[6 5 4 2 3 1]	[3 1 2 5 3 6]

= [2*5*3 2*3*2 2*2*2 2*5*1 2*6*0 2*3*0]
= [30 12 8 10 0 0]
T2[1] = 2 * R2[1] * Abs(XID[2]-PXID[2])
= 2* [1 2 2 5 6 1] * [0 0 0 0 0 0]
= [2*1*0 2*2*0 2*2*0 2*5*0 2*6*0 2*1*0]
= [0 0 0 0 0 0]

Table 12. Details of T1[ ] and T2[ ]

P.No.	T1[ ]	T2[ ]
1	[30 12 8 10 0 0]	[0 0 0 0 0 0]
2	[16 0 8 0 0 0]	[0 0 0 0 0 0]
3	[6 0 12 12 0 48]	[0 0 0 0 0 0]
4	[0 0 0 0 0 0]	[0 0 0 0 0 0]

Step 7: The new velocity of particles (NVID[ ]) and new particle positions (NXID[ ])are calculated based on the revised velocity of the particles (RVID[ ]) and the revised particle positions (RXID[ ]). The revised values are calculated using equations 17 and 18. The weight factor (*W*) is calculated using equation19.

RVID[ ]= (VID[ ] * W) + T1[ ] + T2[ ] (17)

RXID[ ] = XID[ ] + RVID[ ] (18)

$$W = \frac{(0.9 - C)}{(itr + 1)} \tag{19}$$

The value of *C* is assumed as 0.5 after sensitivity analysis and *itr* is the current iteration number. The new velocity of particles (NVID[ ]) and new particle positions (NXID[ ]) is calculated using the following equations 20 and 21.

NXID[ ]= Seq(Rem(RXID[ ]/Number of products)) (20)

NVID[ ] = Seq(Rem(NVID[ ][ ]/ Number of products)) (21)

The value of *W* for *itr* value of 1 is,

$$W = \frac{(0.9 - 0.5)}{(1 + 1)} = 0.2$$

Therefore, RVID[1] = (VID[1] * W) + T1[1] + T2[1]
= ([0 0 0 0 0 0] * 0.2) + [30 12 8 10 0 0]
+ [0 0 0 0 0 0]
= [30 12 8 10 0 0]

RXID[1] = XID[1] + RVID[1]

= [1 3 4 2 5 6] + [30 12 8 10 0 0]
= [1+30 3+12 4+8 2+10 5+0 6+0]
= [31 15 12 12 5 6]

The new velocity of particles (NVID[ ]) and new particle positions (NXID[ ]) are,

NXID[1] = Seq(Rem(RXID[1]/Number of products))
= Seq(Rem([31 15 12 12 5 6]/6))
= Seq([1 3 2 2 5 1])
= [1 3 2 4 5 6]
NVID[1] = Seq(Rem(RVID[1]/ Number of products))
= Seq(Rem([30 12 8 10 0 0]/6))
= Seq([5 2 2 4 0 0]
= [5 2 6 4 1 3]

STAGE IV – Replacement

Step 8: The NXID[ ] is replaced with the existing XID[ ] and NVID[ ] is replaced with the existing VID[ ]. Assume PXID[ ] as GXID[ ].

The best solutions after the first iteration are selected and stored in a file. Stages II to IV are repeated for *nitr=100* number of iterations, with the best solutions selected, or they are selected if there has been no change in the value of average fitness factor for 50 consecutive iterations. The PSO parameters that are assumed in this work are presented in Table 13.

Table 13. PSO parameters value

Particulars	Value / method
Population Size	20
Replacement strategy	Complete replacement
Stopping criteria	100 iterations or no change in the value of average fitness factor in 50 consecutive iteration

COMPUTATIONAL TEST RESULTS AND DISCUSSIONS

Some randomly-generated problems as well as the problems discussed in Pillai et al. (2005), Chen et al. (2001), and Siva Kumar et al. (2011) have also been solved using the proposed methodology. Input data, such as the number of products and their machine type sequences and product demand, are listed in Appendix Table 15. The number of machine types and their duplicate numbers are listed in Appendix Table 16. The cost of individual machine types is listed in Appendix Table 17. The flow time of products between machine to machine for the 10 problems that were solved is listed in Appendix Tables 18-27. The final linear machine sequence, product sequence, total flow distance, total investment cost of machines, total flow time of products, and the total number of machines in the final linear sequence are presented in Table 14.

As compared with the problems and solutions discussed in Pillai et al. (2005), Chen et al. (2001), and Siva Kumar et al. (2011), the algorithm proposed in this paper yields the best linear sequence of machines and minimizes the total flow distance in units, the total number of machines, the total investment cost of machines, and the total flow time for the following reasons.

1. Machines are not assigned based on the descending order of the flow distance (i.e., demand) of a product.
2. The unassigned machine types are incorporated at the beginning of the existing machine sequence.
3. If one of the machine types is assigned and it is available in the existing sequence, its availability in this sequence is verified even if the remaining machine types are unassigned. If any of the remaining machine types are unavailable in the existing sequence and are unassigned, then the machine type is incorporated at the backflow of the existing sequence without affecting the previous product machine type sequences.

CONCLUDING REMARKS

This paper proposed a PSO algorithm for constructing a linear sequence of machines that minimizes four parameters simultaneously, namely: the total flow distance in units, the total investment cost of machines in the final linear sequence, total flow time of products, and the total number of machines in the final linear sequence. Some new problems were generated to test and experiment with the proposed algorithm and the results were compared with previous approaches. Maximum of 19.1% reduction in total flow distance of products, 12.8% reduction in total investment cost of machines, 28.4% reduction in total flow time of products and reduction of 2 number of machines in the final sequence was achieved using the proposed method compared with the previous methods. The reasons for getting good results by the proposed method were discussed in section 7. Hence, the proposed method is highly efficient and it provides better practical support for SRFLP for effective design, minimization of cost of manufacturing and to increase the performance of the system. As the future scope, other optimization tools like simulated annealing (SA) algorithm, artificial bee colony (ABC) algorithm, etc., have to be used to solve the multi-objective SRFLP.

INDUSTRIAL APPLICATION

This work is well suitable for all kinds of mass production industries. Usually, mass production industries are interested in laying types of machinery in a straight line with minimum investment cost on machines by avoiding duplicates vis-a-vis concentration on the reduction of material handling cost and material flow time. The only possibility to attain the above-said objectives is by optimally arranging the machines. The proposed PSO algorithm is used to identify the optimal linear machine sequence among the large possible alternatives.

Table 14. Computational test results

Problem No.	No. of machine types	No. of products	method	Total investment cost of machines in Rs.	Total flow time in min.	Total No. of machines in the sequence	Total flow distance in Units	Product's sequence	Optimal final linear sequence
1	10	5	Proposed	4,51,057	111	10	12800	1-2-3-4-5	5-3-2-7-1-8-9-6-4-10
			Lenin et al. (2014)	4,51,057	111	10	12800	1-2-3-4-5	5-3-2-7-1-8-9-6-4-10
			Lenin et al. (2013)	4,51,057	111	10	12800	1-2-3-4-5	5-3-2-7-1-8-9-6-4-10
			Siva Kumar M et al.	4,51,057	111	10	12800	1-2-3-4-5	5-3-2-7-1-8-9-6-4-10
			Pillai et al.	4,51,057	111	10	12800	1-5-3-4-2	5-3-2-7-1-8-9-6-4-10
2	7	5	Proposed	4,29,000	113	8	8800	1-4-3-2-5	4-1-3-2-6-5-1-7
			Lenin et al. (2014)	4,86,000	126	9	9000	1-2-3-4-5	4-6-1-7-1-3-2-6-5
			Lenin et al. (2013)	4,29,000	113	8	8800	1-4-3-2-5	4-1-3-2-6-5-1-7
			Siva Kumar M et al.	4,86,000	126	9	9000	1-2-3-4-5	4-6-1-7-1-3-2-6-5
			Pillai et al.	4,86,000	126	9	9000	1-2-3-4-5	4-6-1-7-1-3-2-6-5
3	15	4	Proposed	2,16,629	71	12	890	3-4-1-2	2-10-12-14-13-7-11-15-5-3-1-4
			Lenin et al. (2014)	2,16,629	71	12	890	3-4-1-2	2-10-12-14-13-7-11-15-5-3-1-4
			Lenin et al. (2013)	2,16,629	71	12	890	3-4-1-2	2-10-12-14-13-7-11-15-5-3-1-4
			Siva Kumar M et al.	2,16,629	71	12	890	3-4-1-2	2-10-12-14-13-7-11-15-5-3-1-4
			Chen et al.	2,16,629	72	12	989	3-4-1-2	2-10-12-14-13-7-11-15-5-1-4-3
4	14	6	Proposed	2,96,406	194	14	2587	5-2-4-3-6-1	4-2-8-5-3-11-13-14-7-6-12-9-1-10
			Lenin et al. (2014)	2,96,406	210	14	2939	5-3-4-1-6-2	4-2-8-5-3-11-13-14-1-10-7-12-9-6
			Lenin et al. (2013)	2,96,406	204	14	2388	5-3-4-1-6-2	2-4-8-5-3-11-13-14-7-12-9-1-10-6
			Siva Kumar M et al.	2,96,406	204	14	2388	5-3-4-1-6-2	2-4-8-5-3-11-13-14-7-12-9-1-10-6
			Chen et al.	2,96,406	210	14	2939	5-3-4-1-6-2	4-2-8-5-3-11-13-14-1-10-7-12-9-6
5	14	4	Proposed	73,567	179	14	475	1-3-2-4	1-14-2-3-4-6-8-9-7-13-5-10-11-12
			Lenin et al. (2014)	73,567	179	14	475	1-3-2-4	1-14-2-3-4-6-8-9-7-13-5-10-11-12
			Lenin et al. (2013)	73,567	179	14	475	1-3-2-4	1-14-2-3-4-6-8-9-7-13-5-10-11-12
			Siva Kumar M et al.	73,567	179	14	475	1-3-2-4	1-14-2-3-4-6-8-9-7-13-5-10-11-12
			Chen et al.	73,567	183	14	475	1-3-2-4	14-1-2-3-4-6-8-9-7-13-5-10-11-12
6	12	4	Proposed	1,99,931	168	11	438	4-3-1-2	7-9-6-3-4-2-12-5-1-7-3
			Lenin et al. (2014)	2,29,206	180	13	432	4-3-2-1	2-6-3-7-9-6-4-2-12-5-1-7-3
			Lenin et al. (2013)	2,29,206	180	13	432	4-3-2-1	2-6-3-7-9-6-4-2-12-5-1-7-3
			Siva Kumar M et al.	2,14,066	180	12	432	4-3-2-1	6-3-7-9-6-4-2-12-5-1-7-3
			Chen et al.	2,29,206	180	13	432	4-3-2-1	2-6-3-7-9-6-4-2-12-5-1-7-3
7	9	5	Proposed	1,34,198	249	13	912	3-4-5-2-1	9-8-9-1-2-7-3-5-4-8-3-1-6
			Siva Kumar M et al.	1,45,867	348	13	856	5-3-2-1-4	2-9-7-5-8-9-1-4-6-8-3-1-5
			Siva Kumar M et al.	1,45,867	348	13	856	5-3-2-1-4	2-9-7-5-8-9-1-4-6-8-3-1-5
			Siva Kumar M et al.	1,45,867	348	13	856	5-3-2-1-4	2-9-7-5-8-9-1-4-6-8-3-1-5
			Chen et al.	1,45,867	348	13	856	5-3-2-1-4	2-9-7-5-8-9-1-4-6-8-3-1-5
8	11	6	Proposed	1,91,569	237	21	845	5-3-6-1-4-2	1-4-6-9-11-12-8-3-1-2-7-5-3-10-13-12-5-10-7-9-6
			Lenin et al. (2014)	2,10,752	298	22	1044	3-1-6-4-2-5	10-1-4-6-9-11-12-8-7-5-3-8-3-1-2-10-13-12-5-7-9-6
			Lenin et al. (2013)	1,91,569	237	21	845	5-3-6-1-4-2	1-4-6-9-11-12-8-3-1-2-7-5-3-10-13-12-5-10-7-9-6
			Siva Kumar M et al.	2,10,752	298	22	1044	3-1-6-4-2-5	10-1-4-6-9-11-12-8-7-5-3-8-3-1-2-10-13-12-5-7-9-6

REFERENCES

Braglia, M. (1997). Heuristics for single-row layout problems in flexible manufacturing systems. *Production Planning and Control*, *8*(6), 558–567. doi:10.1080/095372897234894

Chen, D. S., Wang, Q., & Chen, H. C. (2001). Linear sequencing for machine layouts by a modified simulated annealing. *International Journal of Production Research*, *39*(8), 1721–1732. doi:10.1080/00207540010023565

Chrysostomos, F., & Vlachos, A. (2005). Optimal solution of linear machine layout problem using ant colony system. *WSEAS Transactions on Information Science and Applications*, *2*(6), 652–662.

Djellab, H., & Gourgand, M. (2001). A new heuristic procedure for the single-row facility layout problem. *International Journal of Computer Integrated Manufacturing*, *14*(3), 270–280. doi:10.1080/09511920010020721

Feng, J., & Che, A. (2018). Novel integer linear programming models for the facility layout problem with fixed-size rectangular departments. *Computers & Operations Research*, *95*, 163–171. doi:10.1016/j.cor.2018.03.013

Heragu, S. S. (2008). *Facilities design*. CRC Press.

Ho, Y. C., & Moodie, C. L. (1998). Machine layout with a linear single-row flow path in an automated manufacturing system. *Journal of Manufacturing Systems*, *17*(1), 1–22. doi:10.1016/S0278-6125(98)80006-X

Houshyar, A., & McGinnis, L. F. (1990). A heuristic for assigning facilities to locations to minimize WIP travel distance in a linear facility. *International Journal of Production Research*, *28*(8), 1485–1498. doi:10.1080/00207549008942807

Kannan, S. M., Sivasubramanian, R., & Jayabalan, V. (2009). Particle swarm optimization for minimizing assembly variation in selective assembly. *International Journal of Advanced Manufacturing Technology*, *42*(7-8), 793–803. doi:10.100700170-008-1638-7

Kouvelis, P., & Chiang, W. C. (1992). A simulated annealing procedure for single row layout problems in flexible manufacturing systems. *International Journal of Production Research*, *30*(4), 717–732. doi:10.1080/00207543.1992.9728452

Kumar, M. S., Islam, M. N., Lenin, N., Vignesh Kumar, D., & Ravindran, D. (2011). A simple heuristic for linear sequencing of machines in layout design. *International Journal of Production Research*, *49*(22), 6749–6768. doi:10.1080/00207543.2010.535860

Lenin, N., Kumar, M. S., Islam, M. N., & Ravindran, D. (2013). Multi-objective optimization in single-row layout design using a genetic algorithm. *International Journal of Advanced Manufacturing Technology*, *67*(5-8), 1777–1790. doi:10.100700170-012-4608-z

Lenin, N., Siva Kumar, M., Ravindran, D., & Islam, M. N. (2014). A tabu search for multi-objective single row facility layout problem. *Journal of Advanced Manufacturing Systems*, *13*(01), 17–40. doi:10.1142/S0219686714500024

Lien, L. C., & Cheng, M. Y. (2012). A hybrid swarm intelligence based particle-bee algorithm for construction site layout optimization. *Expert Systems with Applications*, *39*(10), 9642–9650. doi:10.1016/j.eswa.2012.02.134

Lin, M. T. (2009). The single-row machine layout problem in apparel manufacturing by hierarchical order-based genetic algorithm. *International Journal of Clothing Science and Technology*, *21*(1), 31–43.

Liu, J., Zhang, H., He, K., & Jiang, S. (2018). Multi-objective particle swarm optimization algorithm based on objective space division for the unequal-area facility layout problem. *Expert Systems with Applications*, *102*, 179–192. doi:10.1016/j.eswa.2018.02.035

Nagarajan, L., Mahalingam, S. K., Gurusamy, S., & Dharmaraj, V. K. (2018). Solution for bi-objective single row facility layout problem using artificial bee colony algorithm. *European Journal of Industrial Engineering*, *12*(2), 252–275. doi:10.1504/EJIE.2018.090619

Önüt, S., Tuzkaya, U. R., & Doğaç, B. (2008). A particle swarm optimization algorithm for the multiple-level warehouse layout design problem. *Computers & Industrial Engineering*, *54*(4), 783–799. doi:10.1016/j.cie.2007.10.012

Pillai, V. M., & Gudivada, B. S. (2005, August). A simulated annealing algorithm for linear sequencing of machines for layout design. In *6th international conference on operations and quantitative management* (pp. 9-11). Academic Press.

Şahin, R., & Türkbey, O. (2009). A simulated annealing algorithm to find approximate Pareto optimal solutions for the multi-objective facility layout problem. *International Journal of Advanced Manufacturing Technology*, *41*(9-10), 1003–1018. doi:10.100700170-008-1530-5

Samarghandi, H., & Eshghi, K. (2010). An efficient tabu algorithm for the single row facility layout problem. *European Journal of Operational Research*, *205*(1), 98–105. doi:10.1016/j.ejor.2009.11.034

Samarghandi, H., Taabayan, P., & Jahantigh, F. F. (2010). A particle swarm optimization for the single row facility layout problem. *Computers & Industrial Engineering*, *58*(4), 529–534. doi:10.1016/j.cie.2009.11.015

Satheesh Kumar, R. M., Asokan, P., & Kumanan, S. (2010). An artificial immune system-based algorithm to solve linear and loop layout problems in flexible manufacturing systems. *International Journal of Product Development*, *10*(1-3), 165–179. doi:10.1504/IJPD.2010.029991

Singh, S. P., & Sharma, R. R. (2006). A review of different approaches to the facility layout problems. *International Journal of Advanced Manufacturing Technology*, *30*(5-6), 425–433. doi:10.100700170-005-0087-9

Sivakumar, K., Balamurugan, C., & Ramabalan, S. (2011). Simultaneous optimal selection of design and manufacturing tolerances with alternative manufacturing process selection. *Computer Aided Design*, *43*(2), 207–218. doi:10.1016/j.cad.2010.10.001

Solimanpur, M., Vrat, P., & Shankar, R. (2005). An ant algorithm for the single row layout problem in flexible manufacturing systems. *Computers & Operations Research*, *32*(3), 583–598. doi:10.1016/j.cor.2003.08.005

Suresh, G., & Sahu, S. (1993). Multiobjective facility layout using simulated annealing. *International Journal of Production Economics*, *32*(2), 239–254. doi:10.1016/0925-5273(93)90071-R

Taghavi, A., & Murat, A. (2011). A heuristic procedure for the integrated facility layout design and flow assignment problem. *Computers & Industrial Engineering*, *61*(1), 55–63. doi:10.1016/j.cie.2011.02.011

Teo, Y. T., & Ponnambalam, S. G. (2008, August). A hybrid ACO/PSO heuristic to solve single row layout problem. In *2008 IEEE International Conference on Automation Science and Engineering* (pp. 597-602). IEEE. 10.1109/COASE.2008.4626491

Tompkins, J. A., White, J. A., Bozer, Y. A., & Tanchoco, J. M. A. (2010). *Facilities planning*. John Wiley & Sons.

Wang, T. Y., Lin, H. C., & Wu, K. B. (1998). An improved simulated annealing for facility layout problems in cellular manufacturing systems. *Computers & Industrial Engineering*, *34*(2), 309–319. doi:10.1016/S0360-8352(97)00318-5

APPENDIX

Table 15. Operation sequences and product demand of example problems

Problem no	Products	Operation sequence	Product demand
Literature Problems			
1 Pillai et al.	1	1-8-9-6-4	700
	2	5-3-2-7	600
	3	5-3-2-9	500
	4	3-7-6-4	400
	5	3-2-7-9-10	300
2 Pillai et al.	1	1-3-2-6-5	800
	2	4-6-1-7	400
	3	4-1-6-5	300
	4	4-3-2-5	200
	5	4-1-3-2	100
3 Chen et al.	1	14-13-7-15	34
	2	2-10-12-13	29
	3	11-15-5-3	94
	4	15-5-1-4	89
4 Chen et al.	1	4-5-3-9	69
	2	5-3-7-6	13
	3	13-7-12-9	113
	4	8-5-3-14	72
	5	11-13-14-7	131
	6	2-5-1-10	36
5 Pillai et al. + Chen et al.	1	2-3-4-6-8-9-7	20
	2	14-2-3-4-5-10-11-12	10
	3	2-4-6-8-9-13	15
	4	1-2-3-5-11-12	10
Generated Problems			
6	1	6-3-12-5	6
	2	7-9-6-2	24
	3	4-2-12-3	28
	4	2-12-5-1-7	30
7	1	2-7-4-8	17
	2	9-7-8-3	27
	3	5-4-8-3-1	28
	4	2-7-3-5	5
	5	8-9-1-4-6	44
8	1	8-3-1-2	42
	2	1-4-6-9	21
	3	10-13-12-5	45
	4	11-12-8-13-7	23
	5	10-7-9-6	13
	6	7-5-3-13	40
9	1	8-3-4-12-5	48
	2	9-10-7-11	18
	3	13-10-7-5	32
	4	4-12-2-10-8	41
	5	9-10-7-3-2	1
	6	11-1-10-2-3	3
10	1	3-5-6-9-7	22
	2	1-2-3-4	15
	3	2-3-5-6	17
	4	4-6-7-8	20
	5	2-5-9-8-7	23
	6	2-5-8-3	19
	7	1-9-8-7	16

Table 16. Machine types and their duplicates for the example problems

Problem No.	Machines Types														
	1	**2**	**3**	**4**	**5**	**6**	**7**	**8**	**9**	**10**	**11**	**12**	**13**	**14**	**15**
1	1	1	1	1	1	1	1	1	1	1					
2	2	1	1	1	1	2	1								
3	1	1	1	2	1	2	2	2	2	2	1	2	1	1	1
4	1	1	1	1	1	1	1	1	1	1	1	1	1	1	
5	1	1	1	1	1	1	1	1	1	1	1	1	1	1	
6	2	2	2	1	2	2	2	1	1	2	1	1			
7	2	1	2	1	2	2	1	2	2						
8	2	2	2	1	2	2	2	2	2	2	2	2	1		
9	1	2	2	2	2	1	2	2	1	2	2	1	2		
10	2	2	2	1	2	1	2	2	2						

Table 17. Machine types and their cost for the example literature problems

Machines Types	Problem No.									
	1	**2**	**3**	**4**	**5**	**6**	**7**	**8**	**9**	**10**
1	57,884	72,000	17,552	21,011	8,788	27,400	5,192	1,431	9,570	12,515
2	34,545	37,000	21,657	28,752	6,589	15,140	9,861	16,674	17,415	14,625
3	15,455	42,000	13,255	26,354	3,512	13,616	3,558	13,581	19,323	19,874
4	14,411	35,000	19,558	17,655	6,541	23,146	1,752	10,971	1,180	16,547
5	22,225	52,000	22,668	21,357	3,254	17,863	15,227	2,520	4,951	15,722
6	47,775	57,000	15,478	16,554	9,874	14,135	6,909	14,294	16,465	17,412
7	39,998	62,000	14,192	11,357	6,547	22,890	15,867	8,089	6,091	11,058
8	59,874		12,987	30,699	8,541	13,818	14,483	19,183	15,386	15,302
9	74,325		11,551	19,220	3,256	10,733	19,058	9,974	11,243	18,705
10	84,565		23,490	12,632	1,111	20,851		637	2,676	
11			24,520	10,228	2,222	11,633		3,430	14,342	
12			16,984	24,998	3,333	18,502		15,664	5,793	
13			16,390	27,111	4,445			8,931	17,246	
14			12,402	28,478	5,554					
15			13,961							

Table 18. Material flow time (in min.) between machine to machine for problem no.1

P.No. 1	1	2	3	4	5	6	7	8	9	10
1	0	5	3	4	5	6	5	4	6	5
2	3	0	1	3	4	5	4	5	4	5
3	3	5	0	4	3	3	3	6	5	5
4	5	5	5	0	6	5	4	6	3	4
5	3	3	3	3	0	3	5	5	3	5
6	3	6	5	3	4	0	4	4	5	3
7	4	5	5	4	5	6	0	6	6	6
8	6	3	6	5	6	3	3	0	4	5
9	5	6	5	6	3	4	3	3	0	5
10	3	3	5	6	4	5	4	6	5	0

Table 19. Material flow time (in min.) between machine to machine for problem no. 2

1	2	3	4	5	6	7
0	6	5	3	6	5	4
5	0	4	4	5	5	3
3	4	0	4	5	5	3
4	4	6	0	3	4	3
4	3	5	3	0	3	4
4	3	4	6	6	0	4
3	4	5	5	5	6	0

Table 20. Material flow time (in min.) between machine to machine for problem no. 3

1	2	3	4	5	6	7	8	9	10	11	12	13	14	15
0	5	6	3	5	3	6	6	5	3	4	6	6	5	3
3	0	6	4	5	6	5	4	5	4	5	6	5	5	4
5	6	0	5	3	5	5	6	5	4	6	5	3	5	5
4	6	3	0	4	3	4	5	4	5	4	5	6	4	4
3	5	3	5	0	4	3	3	3	6	5	5	6	3	6
4	5	5	5	6	0	6	5	4	6	3	4	6	4	5
6	4	3	3	5	3	0	3	5	5	3	5	4	6	4
6	3	3	6	5	3	4	0	4	4	5	3	6	6	5
2	6	4	5	5	4	5	6	0	6	6	6	5	5	4
6	5	6	3	4	5	6	3	3	0	4	5	4	3	5
6	3	5	6	3	6	3	4	3	3	0	5	4	5	5
6	5	6	6	4	4	3	5	4	6	5	0	3	5	3
3	5	5	6	6	4	6	6	3	6	4	6	0	3	3
3	4	6	4	3	5	6	6	4	4	5	5	6	0	4
4	5	4	6	6	3	4	4	5	5	6	4	5	6	0

Table 21. Material flow time (in min.) between machine to machine for problem no. 4

M.No.	1	2	3	4	5	6	7	8	9	10	11	12	13	14
1	0	6	6	5	5	4	3	6	5	3	5	5	4	5
2	6	0	3	3	5	3	3	4	6	5	6	5	6	4
3	4	5	0	5	5	3	6	5	3	6	6	6	4	6
4	5	3	4	0	6	5	5	5	4	3	4	6	3	6
5	6	5	3	4	0	3	3	4	5	6	4	4	5	3
6	5	5	4	3	6	0	4	5	6	3	3	6	6	4
7	4	6	5	3	5	3	0	6	3	4	5	6	6	4
8	5	5	4	3	4	5	4	0	3	3	4	3	4	5
9	4	4	5	6	6	5	4	6	0	3	6	6	4	5
10	5	6	4	5	3	3	5	6	4	0	5	4	5	6
11	6	5	5	5	4	5	3	6	5	5	0	6	5	4
12	5	3	6	6	6	4	6	5	4	4	3	0	6	5
13	5	5	4	3	4	6	6	5	3	5	5	3	0	6
14	4	5	4	6	5	4	5	4	5	5	3	3	4	0

Table 22. Material flow time (in min.) between machine to machine for problem no. 5

M.No.	1	2	3	4	5	6	7	8	9	10	11	12	13	14
1	0	7	6	5	5	6	4	6	5	3	5	5	4	7
2	6	0	3	3	5	3	3	4	6	5	6	7	6	4
3	4	7	0	5	6	3	6	5	3	6	6	6	4	6
4	5	3	4	0	6	5	5	5	4	3	4	6	3	6
5	6	5	3	4	0	3	3	4	5	6	4	4	5	3
6	5	5	4	3	6	0	4	5	6	3	3	6	6	4
7	4	6	5	3	5	3	0	6	3	4	5	6	6	7
8	5	5	4	3	4	5	7	0	3	3	4	3	4	5
9	4	4	5	6	6	5	4	6	0	3	6	6	4	5
10	5	6	4	5	3	7	5	6	4	0	5	4	5	6
11	6	5	5	5	4	5	3	7	5	5	0	7	5	4
12	5	3	6	6	6	4	6	5	4	4	3	0	6	5
13	5	5	4	3	4	6	6	5	3	5	5	3	0	6
14	7	5	6	6	5	4	7	4	5	5	3	5	4	0

Table 23. Material flow time (in min.) between machine to machine for problem no. 6

M.No.	1	2	3	4	5	6	7	8	9	10	11	12
1	0	9	8	8	9	7	8	7	8	7	7	9
2	8	0	9	9	9	9	7	9	7	9	8	8
3	7	8	0	8	7	7	8	7	7	8	8	7
4	8	9	7	0	9	7	7	8	7	9	9	7
5	8	7	9	8	0	8	8	9	9	8	8	8
6	8	7	9	7	9	0	8	9	8	8	7	8
7	7	7	9	8	8	8	0	8	8	9	7	7
8	9	7	9	8	8	9	9	0	8	9	9	8
9	8	8	7	9	7	7	7	7	0	9	9	8
10	9	7	9	7	7	8	8	9	9	0	7	9
11	7	8	9	8	7	9	9	9	9	9	0	8
12	7	8	9	9	9	9	9	9	7	9	8	0

Table 24. Material flow time (in min.) between machine to machine for problem no. 7

M.No.	1	2	3	4	5	6	7	8	9
1	0	8	9	7	7	9	7	8	9
2	8	0	8	8	7	7	9	9	9
3	9	8	0	7	7	8	9	9	7
4	9	9	7	0	9	8	8	7	8
5	9	8	9	7	0	8	8	9	7
6	9	8	7	9	9	0	9	8	7
7	7	8	8	7	7	9	0	9	8
8	9	8	9	9	9	8	8	0	9
9	9	8	9	7	8	7	8	7	0

Table 25. Material flow time (in min.) between machine to machine for problem no. 8

M.No.	1	2	3	4	5	6	7	8	9	10	11	12	13
1	0	7	7	7	7	7	8	7	8	7	7	7	7
2	8	0	9	9	7	8	7	9	8	9	8	9	9
3	9	8	0	9	7	8	7	9	9	9	8	9	7
4	8	7	8	0	7	7	7	7	9	7	7	8	7
5	8	7	8	9	0	7	8	7	9	8	7	9	9
6	8	9	7	7	8	0	8	9	8	8	8	8	9
7	8	8	8	8	9	7	0	8	7	9	7	7	7
8	7	9	8	9	7	9	9	0	9	9	9	9	9
9	8	8	8	9	9	8	7	9	0	9	7	8	7
10	7	9	8	7	7	8	7	7	9	0	7	7	8
11	9	7	9	8	7	8	7	9	7	9	0	9	9
12	8	7	7	9	8	9	8	8	8	8	7	0	7
13	8	9	7	9	9	9	9	8	9	8	8	7	0

Table 26. Material flow time (in min.) between machine to machine for problem no. 9

M.No.	1	2	3	4	5	6	7	8	9	10	11	12	13
1	0	9	9	8	8	8	7	8	9	9	7	8	9
2	8	0	8	7	7	8	7	8	9	8	8	7	9
3	8	8	0	7	9	9	9	9	8	9	9	9	8
4	7	9	7	0	7	9	7	7	9	9	7	8	7
5	8	8	8	7	0	7	9	9	9	7	8	8	7
6	9	7	8	7	7	0	9	9	9	7	8	7	8
7	7	8	9	8	9	8	0	7	9	8	8	8	7
8	8	8	7	7	7	9	8	0	8	8	7	9	8
9	7	7	7	8	7	7	7	7	0	8	9	7	7
10	8	7	8	9	7	7	8	8	9	0	9	8	8
11	9	7	9	9	9	7	8	9	7	9	0	7	9
12	8	7	8	9	8	8	8	9	9	9	7	0	9
13	9	8	7	9	9	8	8	7	7	9	9	9	0

Table 27. Material flow time (in min.) between machine to machine for problem no. 10

M.No.	1	2	3	4	5	6	7	8	9
1	0	8	9	5	4	9	7	4	8
2	8	0	7	8	9	9	8	7	8
3	7	7	0	4	7	4	7	8	9
4	8	8	7	0	9	8	8	9	8
5	9	8	7	8	0	7	8	7	6
6	7	6	7	8	9	0	7	8	9
7	8	7	9	6	9	5	0	9	9
8	7	7	7	7	9	7	7	0	7
9	7	8	8	9	9	8	7	8	0

Chapter 15
Progress in Optimization of Physical Vapor Deposition of Thin Films

Fredrick Mwema
https://orcid.org/0000-0001-6116-5587
University of Johannesburg, South Africa

Esther T. Akinlabi
Pan African University for Life and Earth Sciences Institute (PAULESI), Ibadan, Nigeria

Oluseyi Philip Oladijo
https://orcid.org/0000-0003-1650-6244
Botswana International University of Science and Technology, Botswana

ABSTRACT

In this chapter, the current state of the art in optimization of thin film deposition processes is discussed. Based on the reliable and credible published results, the study aims to identify the applications of various optimization techniques in the thin film deposition processes, with emphasis on physical deposition methods. These methods are chosen due to their attractive attributes over chemical deposition techniques for thin film manufacturing. The study identifies the critical parameters and factors, which are significant in designing of the optimization algorithms based on the specific deposition methods. Based on the specific optimization studies, the chapter provides general trends, optimization evaluation criteria, and input-output parameter relationships on thin film deposition. Research gaps and directions for future studies on optimization of physical vapor deposition methods for thin film manufacturing are provided.

INTRODUCTION

There is increasing need for nano-sized materials for applications in various fields such as optical, microelectronics, solar, optoelectronic, dielectrics, biomedical, etc. (Jilani, Abdel-wahab, & Hammad,

DOI: 10.4018/978-1-7998-7206-1.ch015

2017; Mwema, Oladijo, Akinlabi, & Akinlabi, 2018). This need has led to emergence of new class of materials known as thin films (their thickness vary between nanometers to micrometers), which have proved superior in terms of properties and performance over the bulk materials (Jilani et al., 2017). The methods of manufacturing thin film materials are broadly classified into physical and chemical deposition methods. The chemical deposition methods depend on the chemistry of the solutions used and chemical reactions are involved during formation of the thin film materials. There are several chemical methods of depositing thin films, some of which include plating, chemical bath, sol-gel, chemical vapor deposition, CVD (atomic layer deposition, low pressure, etc.) (Sivaram, 1995). Physical deposition methods involve dislodging of material from source and condensing them on a surface of a substrate and the process mostly occurs inside a vacuum. Some of the physical methods include evaporation methods (vacuum thermal, electron beam, ion plating, laser beam, etc.) and sputtering techniques (Mwema et al., 2018; Semaltianos, 2001; Simon, 2018). The preference of the physical methods is based on their capability to produce quality films and flexibility in terms of processing parameters.

There are various parameters governing the deposition of thin films through physical methods. While some of the methods are specific to the methods, the physical methods share common parameters such as time, rate of deposition, vacuum pressure, material deposition yield, temperature, and so on (Mwema et al., 2018). The interrelationships among these parameters and their influence on the quality of the thin films is complex. As such, several studies have reported on the effect of various deposition parameters to the properties of various thin films (Azmand & Kafashan, 2019; Gullu, Isik, & Gasanly, 2018; Khalaf, Al-Taay, & Ali, 2017; Mwema, Oladijo, & Akinlabi, 2018; Mwema, Akinlabi, & Oladijo, 2019; Mwema, Akinlabi, Oladijo, & Dutta Majumdar, 2019; Tondare, Shivaraj, Narasimhamurthy, Krishna, & Subramanyam, 2018). These studies have demonstrated that the choice of deposition parameters is critical for specific performance, quality and applications of the thin films. The quality of the deposited films can be measured by the level of defects such as cracks, porosity, and so forth and these defects significantly depend on the deposition parameters. Readers a referred to a recent review outlining the relationship between porosity and plasma-spray coating process (Odhiambo, Li, Zhao, & Li, 2019). The deposition parameters determine the kinetics of material source removal, transport, diffusion and growth of the thin films.

The purpose of this chapter is to discuss the progress in research on the optimization of physical deposition of thin films. The parameters influencing these methods, optimization methods and summary of the recent studies on the subject are described. The article is an important resource for selection and optimization of parameters during physical deposition of thin films.

OPTIMIZATION IN THIN FILM TECHNOLOGY

Parameters in Physical Deposition of Thin Films

Physical deposition methods involves evaporation and sputtering techniques (Jilani et al., 2017). In general, both techniques involve removal of the coating material from the source (target) and depositing it onto a surface (substrate). The processes involved in physical deposition of thin films can be summarized in Figure 1. In evaporation processes, the source material is usually heated until it boils and vaporizes. The source of heat maybe resistive heater or e-beam evaporator (Quero, Perdigones, & Aracil, 2018). The evaporated material is transported or accelerated, through vacuum chamber, to the substrate due to the potential difference between the substrate and target. In sputtering, the material is removed by the ion (from the plasma) bombardment onto the target. The energy of the ions must be enough to dislodge atoms from the target (which is usually cathode). The vacuum chamber, which acts as anode, composes of inert gases (mostly argon) to transport the source atoms onto the substrate. As shown in Figure 1, physical processes occurring on the substrate include diffusion, condensation, nucleation and grain growth, which eventually lead to film formation. These processes are detailed in literature on film growth models (Thornton, 1986; To, de Sousa, & Aarão Reis, 2018). Based on Figure 1, there are various parameters, which influence the physical deposition of thin films.

Figure 1. Processes in physical vapor deposition methods

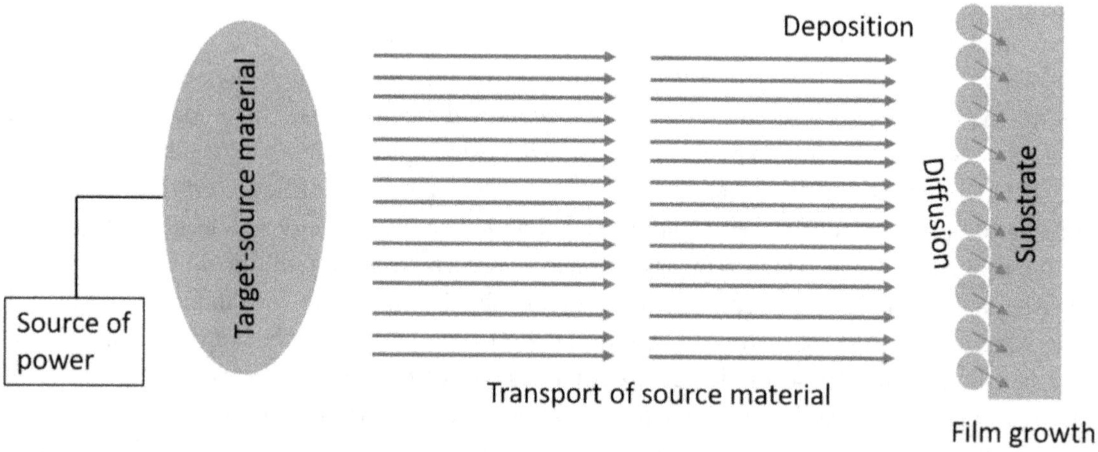

The quality, composition and type of the source and substrate materials is one of the obvious parameters (Chang & Yang, 2018; Gaines, 2019; Khan, Wang, & Jin, 2018). Of course, the quality of the target in terms of stoichiometry and purity determines the performance and properties of the thin films (J. H. Kim, Lee, & Im, 1999). The power (type and amount) applied on the target either in sputtering or evaporation also influences the deposition process. For instance in DC sputtering process, the increase in target power increases the deposition rate and hence the thickness of the films (Kavitha, Kannan, & Rajashabala, 2017). The voltage applied on the target determines the heating rate and hence the evaporation of the source material in thermal evaporation process of thin films. The temperature of the target in physical deposition process also influence the kinetics of material removal, transport and diffusion onto the substrate (Bondarenko, Kolomiytsev, & Shapovalov, 2016). It has been reported that increase in target temperature results in high rate of deposition during magnetron sputtering (Caillard, El'Mokh, Lecas, & Thomann, 2018). The temperature of the substrate during physical deposition of thin films is another general parameter, which greatly influences the process and the properties of the thin films. The substrate temperature influences the process of diffusion, grain growth and nucleation of the target atoms onto the substrate. It has been shown that the higher the substrate temperature, the higher the diffusion, nucleation and grain growth, as such films deposited at higher temperatures are likely to exhibit large grain sizes (Higo, Fujita, Mitsushio, Yoshidome, & Kakoi, 2007; Mwema et al., 2018; Mwema, Akinlabi, Oladijo, & Majumdar, 2019; Mwema, Akinlabi, & Oladijo, 2019b; Yoon, Kim, Kim, Lee, &

Yoon, 2005). Figure 2 summarizes the general parameters associated with evaporation and sputtering deposition techniques of thin films.

Figure 2. Illustrating general parameters influencing evaporation and sputtering techniques of thin film deposition

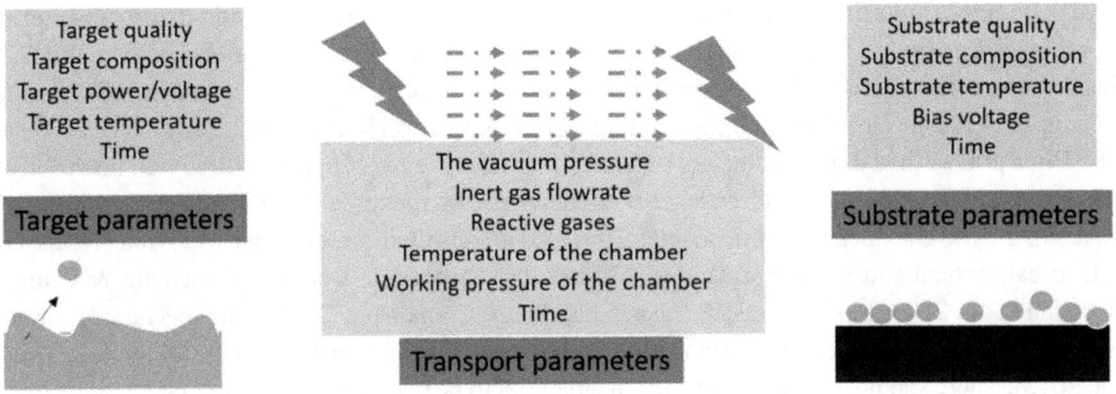

As shown in Figure 2, the composition and quality of the substrate influences the growth of thin films in either of the processes. It has been reported that pure aluminium thin films grown on glass, silicon and steel substrates through sputtering technology have different deposition rates and growth mechanisms (Khachatryan, Lee, Kim, Kim, & Kim, 2018; Mwema, Akinlabi, & Oladijo, 2019). The films deposited on metallic substrates usually exhibit developed microstructure as compared to those deposited on glass and silicon substrates, and the results have been directly linked to the quality and composition of the substrate. Additionally, substrate with relatively higher roughness (such as metals) have been shown to enhance adhesion of the deposited films, although they contribute to surface defects of the films (Kim et al., 2018). The surface roughness of the substrate has also been shown to significantly influence the superhydrophobic characteristics of sputtered PTFE thin films (Barshilia, Mohan, Selvakumar, & Rajam, 2009), optical properties (Hamada, Ogawa, Okumura, & Ishihara, 2019) and the magnetic properties of thin fcc Co(001) films (Steinmuller et al., 2007). The roughness of the substrate greatly affect the electrical transport characteristics of grown thin films-it influences the mobility of the electrons, resistivity, hall mobility in semiconductors and electron scattering (Tóth, 1987).

The substrate bias voltage has been reported to influence the properties of thin films prepared through sputtering technology (Wang et al., 2014). Increasing the bias voltage and decreasing target power was shown to slower the films deposition rates by magnitudes of more than 0.5 (Wang et al., 2014). Increasing substrate bias voltage for Al-Sc thin films was shown to enhance coarsening of the columnar structure and increase hardness and surface roughness (Kovac, Stock, & Zoch, 2012). At higher voltage however, higher defects were reported for sputtered films with higher sheet resistance of the films (Kovac et al., 2012). Contrary results were reported for sputtered Cu thin films-very high voltage bias enhanced the electrical properties and reduced the coarsening of the columnar structures (Bae et al., 2009). The adhesion strength of sputtered thin films of Cu on steel sheets increased with the substrate bias voltage as reported by Ouis and Cailler (Ouis & Cailler, 2013). The substrate bias voltage increases the deposition rates during sputtering and therefore has a direct relationship with the film thickness and growth mechanism of the structures. Besides, the bias voltage strongly determines the preferential orientation of sputtered thin films (Lin, Sproul, Moore, Wu, & Lee, 2011; Priyadarshini, Aich, & Chakraborty, 2014). An extensive literature detailing influence of the bias voltage on the growth of thin films is available in various databases (Ding et al., 2018; Lee & Lee, 1997; Lim, Kang, Kwon, Park, & Kwak, 2007; Low et al., 2014; Shieh, Kryder, & Lee, 1988; Stephens, 1976; Zeng et al., 2015).

Time of deposition is another important parameter for both evaporation and sputtering processes as shown in Figure 2. Time influences the material removal from the target, its transport to the substrate and the processes involved in film growth. It is directly related to the rate of deposition and film thickness (Bhatia, Verma, & Aggarwal, 2018). In sputtering evaporation technologies, a longer time of deposition results in thicker films as illustrated for sputtered ZnO thin films (Bhatia et al., 2018). The thickness of the films affects the lattice parameters but does not affect the *hkl* orientation for ZnO, however, the preferential orientation was seen to increase with thickness for sputtered ITO (Bhatia et al., 2018; J.-H. Kim et al., 2018). The thickness of deposited films also affects their performance in various fields such as electrical, optical and surface protection (Guilemany, Fernández, Delgado, Benedetti, & Climent, 2002; Han et al., 2009; Mishra, Prasad, Dave, Chandra, & Choudhary, 2015). The mechanical properties such as hardness and wear have been shown to enhance with film thickness (Kucukrende & Hakan Yetg, 2013). There should be enough time for the material to be dislodged or evaporated from the target, transported to the substrate and allowed to diffuse and form films on the substrate surface. Therefore, the choice of time of deposition should not only be based on the required film thickness but also on the desired properties and performance of the films.

Common Optimization Techniques in Manufacturing

In modern manufacturing processes, there is increased emphasis on low cost manufacturing for high quality products and achieving the balance between the two factors is critical. Within a manufacturing environment or system, there are several factors affecting the quality and rates of manufacturing some of which include generally, condition of the machines, process parameters, surrounding environment, desired quality and workmanship. For instance, in a machining process, the goal is to machine products at low cost and high-quality surface finishes (Bharathi Raja & Baskar, 2010). Optimization is aimed at selecting the appropriate factors for quality manufacturing of products at reasonable costs. In this section, a summary of the most commonly used techniques is presented in the chart shown in Figure 3.

Taguchi approach is one of the oldest techniques for optimization of manufacturing processes. The technique was introduced into American manufacturing in late 1970's by Dr. Taguchi (Lindeke & Andrew Liou, 1989). The technique involves selecting the most appropriate parameters affecting a certain process, these parameters are both dependent and independent. Dependent variables are usually the direct measurement of the manufacturing process, for instance the surface roughness of a machined part. These parameters show the system performance characteristics, which has three possibilities: smaller-is-better, larger-is-better and nominal-is-the-best (Mwema, Akinlabi, & Oladijo, 2019a). The independent variables on the other hand are those factors of the process, which significantly determine the dependent variables, for instance, the machine settings or the manufacturing parameters. Usually in a Taguchi optimization method, identification of these parameters and choosing their appropriate levels are most important aspects. For example, in a milling operation of a hard material, the range of the milling speed should be carefully considered if it falls within the machining range of that material (Kivak, 2014). The other important aspect of the Taguchi is the choice of orthogonal array for the design of experiment, which is chosen based on the independent variables and their levels. The results of the experimentation are analyzed, and an optimization algorithm proposed. Details of Taguchi method on manufacturing processes has been extensively explored in literature (Ronoh et al., 2019). Response surface methodology is another important technique for optimization of manufacturing processes. The method involves mapping responses of a process to independent decision variables and a decision on optimization is reached based on the model of the dependent and independent variables (Mukherjee & Ray, 2006).

In the chart in Figure 3, optimization techniques have been classified as traditional and modern for manufacturing processes. As shown, design of experiments and mathematically based methods are classified as traditional techniques whereas heuristic search methods are known as non-traditional techniques.

Application of optimization techniques to machining processes is quite advanced and widely researched. As such, there are many studies utilizing various techniques to optimize machining operations (Bharathi Raja & Baskar, 2010). A multi-pass turning operation has been optimized using Taguchi and artificial bee colony methods (Yildiz, 2013), particle swarm and genetic algorithm methods have been used to optimize roughness of milled surfaces (Bharathi Raja & Baskar, 2012; Zain, Haron, & Sharif, 2010), genetic algorithm and Taguchi have been used to optimize micro-turning processes of metals, and so forth (Mia et al., 2018; Peña-Parás et al., 2019; Viswanathan, Ramesh, & Subburam, 2018).

State of the Art of Studies

Extensive studies have been undertaken on optimization of thin film deposition including Taguchi, genetic algorithms, RSM, etc. (Li & Watson, 1998). In general, the focus has been on achieving the optimal properties and performance of various films for applications such as optical, electrical, solar cells, energy, microelectronics, and so forth. Genetic algorithm and transfer mixed methods were used to optimize the thickness of multilayer nanostructured thin films of SiO_2 and ITO for antireflective applications (Jalali, Jafari, & Mohammadi, 2019). The optimization in this study revealed that antireflective coatings consisting of low-n SiO_2 and high-n ITO on Si substrate under room temperature conditions were the best for enhancement of efficiency in solar cells. The optimization was important for enhancing the solar absorption capacity of solar cell systems.

Taguchi optimization of vanadium nitride (VN) thin films grown through unbalanced magnetron sputtering process was reported (Wu, Huang, & Yu, 2019). An L9 Taguchi orthogonal array was applied in which four independent variables, each with three levels were utilized. The variables considered were substrate bias voltage (-40, -65, -90 V), nitrogen flow rate (1, 2, 3 sccm), substrate temperature (150, 300, 450°C) and substrate rotation speed (10, 15, 20 rpm) (Wu et al., 2019). The dependent variables used to measure the influence of the independent parameters to VN thin films were surface roughness, mechanical properties (hardness and Young's modulus), grain size, film's thickness, chemical constituents (ratio of N to V and presence of O), lattice parameter, residual stress and electrical resistivity. The optimization was aimed at achieving optimal electrical performance of the VN thin films and it was seen that the best films for such applications were obtained at voltage of -90 V, temperature of 300°C, substrate rotation of 20 rpm and nitrogen flow rate of 3 sccm. A similar study was undertaken on ZnO thin films based on single variable optimization technique (Subramanyam, Goutham, Pavan Kumar, Yadhuraj, & Geetha, 2018).

Figure 3. Classification of optimization techniques used in various manufacturing processes (Mukherjee & Ray, 2006)

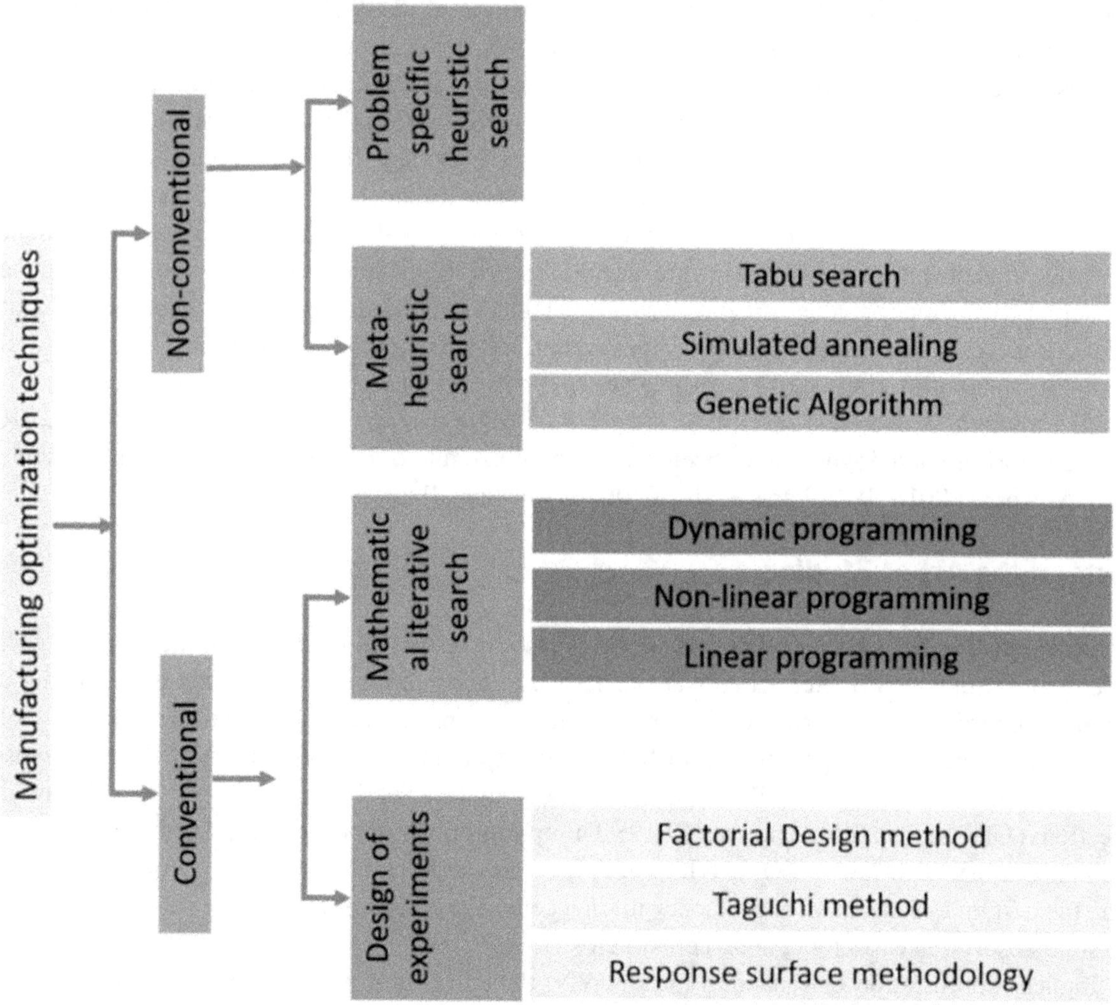

A design of experiment was undertaken to optimize the DC magnetron sputtering of Ni-Cr thin films on thermally sprayed alumina for sensor application by Tillman, Kokaji and Stangier (Tillmann, Kokalj, & Stangier, 2018). The independent parameters considered were bias-voltage, power and working pressure in the chamber whereas the dependent parameters were electrical resistance, scratch resistance and hardness/Young's modulus (Tillmann et al., 2018). The substrate bias voltage was shown to be the most significant parameter influencing deposition of Ni-Cr thin films on thermally sprayed alumina. The optimized films were obtained at voltage of -150 V, power of 1200 W and pressure of 250 mPa and a confirmatory study revealed that the films prepared at the optimized parameters exhibited high resistance to scratch, high mechanical properties and better electrical properties.

In another study, a hybrid approach involving investigated design and particle swarm optimization method was used to optimize the ITO/Ag/ITO tri-layered thin films to overcome the balance between the electrical and optical performances (Ferhati, Djeffal, & Benhaya, 2019). A figure of merit (FOG) has also been used to optimize thin films during sputtering technique (Chee, Meng, Lai, & Huang, 2018; Fekkai, 2014). An optimization method involving inspection of the set of RF magnetron sputtering of ITO films for solar cell applications deposited at low temperature has been presented by Sousa and Cunha (Sousa & da Cunha, 2019). The independent parameters used were the temperature, working pressure, plasma irradiation, gas flow rate and content of hydrogen in the working gas and their effect to the structural, optical and electrical properties of the ITO films investigated. The optimization revealed that low working pressure with partial pressure of hydrogen and plasma irradiation of the target enhances deposition ITO films with low resistivity. An ITO thin film of 248 nm, 90% transmittance in visible range and 23 Ω was produced within the optimization window.

Other studies involving optimization of thin films include zirconia (Usmani, Vijay, Chhibber, & Dixit, 2017), NiO_x (Yan et al., 2018), nickel-nickel oxide film (Zhao, Avendano, Gelin, Lu, & Wackelgard, 2006), Bi_2Te_3 thin films (Somdock, Kianwimol, Harnwunggmoung, Sakulkalavek, & Sakdanuphab, 2019), AlN thin films (Jinno et al., 2017), AZO and GZO thin films (Zhu et al., 2012; Zhu et al., 2014), using Taguchi, neural networks, desirability functions, genetic algorithms, etc. (Bushroa, Masjuki, Muhamad, & Beake, 2011; H. C. Lin, Su, Wang, Chang, & Juang, 2012; Venkata Ramana, Saravanan, Kamat, & Aparna, 2012). Similarly, studies on optimization of thermal evaporation include Gromov et al. (Gromov et al., 2019), Kumarasinge et al. (Kumarasinghe, Dissanayake, Pemasiri, & Dassanayake, 2017), Chander et al.(Chander & Dhaka, 2015), Ramarajan et al. (Ramarajan et al., 2018), Alex et al. (Alex, Kumar P, Chattopadhyay, Barshilia, & Basu, 2019), etc. In all these studies, interrelationships between the independent and dependent parameters and their influence on the properties of the thin films. The interrelationships are complex and through optimization/design of experiments the most significant parameters of the process can be identified and the single-parameter influence into the properties and functionality of the films can further be deduced.

CONCLUSION AND FUTURE STUDIES

In this article, an overview of the optimization of physical vapor deposition methods (evaporation and sputtering technologies) has been provided. A background on the general deposition parameters and techniques of optimization in manufacturing processes is also provided. There are various parameters affecting thin film deposition processes through physical methods, some of which include, the quality, composition and condition of the target (source) material, the distance between the target and substrate, the transport parameters and the properties of the substrate surface. Most of the studies on optimization of physical deposition of thin films have been on sputtering technology, which could be attributed to the attributes offered by the process over the evaporation technique. It has also been observed that most of the optimization studies in sputtering technology has been on the transport properties (plasma characteristics), multilayer properties and the substrate temperature and bias voltage. The future efforts in this subject should also incorporate the optimization of the target properties. Additionally, comparisons of various optimization techniques for physical deposition of thin films should be undertaken to determine the most appropriate techniques for these applications. There is need for more studies on optimization to cover a large resource of thin film materials for various applications.

REFERENCES

Alex, S., Kumar, P. R., Chattopadhyay, K., Barshilia, H. C., & Basu, B. (2019). Thermally evaporated Cu–Al thin film coated flexible glass mirror for concentrated solar power applications. *Materials Chemistry and Physics*, *232*(April), 221–228. doi:10.1016/j.matchemphys.2019.04.078

Azmand, A., & Kafashan, H. (2019). Al-doped ZnS thin films: Physical and electrochemical characterizations. *Journal of Alloys and Compounds*, *779*, 301–313. doi:10.1016/j.jallcom.2018.11.268

Bae, J. W., Lim, J. W., Mimura, K., Uchikoshi, M., Wada, M., Ikeda, M., & Isshiki, M. (2009). Influence of substrate bias voltage on properties of Pt thin films deposited by non-mass separated ion beam deposition method. *Materials Letters*, *63*(26), 2181–2184. doi:10.1016/j.matlet.2009.07.015

Barshilia, H. C., Mohan, D. K., Selvakumar, N., & Rajam, K. S. (2009). Effect of substrate roughness on the apparent surface free energy of sputter deposited superhydrophobic polytetrafluoroethylene thin films. *Applied Physics Letters*, *95*(3), 033116. Advance online publication. doi:10.1063/1.3186079

Bharathi Raja, S., & Baskar, N. (2010). Optimization techniques for machining operations: A retrospective research based on various mathematical models. *International Journal of Advanced Manufacturing Technology*, *48*(9–12), 1075–1090. doi:10.100700170-009-2351-x

Bharathi Raja, S., & Baskar, N. (2012). Application of Particle Swarm Optimization technique for achieving desired milled surface roughness in minimum machining time. *Expert Systems with Applications*, *39*(5), 5982–5989. doi:10.1016/j.eswa.2011.11.110

Bhatia, S., Verma, N., & Aggarwal, M. (2018). Effect of deposition time on sputtered ZnO thin films and their gas sensing application. *Journal of Materials Science Materials in Electronics*, *29*(21), 18136–18143. doi:10.100710854-018-9925-z

Bondarenko, A., Kolomiytsev, A., & Shapovalov, V. (2016). The target heating influence on the reactive magnetron sputtering process. *Journal of Physics: Conference Series*, *729*(1), 012006. doi:10.1088/1742-6596/729/1/012006

Bushroa, A. R., Masjuki, H. H., Muhamad, M. R., & Beake, B. D. (2011). Optimized scratch adhesion for TiSiN coatings deposited by a combination of DC and RF sputtering. *Surface and Coatings Technology*, *206*(7), 1837–1844. doi:10.1016/j.surfcoat.2011.07.048

Caillard, A., El'Mokh, M., Lecas, T., & Thomann, A. L. (2018). Effect of the target temperature during magnetron sputtering of Nickel. *Vacuum*, *147*, 82–91. doi:10.1016/j.vacuum.2017.10.016

Chander, S., & Dhaka, M. S. (2015). Optimization of physical properties of vacuum evaporated CdTe thin films with the application of thermal treatment for solar cells. *Materials Science in Semiconductor Processing*, *40*, 708–712. doi:10.1016/j.mssp.2015.07.063

Chang, C.-L., & Yang, F.-C. (2018). Effect of target composition on the microstructural, mechanical, and corrosion properties of TiAlN thin films deposited by high-power impulse magnetron sputtering. *Surface and Coatings Technology*, *352*, 330–337. doi:10.1016/j.surfcoat.2018.08.023

Chee, K. W. A., Meng, F., Lai, D. C., & Huang, F. (2018). Measurement-based optimization and analysis of α-IGZO/Ag/α-IGZO transparent conducting electrodes fabricated using DC magnetron sputter deposition. *Ceramics International, 44*(17), 20939–20946. doi:10.1016/j.ceramint.2018.06.277

Ding, J., Zhang, T., Mei, H., Yun, J. M., Jeong, S. H., Wang, Q., & Kim, K. H. (2018). Effects of negative bias voltage and ratio of nitrogen and argon on the structure and properties of NbN coatings deposited by HiPIMS deposition system. *Coatings, 8*(1), 10. Advance online publication. doi:10.3390/coatings8010010

Fekkai, Z. (2014). Optical, Morphological and Electrical Properties of Silver and Aluminium Metallization Contacts for Solar Cells. *American Journal of Modern Physics, 3*(2), 45. doi:10.11648/j.ajmp.20140302.13

Ferhati, H., Djeffal, F., & Benhaya, A. (2019). Optimized high-performance ITO/Ag/ITO multilayer transparent electrode deposited by RF magnetron sputtering. *Superlattices and Microstructures, 129*(March), 176–184. doi:10.1016/j.spmi.2019.03.027

Gaines, J. R. (2019). The Effect of Sputter Cathode Design on Deposition Parameters. *Vakuum in Forschung Und Praxis, 31*(5), 14–19. doi:10.1002/vipr.201900718

Gromov, D. G., Dubkov, S. V., Savitskiy, A. I., Shaman, Y. P., Polokhin, A. A., Belogorokhov, I. A., & Trifonov, A. Y. (2019). Optimization of nanostructures based on Au, Ag, Au Ag nanoparticles formed by thermal evaporation in vacuum for SERS applications. *Applied Surface Science, 489*(November), 701–707. doi:10.1016/j.apsusc.2019.05.286

Guilemany, J. M., Fernández, J., Delgado, J., Benedetti, A. V., & Climent, F. (2002). Effects of thickness coating on the electrochemical behaviour of thermal spray Cr3C2–NiCr coatings. *Surface and Coatings Technology, 153*(2–3), 107–113. doi:10.1016/S0257-8972(01)01679-6

Gullu, H. H., Isik, M., & Gasanly, N. M. (2018). Structural and optical properties of thermally evaporated Cu-Ga-S (CGS) thin films. *Physica B, Condensed Matter, 547*, 92–96. doi:10.1016/j.physb.2018.08.015

Hamada, K., Ogawa, T., Okumura, H., & Ishihara, K. N. (2019). The effect of substrate roughness on the properties of RF sputtered AZO thin film. *MRS Communications, 9*(2), 697–701. doi:10.1557/mrc.2019.66

Han, M. S., Woo, Y., Ko, S.-C., Jeong, Y.-J., Jang, S.-K., & Kim, S.-J. (2009). Effects of thickness of Al thermal spray coating for STS 304. *Transactions of Nonferrous Metals Society of China, 19*(4), 925–929. doi:10.1016/S1003-6326(08)60379-9

Higo, M., Fujita, K., Mitsushio, M., Yoshidome, T., & Kakoi, T. (2007). Epitaxial growth and surface morphology of aluminum films deposited on mica studied by transmission electron microscopy and atomic force microscopy. *Thin Solid Films, 516*(1), 17–24. doi:10.1016/j.tsf.2007.04.046

Jalali, T., Jafari, M., & Mohammadi, A. (2019). Genetic algorithm optimization of antireflection coating consisting of nanostructured thin films to enhance silicon solar cell efficacy. *Materials Science and Engineering: B, 247*(January), 114354. doi:10.1016/j.mseb.2019.05.016

Jilani, A., Abdel-wahab, M. S., & Hammad, A. H. (2017). Advance Deposition Techniques for Thin Film and Coating. In *Modern Technologies for Creating the Thin-film Systems and Coatings* (Vol. 2, pp. 137–149). InTech., doi:10.5772/65702

Jinno, D., Otsuki, S., Sugimori, S., Daicho, H., Iwaya, M., Takeuchi, T., Kamiyama, S., & Akasaki, I. (2017). Characterization and optimization of sputtered AlN buffer layer on r -plane sapphire substrate to improve the crystalline quality of nonpolar a -plane GaN. *Journal of Crystal Growth*, *480*, 90–95. doi:10.1016/j.jcrysgro.2017.10.018

Kavitha, A., Kannan, R., & Rajashabala, S. (2017). Effect of target power on the physical properties of Ti thin films prepared by DC magnetron sputtering with supported discharge. *Materials Science Poland*, *35*(1), 173–180. doi:10.1515/msp-2017-0022

Khachatryan, H., Lee, S., Kim, K.-B., Kim, H.-K., & Kim, M. (2018). Al thin film: The effect of substrate type on Al film formation and morphology. *Journal of Physics and Chemistry of Solids*, *122*(May), 109–117. doi:10.1016/j.jpcs.2018.06.018

Khalaf, M. K., Al-Taay, H. F., & Ali, D. S. (2017). Effect of radio frequency magnetron sputtering power on structural and optical properties of Ti6Al4V thin films. *Photonic Sensors*, *7*(2), 163–170. doi:10.100713320-017-0390-8

Khan, W., Wang, Q., & Jin, X. (2018). Effect of target composition and sputtering deposition parameters on the functional properties of nitrogenized Ag-Permalloy flexible thin films deposited on polymer substrates. *Materials (Basel)*, *11*(3), 439. Advance online publication. doi:10.3390/ma11030439 PMID:29562603

Kim, D., Lee, H., Bae, J., Jeong, H., Choi, B., Nam, T., & Noh, J. (2018). Effect of Substrate Roughness on Adhesion and Structural Properties of Ti–Ni Shape Memory Alloy Thin Film. *Journal of Nanoscience and Nanotechnology*, *18*(9), 6201–6205. doi:10.1166/jnn.2018.15636 PMID:29677767

Kim, J. H., Lee, S., & Im, H. S. (1999). Effect of target density and its morphology on TiO2 thin films grown on Si(100) by PLD. *Applied Surface Science*, *151*(1), 6–16. doi:10.1016/S0169-4332(99)00269-X

Kim, J.-H., Seong, T.-Y., Ahn, K.-J., Chung, K.-B., Seok, H.-J., Seo, H.-J., & Kim, H.-K. (2018). The effects of film thickness on the electrical, optical, and structural properties of cylindrical, rotating, magnetron-sputtered ITO films. *Applied Surface Science*, *440*, 1211–1218. doi:10.1016/j.apsusc.2018.01.318

Kivak, T. (2014). Optimization of surface roughness and flank wear using the Taguchi method in milling of Hadfield steel with PVD and CVD coated inserts. *Measurement: Journal of the International Measurement Confederation*, *50*(1), 19–28. doi:10.1016/j.measurement.2013.12.017

Kovac, J., Stock, H.-R., & Zoch, H.-W. (2012). Influence of Substrate Bias Voltage on the Properties of Sputtered Aluminum-Scandium Thin Sheets. *Journal of Surface Engineered Materials and Advanced Technology*, *02*(02), 115–119. doi:10.4236/jsemat.2012.22018

Kucukrende, I., & Hakan Yetg, S. (2013). Coating Parameters Influences on Mechanical Properties of Coating. *Journal of Applied Sciences (Faisalabad)*, *13*(4), 645–649. doi:10.3923/jas.2013.645.649

Kumarasinghe, P. K. K., Dissanayake, A., Pemasiri, B. M. K., & Dassanayake, B. S. (2017). Thermally evaporated CdTe thin films for solar cell applications: Optimization of physical properties. *Materials Research Bulletin*, *96*, 188–195. doi:10.1016/j.materresbull.2017.04.026

Lee, H. C., & Lee, J. Y. (1997). Effect of negative bias voltage on the microstructures of AIN thin films fabricated by reactive r.f. magnetron sputtering. *Journal of Materials Science Materials in Electronics*, *8*(6), 385–390. doi:10.1023/A:1018551726015

Li, D. G., & Watson, A. C. (1998). Optical thin film optimization design using genetic algorithms. In *1997 IEEE International Conference on Intelligent Processing Systems (Cat. No.97TH8335)* (Vol. 1, pp. 132–136). IEEE. 10.1109/ICIPS.1997.672752

Lim, D. G., Kang, G. S., Kwon, S. I., Park, M. W., & Kwak, D. J. (2007). Influence of Positive Substrate Bias on the Electrical Properties of ZnO:Al Films Prepared by DC Magnetron Sputtering. *Journal of the Korean Physical Society*, *50*(6), 1697. doi:10.3938/jkps.50.1697

Lin, H. C., Su, C. T., Wang, C. C., Chang, B. H., & Juang, R. C. (2012). Parameter optimization of continuous sputtering process based on Taguchi methods, neural networks, desirability function, and genetic algorithms. *Expert Systems with Applications*, *39*(17), 12918–12925. doi:10.1016/j.eswa.2012.05.032

Lin, J., Sproul, W. D., Moore, J. J., Wu, Z. L., & Lee, S. L. (2011). Effect of negative substrate bias voltage on the structure and properties of CrN films deposited by modulated pulsed power (MPP) magnetron sputtering. *Journal of Physics. D, Applied Physics*, *44*(42), 425305. Advance online publication. doi:10.1088/0022-3727/44/42/425305

Lindeke, R. R., & Andrew Liou, Y.-H. (1989). Methods for optimization in the manufacturing system - the taguchi method: An engineering approach to its implementation. *Journal of Mechanical Working Technology*, *20*(C), 205–218. doi:10.1016/0378-3804(89)90031-4

Low, J. W., Nafarizal, N., Sahdan, M. Z., Abd Kadir, M., bin Ahmad, M. K., Yeon, M. S. A., Ammar, Z., Ahmad Saad, F. S., & Mohd Zain, A. F. (2014). Effect of Substrate Bias in Copper Sputtering Plasma Measured by Langmuir Probe. *Advanced Materials Research*, *925*, 238–242. . doi:10.4028/www.scientific.net/AMR.925.238

Mia, M., Morshed, M. S., Kharshiduzzaman, M., Razi, M. H., Mostafa, M. R., Rahman, S. M. S., Ahmad, I., Hafiz, M. T., & Kamal, A. M. (2018). Prediction and optimization of surface roughness in minimum quantity coolant lubrication applied turning of high hardness steel. *Measurement*, *118*(January), 43–51. doi:10.1016/j.measurement.2018.01.012

Mishra, P. K., Prasad, J. N., Dave, V., Chandra, R., & Choudhary, A. K. (2015). The significant effect of film thickness on the properties of chalcopyrite thin absorbing films deposited by RF magnetron sputtering. *Materials Science in Semiconductor Processing*, *34*, 350–358. doi:10.1016/j.mssp.2015.02.047

Mukherjee, I., & Ray, P. K. (2006). A review of optimization techniques in metal cutting processes. *Computers & Industrial Engineering*, *50*(1–2), 15–34. doi:10.1016/j.cie.2005.10.001

Mwema, Oladijo, & Akinlabi. (2018). Effect of Substrate Temperature on Aluminium Thin Films Prepared byRF-Magnetron Sputtering. *Materials Today: Proceedings*, *5*(9), 20464–20473. doi:10.1016/j.matpr.2018.06.423

Mwema, F. M., Akinlabi, E. T., & Oladijo, O. P. (2019). *Two-Dimensional Fast Fourier Transform Analysis of Surface Microstructures of Thin Aluminium Films Prepared by Radio-Frequency (RF) Magnetron Sputtering. In Advances in Material Sciences and Engineering. Lecture Notes in Mechanical Engineering*. Springer. doi:10.1007/978-981-13-8297-0_27

Mwema, F. M., Akinlabi, E. T., & Oladijo, O. P. (2019). Effect of Substrate Type on the Fractal Characteristics of AFM Images of Sputtered Aluminium Thin Films. *Materials Science*, *26*(1), 49–57. doi:10.5755/j01.ms.26.1.22769

Mwema, F. M., Akinlabi, E. T., & Oladijo, O. P. (2019a). Correction of Artifacts and Optimization of Atomic Force Microscopy Imaging. In K. Kumar & J. Paulo Davim (Eds.), *Title: Design, Development, and Optimization of Bio-Mechatronic Engineering Products* (pp. 158–179). IGI Global. doi:10.4018/978-1-5225-8235-9.ch007

Mwema, F. M., Akinlabi, E. T., & Oladijo, O. P. (2019b). Fractal analysis of hillocks: A case of RF sputtered aluminum thin films. *Applied Surface Science*, *489*(May), 614–623. doi:10.1016/j.apsusc.2019.05.340

Mwema, F. M., Akinlabi, E. T., Oladijo, O. P., & Dutta Majumdar, J. (2019). Evolution of microstructure and wear properties of aluminum thin films with sputtering substrate temperature. In *2019 IEEE 10th International Conference on Mechanical and Intelligent Manufacturing Technologies (ICMIMT)* (pp. 31–36). IEEE. 10.1109/ICMIMT.2019.8712046

Mwema, F. M., Akinlabi, E. T., Oladijo, O. P., & Majumdar, J. D. (2019). Effect of varying low substrate temperature on sputtered aluminium films. *Materials Research Express*, *6*(5), 056404. doi:10.1088/2053-1591/ab014a

Mwema, F. M., Oladijo, O. P., Akinlabi, S. A., & Akinlabi, E. T. (2018). Properties of physically deposited thin aluminium film coatings: A review. *Journal of Alloys and Compounds*, *747*, 306–323. doi:10.1016/j.jallcom.2018.03.006

Odhiambo, J. G., Li, W., Zhao, Y., & Li, C. (2019). Porosity and Its Significance in Plasma-Sprayed Coatings. *Coatings*, *9*(7), 460. doi:10.3390/coatings9070460

Ouis, A., & Cailler, M. (2013). Effects of substrate bias voltage on adhesion of DC magnetron-sputtered copper films on E24 carbon steel: Investigations by Auger electron spectroscopy. *Journal of Adhesion Science and Technology*, *27*(21), 2367–2386. doi:10.1080/01694243.2013.777320

Peña-Parás, L., Maldonado-Cortés, D., Rodríguez-Villalobos, M., Romero-Cantú, A. G., Montemayor, O. E., & Herrera, M. (2018, September). … Hugler, W. (2019). Optimization of milling parameters of 1018 steel and nanoparticle additive concentration in cutting fluids for enhancing multi-response characteristics. *Wear*, *426–427*, 877–886. doi:10.1016/j.wear.2019.01.078

Priyadarshini, B. G., Aich, S., & Chakraborty, M. (2014). Substrate bias voltage and deposition temperature dependence on properties of rf-magnetron sputtered titanium films on silicon (100). *Bulletin of Materials Science*, *37*(7), 1691–1700. doi:10.100712034-014-0722-x

Quero, J. M., Perdigones, F., & Aracil, C. (2018). Microfabrication technologies used for creating smart devices for industrial applications. In *Smart Sensors and MEMs* (pp. 291–311). Elsevier., doi:10.1016/B978-0-08-102055-5.00011-5

Ramarajan, R., Kovendhan, M., Phan, D. T., & Thangaraju, K. (2018). Optimization of Zn 2 SnO 4 thin film by post oxidation of thermally evaporated alternate Sn and Zn metallic multi-layers. *Applied Surface Science*, *449*, 68–76. doi:10.1016/j.apsusc.2018.01.029

Ronoh, N. K., Mwema, F. M., Akinlabi, S. A., Akinlabi, E. T., Karuri, N. W., & Ngetha, H. T. (2019). Effects of cooling conditions and grinding depth on sustainable surface grinding of Ti-6Al-4V: Taguchi approach. *AIMS Materials Science*, *6*(5), 697–712. doi:10.3934/matersci.2019.5.697

Semaltianos, N. G. (2001). Thermally evaporated aluminium thin films. *Applied Surface Science*, *183*(3–4), 223–229. doi:10.1016/S0169-4332(01)00565-7

Shieh, H. P. D., Kryder, M. H., & Lee, J. W. (1988). Effects of bias-sputtering on magnetron-sputtered magneto-optical recording media. *IEEE Transactions on Magnetics*, *24*(6), 2446–2448. doi:10.1109/20.92136

Simon, A. H. (2018). Sputter Processing. In *Handbook of Thin Film Deposition* (pp. 195–230). Elsevier. doi:10.1016/B978-0-12-812311-9.00007-4

Sivaram, S. (1995). *Chemical Vapor Deposition. Appl. Phys. A* (Vol. 73). Boston, MA: Springer US. doi:10.1007/978-1-4757-4751-5

Somdock, N., Kianwimol, S., Harnwunggmoung, A., Sakulkalavek, A., & Sakdanuphab, R. (2019). Simultaneous stoichiometric composition and highly (00l) orientation of flexible Bi2Te3thin films via optimising the DC magnetron sputter-deposition process. *Journal of Alloys and Compounds*, *773*, 78–85. doi:10.1016/j.jallcom.2018.09.216

Sousa, M. G., & da Cunha, A. F. (2019). Optimization of low temperature RF-magnetron sputtering of indium tin oxide films for solar cell applications. *Applied Surface Science*, *484*, 257–264. doi:10.1016/j.apsusc.2019.03.275

Steinmuller, S. J., Vaz, C. A. F., Ström, V., Moutafis, C., Tse, D. H. Y., Gürtler, C. M., Kläui, M., Bland, J. A. C., & Cui, Z. (2007). Effect of substrate roughness on the magnetic properties of thin fcc Co films. *Physical Review B: Condensed Matter and Materials Physics*, *76*(5), 1–8. doi:10.1103/PhysRevB.76.054429

Stephens, A. W., Vossen, J. L., & Kern, W. (1976). The Effect of Substrate Bias on the Properties of Reactively Sputtered Silicon Nitride. *Journal of the Electrochemical Society*, *123*(2), 303–305. doi:10.1149/1.2132809

Subramanyam, T. K., Goutham, P., Pavan Kumar, S., Yadhuraj, S. R., & Geetha, K. S. (2018). Optimization of Sputtered AZO Thin Films for Device Application. *Materials Today: Proceedings*, *5*(4), 10851–10859. doi:10.1016/j.matpr.2017.12.373

Thornton, J. A. (1986). The microstructure of sputter-deposited coatings. *Journal of Vacuum Science & Technology. A, Vacuum, Surfaces, and Films*, *4*(6), 3059–3065. doi:10.1116/1.573628

Tillmann, W., Kokalj, D., & Stangier, D. (2018). Optimization of the deposition parameters of Ni-20Cr thin films on thermally sprayed Al2O3 for sensor application. *Surface and Coatings Technology*, *344*(October), 223–232. doi:10.1016/j.surfcoat.2018.03.029

To, T. B. T., de Sousa, V. B., & Aarão Reis, F. D. A. (2018). Thin film growth models with long surface diffusion lengths. *Physica A, 511*, 240–250. Advance online publication. doi:10.1016/j.physa.2018.07.024

Tondare, R. S., Shivaraj, B. W., Narasimhamurthy, H. N., Krishna, M., & Subramanyam, T. K. (2018). Effect of deposition time on structural, electrical and optical properties of Aluminium doped ZnO thin films by RF magnetron sputtering. *Materials Today: Proceedings, 5*(1), 2710–2715. doi:10.1016/j.matpr.2018.01.052

Tóth, L. (1987). A model of substrate surface roughness effect on the electrical properties of thin films. *Vacuum, 37*(1–2), 103–106. doi:10.1016/0042-207X(87)90094-7

Usmani, B., Vijay, V., Chhibber, R., & Dixit, A. (2017). Optimization of sputtered zirconium thin films as an infrared reflector for use in spectrally-selective solar absorbers. *Thin Solid Films, 627*, 17–25. doi:10.1016/j.tsf.2017.02.055

Venkata Ramana, G., Saravanan, P., Kamat, S. V., & Aparna, Y. (2012). Optimization of sputtering parameters for SmCo thin films using design of experiments. *Applied Surface Science, 261*, 110–117. doi:10.1016/j.apsusc.2012.07.109

Viswanathan, R., Ramesh, S., & Subburam, V. (2018). Measurement and optimization of performance characteristics in turning of Mg alloy under dry and MQL conditions. *Measurement, 120*(November), 107–113. doi:10.1016/j.measurement.2018.02.018

Wang, P., Takeno, T., Fontaine, J., Aono, M., Adachi, K., Miki, H., & Takagi, T. (2014). Effects of substrate bias voltage and target sputtering power on the structural and tribological properties of carbon nitride coatings. *Materials Chemistry and Physics, 145*(3), 434–440. doi:10.1016/j.matchemphys.2014.02.032

Wu, C.-K., Huang, J.-H., & Yu, G.-P. (2019). Optimization of deposition processing of VN thin films using design of experiment and single-variable (nitrogen flow rate) methods. *Materials Chemistry and Physics, 224*, 246–256. doi:10.1016/j.matchemphys.2018.12.038

Yan, X., Zheng, J., Zheng, L., Lin, G., Lin, H., Chen, G., Du, B., & Zhang, F. (2018). Optimization of sputtering NiO films for perovskite solar cell applications. *Materials Research Bulletin, 103*, 150–157. doi:10.1016/j.materresbull.2018.03.027

Yildiz, A. R. (2013). Optimization of cutting parameters in multi-pass turning using artificial bee colony-based approach. *Information Sciences, 220*, 399–407. doi:10.1016/j.ins.2012.07.012

Yoon, S. G., Kim, H. K., Kim, M. J., Lee, H. M., & Yoon, D. H. (2005). Effect of substrate temperature on surface roughness and optical properties of Ta2O5 using ion-beam sputtering. In Thin Solid Films (Vol. 475, pp. 239–242). doi:10.1016/j.tsf.2004.07.043

Zain, A. M., Haron, H., & Sharif, S. (2010). Application of GA to optimize cutting conditions for minimizing surface roughness in end milling machining process. *Expert Systems with Applications, 37*(6), 4650–4659. doi:10.1016/j.eswa.2009.12.043

Zeng, Y., Tan, Z., Zhou, L., Jiang, M., Qiu, Y., Fang, F., Huang, H., Zhang, X., & Jiang, J. (2015). Effects of Bias Voltage on Fen Films Prepared by Magnetron Sputtering. *Materials Research, 18*(suppl 1), 115–119. doi:10.1590/1516-1439.334614

Zhao, S., Avendano, E., Gelin, K., Lu, J., & Wackelgard, E. (2006). Optimization of an industrial DC magnetron sputtering process for graded composition solar thermal absorbing layer. *Solar Energy Materials and Solar Cells*, *90*(3), 308–328. doi:10.1016/j.solmat.2005.03.018

Zhu, B. L., Wang, J., Zhu, S. J., Wu, J., Zeng, D. W., & Xie, C. S. (2012). Optimization of sputtering parameters for deposition of Al-doped ZnO films by rf magnetron sputtering in Ar + H 2 ambient at room temperature. *Thin Solid Films*, *520*(23), 6963–6969. doi:10.1016/j.tsf.2012.07.049

Zhu, D. L., Wang, Q., Han, S., Cao, P. J., Liu, W. J., Jia, F., Zeng, Y. X., Ma, X. C., & Lu, Y. M. (2014). Optimization of process parameters for the electrical properties in Ga-doped ZnO thin films prepared by r.f. magnetron sputtering. *Applied Surface Science*, *298*, 208–213. doi:10.1016/j.apsusc.2014.01.163

ADDITIONAL READING

Abegunde, O. O., Akinlabi, E. T., & Oladijo, O. P. (2020). Developing an empirical relationship for optimizing surface roughness of TiC thin film grown by magnetron sputtering using Taguchi analysis. *Materials Today Proceedings*, *26*(2), 3282–3287. doi:10.1016/j.matpr.2020.02.253

Abegunde, O. O., Akinlabi, E. T., & Oladijo, O. P. (2020). Dataset on microstructural, structural and tribology characterization of TiC thin film on CpTi substrate grown by RF magnetron sputtering. *Data in Brief*, *29*, 105205. doi:10.1016/j.dib.2020.105205 PMID:32055672

Abegunde, O. O., Akinlabi, E. T., & Oladijo, O. P. (2020). Influence of sputtering parameters on the structural and mechanical properties of TiC thin film coating. *Applied Surface Science*, *520*, 146323. doi:10.1016/j.apsusc.2020.146323

Ghadai, R. K., Kalita, K., Mondal, S. C., & Swain, B. P. (2018). PECVD process parameter optimization: Towards increased hardness of diamond-like carbon thin films. *Materials and Manufacturing Processes*, *33*(16), 1905–1913. doi:10.1080/10426914.2018.1512114

Pulker, H. K., & Girardet, E. (1969). Problems Encountered During Deposition of Optical Thin Films with Reproducible Characteristics in an Automatic Coater. *Journal of Vacuum Science and Technology*, *6*(1), 131–134. doi:10.1116/1.1492643

Wu, Y. J., Hsu, S.-C., Lin, Y.-C., Xu, Y., Chuang, T.-H., & Chen, S.-C. (2020). Study on thermoelectric property optimization of mixed-phase bismuth telluride thin films deposited by co-evaporation process. *Surface and Coatings Technology*, *394*, 125694. doi:10.1016/j.surfcoat.2020.125694

KEY TERMS AND DEFINITIONS

Physical Vapor Deposition (PVD): These are thin films manufacturing processes which involves removal of atoms from the source material (target) through atomization in a vacuum and then the atoms are transported to the surface substrate for condensation.

Process Parameters: These are the factors which affect a process such as PVD.

Substrate: The surface on which the thin films are deposited/grown.

Target: The source of the material for formation of the thin films.

Thin Films: These are films with thickness less than 1 μm and are usually deposited on substrate surfaces.

Compilation of References

Abd-El-Wahed, W. F., El-Shorbagy, M. A., & Mousa, A. (2012). A local search based hybrid particle swarm optimization algorithm for multi objective optimization. *Swarm and Evolutionary Computation*, *3*, 1–14. doi:10.1016/j.swevo.2011.11.005

Abhishek, K., Datta, S., & Mahapatra, S. S. (2016). Multi-objective optimization in drilling of CFRP (polyester) composites: Application of a fuzzy embedded harmony search (HS) algorithm. *Measurement*, *77*, 222–239. doi:10.1016/j.measurement.2015.09.015

Acharya, S. G., Vadher, J. A., Sheladiya, M. V., & Madhnani, M. (2016). Quality casting of motor body using design of experiment and casting simulation. *International Journal of Manufacturing Research*, *11*(2), 111–125. doi:10.1504/IJMR.2016.078243

Adalı, E. A., & Işık, A. T. (2017). CRITIC and MAUT methods for the contract manufacturer selection problem. *European Journal of Multidisciplinary Studies*, *2*(5), 93–101. doi:10.26417/ejms.v5i1.p93-101

Adke, M. N., & Shrikant, V. K. (2014). Optimization of die-casting process parameters to identify optimized level for cycle time using Taguchi method. *Int. J. Innov. Eng. Technol*, *4*, 365–375.

Afzal, A. (2014). Implantable zirconia bioceramics for bone repair and replacement: A chronological review. *Materials Express*, *4*(1), 1–12. doi:10.1166/mex.2014.1148

Afzulpurkar, N. V., & Navalertporn, T. (2011). Optimization of tile manufacturing process using particle swarm optimization. *Swarm and Evolutionary Computation*, *1*(2), 97–109. doi:10.1016/j.swevo.2011.05.003

Akao, Y. (1990). *Quality function deployment*. Productivity Press.

Alex, S., Kumar, P. R., Chattopadhyay, K., Barshilia, H. C., & Basu, B. (2019). Thermally evaporated Cu–Al thin film coated flexible glass mirror for concentrated solar power applications. *Materials Chemistry and Physics*, *232*(April), 221–228. doi:10.1016/j.matchemphys.2019.04.078

Ambrogio, G., Cozza, V., Filice, L., & Micari, F. (2007). An analytical model for improving precision in single point incremental forming. *Journal of Materials Processing Technology*, *191*(1-3), 92–95. doi:10.1016/j.jmatprotec.2007.03.079

Anastasiou, K. S. (2002). Optimization of the aluminium die casting process based on the Taguchi method. *Proceedings of the Institution of Mechanical Engineers. Part B, Journal of Engineering Manufacture*, *216*(7), 969–977. doi:10.1243/09544050260174175

Arbabi, V., & Ebrahimzadeh, I. (2010). Effects of wall thickness on microstructures and properties of α/β brasses pipes produced by horizontal continuous casting. *International Journal of Cast Metals Research*, *23*(3), 150–157. doi:10.1179/136404609X12490478029353

Asiltürk, I., & Akkuş, H. (2011). Determining the effect of cutting parameters on surface roughness in hard turning using the Taguchi method. *Measurement, 44*(9), 1697–1704.

Aversa, R., Petrescu, R. V., Petrescu, F. I., & Apicella, A. (2016). Smart-factory: Optimization and process control of composite centrifuged pipes. *American Journal of Applied Sciences, 13*(11), 1330–1341. doi:10.3844/ajassp.2016.1330.1341

Azmand, A., & Kafashan, H. (2019). Al-doped ZnS thin films: Physical and electrochemical characterizations. *Journal of Alloys and Compounds, 779*, 301–313. doi:10.1016/j.jallcom.2018.11.268

Babu, B., Karthikeyan, S. K., & Adithya, V. (2014). Experimental Investigation and Analysis of Corrosion and Hardness using Aluminium Composites. *International Journal of Latest Trends in Engineering and Technology.*

Bacaicoa, I., Wicke, M., Luetje, M., Zeismann, F., Brueckner-Foit, A., Geisert, A., & Fehlbier, M. (2017). Characterization of casting defects in a Fe-rich Al-Si-Cu alloy by microtomography and finite element analysis. *Engineering Fracture Mechanics, 183*, 159–169. doi:10.1016/j.engfracmech.2017.03.015

Badi, I., & Pamucar, D. (2020). Supplier selection for steelmaking company by using combined Grey-MARCOS methods. *Decision Making: Applications in Management and Engineering, 3*(2), 37–48.

Bae, J. W., Lim, J. W., Mimura, K., Uchikoshi, M., Wada, M., Ikeda, M., & Isshiki, M. (2009). Influence of substrate bias voltage on properties of Pt thin films deposited by non-mass separated ion beam deposition method. *Materials Letters, 63*(26), 2181–2184. doi:10.1016/j.matlet.2009.07.015

Bagudancha, I., Romeua, M. L., Centenob, G., & Zuniga, A. E. (2015). Forming force and temperature effects on single point incremental forming of polyvinylchloride. *Journal of Materials Processing Technology, 219*, 221–229. doi:10.1016/j.jmatprotec.2014.12.004

Bahoul, R., Arfa, H., & Belhadj, H. S. (2014). A study on optimal design of process parameters in single point incremental forming of sheet metal by combining Box–Behnken design of experiments, response surface methods and genetic algorithms. *Int J Adv Manuf, 74,* 163-185.

Barshilia, H. C., Mohan, D. K., Selvakumar, N., & Rajam, K. S. (2009). Effect of substrate roughness on the apparent surface free energy of sputter deposited superhydrophobic polytetrafluoroethylene thin films. *Applied Physics Letters, 95*(3), 033116. Advance online publication. doi:10.1063/1.3186079

Bauer, S., Schmuki, P., von der Mark, K., & Park, J. (2013). Engineering biocompatible implant surfaces Part I: Materials and surfaces. *Progress in Materials Science, 58*(3), 261–326. doi:10.1016/j.pmatsci.2012.09.001

Bendell, T. (Ed.). (1989). *Taguchi Methods: Proceedings of the 1988 European Conference; the 1. European Conference on Taguchi Methods, the Cafe Royal.* Elsevier.

Bhandare, R. G., & Sonawane, P. M. (2013). Preparation of aluminium matrix composite by using stir casting method. *International Journal of Engineering and Advanced Technology, 3*(3), 61–65.

Bharathi Raja, S., & Baskar, N. (2010). Optimization techniques for machining operations: A retrospective research based on various mathematical models. *International Journal of Advanced Manufacturing Technology, 48*(9–12), 1075–1090. doi:10.100700170-009-2351-x

Bharathi Raja, S., & Baskar, N. (2012). Application of Particle Swarm Optimization technique for achieving desired milled surface roughness in minimum machining time. *Expert Systems with Applications, 39*(5), 5982–5989. doi:10.1016/j.eswa.2011.11.110

Bhatia, S., Verma, N., & Aggarwal, M. (2018). Effect of deposition time on sputtered ZnO thin films and their gas sensing application. *Journal of Materials Science Materials in Electronics, 29*(21), 18136–18143. doi:10.100710854-018-9925-z

Bhattacharya, A., Das, S., Majumder, P., & Batish, A. (2009). Estimating the effect of cutting parameters on surface finish and power consumption during high speed machining of AISI 1045 steel using Taguchi design and ANOVA. *Production Engineering*, *3*(1), 31–40. doi:10.100711740-008-0132-2

Bhutia, P. W., & Phipon, R. (2012). Application of AHP and TOPSIS method for supplier selection problem. *IOSR Journal of Engineering*, *2*(10), 43–50. doi:10.9790/3021-021034350

Bodunrin, M. O., Alaneme, K. K., & Chown, L. H. (2015). Aluminium matrix hybrid composites: A review of reinforcement philosophies; mechanical, corrosion and tribological characteristics. *Journal of Materials Research and Technology*, *4*(4), 434–445. doi:10.1016/j.jmrt.2015.05.003

Bondarenko, A., Kolomiytsev, A., & Shapovalov, V. (2016). The target heating influence on the reactive magnetron sputtering process. *Journal of Physics: Conference Series*, *729*(1), 012006. doi:10.1088/1742-6596/729/1/012006

Borowiecki, B., Borowiecka, O., & Szkodzińska, E. (2011). Casting defects analysis by the Pareto method. *Archives of Foundry Engineering, 11*.

Bosco, M. A., Palanikumar, K., Prasad, B. D., & Velayudham, A. (2013). Influence of machining parameters on delamination in drilling of GFRP-armour steel sandwich composites. *Procedia Engineering*, *51*, 758–763. doi:10.1016/j.proeng.2013.01.108

Bose, G. K., & Pain, P. (2020). 9 Optimization of EDM process through evolutionary computing and fuzzy MCDM techniques. *Soft Computing: Techniques in Engineering Sciences*, *1*, 165.

Bose, P. K., Deb, M., Banerjee, R., & Majumder, A. (2013). Multi objective optimization of performance parameters of a single cylinder diesel engine running with hydrogen using a Taguchi-fuzzy based approach. *Energy*, *63*, 375–386. doi:10.1016/j.energy.2013.10.045

Braglia, M. (1997). Heuristics for single-row layout problems in flexible manufacturing systems. *Production Planning and Control*, *8*(6), 558–567. doi:10.1080/095372897234894

Bushroa, A. R., Masjuki, H. H., Muhamad, M. R., & Beake, B. D. (2011). Optimized scratch adhesion for TiSiN coatings deposited by a combination of DC and RF sputtering. *Surface and Coatings Technology*, *206*(7), 1837–1844. doi:10.1016/j.surfcoat.2011.07.048

Caceres, C. H., & Selling, B. I. (1996). Casting defects and the tensile properties of an AlSiMg alloy. *Materials Science and Engineering A*, *220*(1-2), 109–116. doi:10.1016/S0921-5093(96)10433-0

Caillard, A., El'Mokh, M., Lecas, T., & Thomann, A. L. (2018). Effect of the target temperature during magnetron sputtering of Nickel. *Vacuum*, *147*, 82–91. doi:10.1016/j.vacuum.2017.10.016

Chakladar, N. D., & Chakraborty, S. (2008). A combined TOPSIS- AHP method based approach for non-traditional machining pro- cesses selection. *Proceedings of the Institution of Mechanical Engineers. Part B, Journal of Engineering Manufacture*, *222*(12), 1613–1623. doi:10.1243/09544054JEM1238

Chakladar, N. D., Das, R., & Chakraborty, S. (2009). A digraph-based expert system for non-traditional machining processes selection. *International Journal of Advanced Manufacturing Technology*, *43*(3-4), 226–237. doi:10.100700170-008-1713-0

Chakraborty, S. (2011). Applications of the MOORA method for decision making in manufacturing environment. *International Journal of Advanced Manufacturing Technology*, *54*(9-12), 1155–1166. doi:10.100700170-010-2972-0

Chakraborty, S., Chattopadhyay, R., & Chakraborty, S. (2020). An integrated D-MARCOS method for supplier selection in an iron and steel industry. *Decision Making: Applications in Management and Engineering*, *3*(2), 49–69.

Chakraborty, S., Dandge, S. S., & Agarwal, S. (2020). Non-traditional machining processes selection and evaluation: A rough multi-attributive border approximation area comparison approach. *Computers & Industrial Engineering*, *139*, 106201. doi:10.1016/j.cie.2019.106201

Chakraborty, S., & Dey, S. (2006). Design of an analytic- hierarchy-process-based expert system for non-traditional machining process selection. *International Journal of Advanced Manufacturing Technology*, *31*(5-6), 490–500. doi:10.100700170-005-0216-5

Chakraborty, S., & Dey, S. (2007). QFD-based expert system for non-traditional machining processes selection. *Expert Systems with Applications*, *32*(4), 1208–1217. doi:10.1016/j.eswa.2006.02.010

Chander, S., & Dhaka, M. S. (2015). Optimization of physical properties of vacuum evaporated CdTe thin films with the application of thermal treatment for solar cells. *Materials Science in Semiconductor Processing*, *40*, 708–712. doi:10.1016/j.mssp.2015.07.063

Chandraseelan, E. R., Jehadeesan, R., & Raajenthiren, M. (2008). Web-based knowledge base system for selection of non-traditional machining processes. *Malaysian Journal of Computer Science*, *21*(1), 45–56. doi:10.22452/mjcs.vol21no1.5

Chang, C.-L., & Yang, F.-C. (2018). Effect of target composition on the microstructural, mechanical, and corrosion properties of TiAlN thin films deposited by high-power impulse magnetron sputtering. *Surface and Coatings Technology*, *352*, 330–337. doi:10.1016/j.surfcoat.2018.08.023

Chang, M. S., & ... (2004). Use of Taguchi method to develop a robust design for the magnesium alloy die casting process. *Materials Science and Engineering A*, *379*(1-2), 366–371. doi:10.1016/j.msea.2004.03.006

Chan, L. K., & Wu, M. L. (2002). Quality function deployment: A literature review. *European Journal of Operational Research*, *143*(3), 463–497. doi:10.1016/S0377-2217(02)00178-9

Chatterjee, P., & Chakraborty, S. (2013). Non-traditional machining processes selection using evaluation of mixed data method. *International Journal of Advanced Manufacturing Technology*, *68*(5-8), 1613–1626. doi:10.100700170-013-4958-1

Chaudhary, G., Kumar, M., Verma, S., & Srivastav, A. (2014). Optimization of drilling parameters of hybrid metal matrix composites using response surface methodology. *Procedia Materials Science, 6*, 229-237.

Chee, K. W. A., Meng, F., Lai, D. C., & Huang, F. (2018). Measurement-based optimization and analysis of α-IGZO/Ag/α-IGZO transparent conducting electrodes fabricated using DC magnetron sputter deposition. *Ceramics International*, *44*(17), 20939–20946. doi:10.1016/j.ceramint.2018.06.277

Chen, D. S., Wang, Q., & Chen, H. C. (2001). Linear sequencing for machine layouts by a modified simulated annealing. *International Journal of Production Research*, *39*(8), 1721–1732. doi:10.1080/00207540010023565

Chhabra, M., & Singh, R. (2012). Obtaining desired surface roughness of castings produced using ZCast direct metal casting process through Taguchi's experimental approach. *Rapid Prototyping Journal*, *18*(6), 458–471. doi:10.1108/13552541211272009

Choudhury, T., Das, P. P., Roy, M. K., Shivakoti, I., Ray, A., & Pradhan, B. B. (2013). Selection of non-traditional machining process- a distance based approach. In *IEEE International Conference on Industrial Engineering and Engineering Management* (pp. 852-856). IEEE. 10.1109/IEEM.2013.6962532

Christy, J. V., Mourad, I. A.-H., & Arunachalam, R. (2019). Mechanical and Tribological Evaluation of Aluminum Metal Matrix Composite Pipes Fabricated by Gravity and Squeeze Stir Casting. *Pressure Vessels and Piping Conference, 58974*, V06AT06A018. 10.1115/PVP2019-93857

Chrysostomos, F., & Vlachos, A. (2005). Optimal solution of linear machine layout problem using ant colony system. *WSEAS Transactions on Information Science and Applications*, *2*(6), 652–662.

Chuang, P. T. (2001). Combining the analytic hierarchy process and quality function deployment for a location decision from a requirement perspective. *International Journal of Advanced Manufacturing Technology*, *18*(11), 842–849. doi:10.1007001700170010

Clarke, S. A., Walsh, P., Maggs, C. A., & Buchanan, F. (2011). Designs from the deep: Marine organisms for bone tissue engineering. *Biotechnology Advances*, *29*(6), 610–617. doi:10.1016/j.biotechadv.2011.04.003 PMID:21527337

Coelho, L. S., & Lee, C. (2008). Solving economic load dispatch problems in power systems using chaotic and Gaussian particle swarm optimization approaches. *Electrical Power and Energy Systems*, *30*(5), 297–307. doi:10.1016/j.ijepes.2007.08.001

Cogun, C. (1993). Computer-aided system for the selection of nontraditional machining operations. *Computers in Industry*, *22*(2), 169–179. doi:10.1016/0166-3615(93)90063-7

Cogun, C. (1994). Computer-aided preliminary selection of non- traditional machining processes. *International Journal of Machine Tools & Manufacture*, *34*(3), 315–326. doi:10.1016/0890-6955(94)90002-7

Cohen, L. (1995). *Quality Function Deployment- how to make QFD work for you.* Addison-Wesley.

Cunningham, E., Dunne, N., Walker, G., Maggs, C., Wilcox, R., & Buchanan, F. (2010). Hydroxyapatite bone substitutes developed via replication of natural marine sponges. *Journal of Materials Science*, *21*, 2255–2261. PMID:20012771

Dabade, U. A., & Bhedasgaonkar, R. C. (2013). Casting defect analysis using design of experiments (DoE) and computer aided casting simulation technique. *Procedia Cirp*, *7*, 616–621. doi:10.1016/j.procir.2013.06.042

Das, P. P., & Chakraborty, S. (2020a). Lexicographic method-based parametric optimization of non-traditional machining processes for ceramic materials. *OPSEARCH*, *57*(3), 700–715. doi:10.100712597-020-00439-8

Das, P. P., & Chakraborty, S. (2020b). Application of grey correlation-based EDAS method for parametric optimization of non-traditional machining processes. *Scientia Iranica*. Advance online publication. doi:10.24200/SCI.2020.53943.3499

Das, S., & Chakraborty, S. (2011). Selection of non-traditional machining processes using analytic network process. *Journal of Manufacturing Systems*, *30*(1), 41–53. doi:10.1016/j.jmsy.2011.03.003

Davim, J. P., & Figueira, L. (2007). Machinability evaluation in hard turning of cold work tool steel (D2) with ceramic tools using statistical techniques. *Materials & Design*, *28*(4), 1186–1191. doi:10.1016/j.matdes.2006.01.011

Davim, J. P., & Reis, P. (2003). Drilling carbon fiber reinforced plastics manufactured by autoclave_experimental and statistical study. *Materials & Design*, *24*(5), 315–324. doi:10.1016/S0261-3069(03)00062-1

Davim, J. P., & Reis, P. (2003). Study of delamination in drilling carbon fiber reinforced plastics (CFRP) using design experiments. *Composite Structures*, *59*(4), 481–487. doi:10.1016/S0263-8223(02)00257-X

Deepa, S. N., & Sugumaran, G. (2011). Model order formulation of a multivariable discrete system using a modified particle swarm optimization approach. *Swarm and Evolutionary Computation*, *1*(4), 204–212. doi:10.1016/j.swevo.2011.06.005

Deng, H., Yeh, C. H., & Willis, R. J. (2000). Inter-company comparison using modified TOPSIS with objective weights. *Computers & Operations Research*, *27*(10), 963–973. doi:10.1016/S0305-0548(99)00069-6

Dhakar, K., & Dvivedi, A. (2016). Parametric evaluation on near-dry electric discharge machining. *Materials and Manufacturing Processes*, *31*(4), 413–421. doi:10.1080/10426914.2015.1037905

Dhinakaran, S., & Moorthy, T. V. (2014). Fabrication and Characteristic of Boron Carbide Particulate Reinforced Aluminum Metal Matrix Composites. *International Journal of Engineering Research Science*, *3*(7), 1051–1078.

Diakoulaki, D., Mavrotas, G., & Papayannakis, L. (1995). Determining objective weights in multiple criteria problems: The critic method. *Computers & Operations Research*, *22*(7), 763–770. doi:10.1016/0305-0548(94)00059-H

Ding, J., Zhang, T., Mei, H., Yun, J. M., Jeong, S. H., Wang, Q., & Kim, K. H. (2018). Effects of negative bias voltage and ratio of nitrogen and argon on the structure and properties of NbN coatings deposited by HiPIMS deposition system. *Coatings*, *8*(1), 10. Advance online publication. doi:10.3390/coatings8010010

Diyaley, S., Shilal, P., Shivakoti, I., Ghadai, R. K., & Kalita, K. (2017). PSI and TOPSIS based selection of process parameters in WEDM. *Periodica Polytechnica Mechanical Engineering*, *61*(4), 255–260. doi:10.3311/PPme.10431

Diyaley, S., Shilal, P., Shivakoti, I., Ghadai, R. K., & Kalita, K. (2017). PSI and TOPSIS based selection of process parameters in WEDM. *PeriodicaPolytechnica Mechanical Engineering*, *61*(4), 255–260.

Djellab, H., & Gourgand, M. (2001). A new heuristic procedure for the single-row facility layout problem. *International Journal of Computer Integrated Manufacturing*, *14*(3), 270–280. doi:10.1080/09511920010020721

Durante, M., Formisano, A., & Langella, A. (2010). Comparison between analytical and experimental roughness values of components created by incremental forming. *Journal of Materials Processing Technology*, *210*(14), 1934–1941. doi:10.1016/j.jmatprotec.2010.07.006

Dvorkin, E. N., & Toscano, R. G. (2003). Finite element models in the steel industry: Part II: Analyses of tubular products performance. *Computers & Structures*, *81*(8-11), 575–594. doi:10.1016/S0045-7949(02)00403-0

Elsayed, E. A., Dawood, A. S., & Karthikeyan, R. (2017). Evaluating alternatives through the application of TOPSIS method with entropy weight. *Int. J. Eng. Trends Technol*, *46*(2), 60–66. doi:10.14445/22315381/IJETT-V46P211

El-Zonkoly, A. M. (2011). Optimal placement of multi-distributed generation units including different load models using particle swarm optimization. *Swarm and Evolutionary Computation*, *1*(1), 50–59. doi:10.1016/j.swevo.2011.02.003

Enemuoh, E. U., El-Gizawy, A. S., & Okafor, A. C. (2001). An approach for development of damage-free drilling of carbon fiber reinforced thermosets. *International Journal of Machine Tools & Manufacture*, *41*(12), 1795–1814. doi:10.1016/S0890-6955(01)00035-9

Falamaki, C., & Veysizadeh, J. (2008). Taguchi design of experiments approach to the manufacture of one-step alumina microfilter/membrane supports by the centrifugal casting technique. *Ceramics International*, *34*(7), 1653–1659. doi:10.1016/j.ceramint.2007.07.020

Fatile, O. B., Akinruli, J. I., & Amori, A. A. (2014). Microstructure and mechanical behaviour of stir-cast Al-Mg-Sl alloy matrix hybrid composite reinforced with corn cob ash and silicon carbide. *International Journal of Engineering and Technology Innovation*, *4*(4), 251.

Fekkai, Z. (2014). Optical, Morphological and Electrical Properties of Silver and Aluminium Metallization Contacts for Solar Cells. *American Journal of Modern Physics*, *3*(2), 45. doi:10.11648/j.ajmp.20140302.13

Fenggou, C., & Dayong, Y. (2004). The study of high efficiency and intelligent optimization system in EDM sinking process. *Journal of Materials Processing Technology*, *149*(1-3), 83–87. doi:10.1016/j.jmatprotec.2003.10.059

Feng, J., & Che, A. (2018). Novel integer linear programming models for the facility layout problem with fixed-size rectangular departments. *Computers & Operations Research*, *95*, 163–171. doi:10.1016/j.cor.2018.03.013

Ferhati, H., Djeffal, F., & Benhaya, A. (2019). Optimized high-performance ITO/Ag/ITO multilayer transparent electrode deposited by RF magnetron sputtering. *Superlattices and Microstructures*, *129*(March), 176–184. doi:10.1016/j.spmi.2019.03.027

Gadakh, V. S. (2011). Application of MOORA method for parametric optimization of milling process. *International Journal of Advanced Engineering Research*, *1*(4), 743–758.

Gadalla, A. M., & Tsai, W. (1989). Machining of WC-Co composites. *Material and Manufacturing Process*, *4*(3), 411–423. doi:10.1080/10426918908956301

Gaines, J. R. (2019). The Effect of Sputter Cathode Design on Deposition Parameters. *Vakuum in Forschung Und Praxis*, *31*(5), 14–19. doi:10.1002/vipr.201900718

Galvez, A., & Iglesias, A. (2012). Particle swarm optimization for non-uniform rational B-spline surface reconstruction from clouds of 3D data points. *Information Sciences*, *192*, 174–192. doi:10.1016/j.ins.2010.11.007

Ganai, A. R., & Singh, B. (2020). Study of Microstructure, Hardness and Dimensional Accuracy in Al-6061 Centrifugally Cast Pipe. In Manufacturing Engineering (pp. 51–60). Springer.

Ganesh, N., Ghadai, R. K., Bhoi, A. K., Kalita, K., & Gao, X. Z. (2020). An Intelligent Predictive Model-Based Multi-Response Optimization of EDM Process. *Computer Modeling in Engineering & Sciences*, *124*(2), 459–476. doi:10.32604/cmes.2020.09645

Gangwar, V., Kumar, S., Singh, V., & Singh, H. (2020). Effect of Process Parameters on Hardness of AA-6063 In-Situ Microwave Casting by Using Taguchi Method. *IOP Conference Series. Materials Science and Engineering*, *804*, 012019. doi:10.1088/1757-899X/804/1/012019

Garg, A., & Lam, J. S. (2016). Modeling multiple-response environmental and manufacturing characteristics of EDM process. *Journal of Cleaner Production*, *137*, 1588–1601. doi:10.1016/j.jclepro.2016.04.070

Ghabezila, P., & Khoran, M. (2014). Optimization of drilling parameters in composite sandwich structures (PVC core). *Indian J. Sci. Res*, *2*, 173-179.

Gholipoor, A., Baseri, H., & Shabgard, M. R. (2015). Investigation of near dry EDM compared with wet and dry EDM processes. *Journal of Mechanical Science and Technology*, *29*(5), 2213–2218. doi:10.100712206-015-0441-2

Ghorabaee, M. K., Amiri, M., Zavadskas, E. K., & Antucheviciene, J. (2018). A new hybrid fuzzy MCDM approach for evaluation of construction equipment with sustainability considerations. *Archives of Civil and Mechanical Engineering*, *18*(1), 32–49. doi:10.1016/j.acme.2017.04.011

Ghosh, A., & Mallik, A. K. (2008). *Manufacturing Science*. East West Press Private Limited.

Ghosh, G., Mandal, P., & Mondal, S. C. (2019). Modeling and optimization of surface roughness in keyway milling using ANN, genetic algorithm, and particle swarm optimization. *International Journal of Advanced Manufacturing Technology*, *100*(5), 1223–1242. doi:10.100700170-017-1417-4

Ghosh, S., Mandal, P., & Mondal, S. C. (2017). Application of simulated annealing for the optimization of process parameters in WEDM process for machining 201LN stainless steel. In *2017 International Conference on Advances in Mechanical, Industrial, Automation and Management Systems (AMIAMS)* (pp. 24-29). IEEE. 10.1109/AMIAMS.2017.8069183

Giridharan, A., & Samuel, G. L. (2016). Analysis on the effect of discharge energy on machining characteristics of wire electrical discharge turning process. *Proceedings of the Institution of Mechanical Engineers. Part B, Journal of Engineering Manufacture*, *230*(11), 2064–2081. doi:10.1177/0954405415615732

Gopal Krishna, U. B., Sreenivas Rao, K. V., & Vasudeva, B. (2013). Effect of boron carbide reinforcement on aluminium matrix composites. *International Journal of Metallurgical & Materials Science and Engineering (IJMMSE)*, *3*, 41–48.

Gopalsamy, B. M., Mondal, B., & Ghosh, S. (2009). *Taguchi method and ANOVA: An approach for process parameters optimization of hard machining while machining hardened steel*. Academic Press.

Gostimirovic, M., Kovac, P., Sekulic, M., & Skoric, B. (2012). Influence of discharge energy on machining characteristics in EDM. *Journal of Mechanical Science and Technology, 26*(1), 173-179.

Gostimirovic, M., Pucovsky, V., Sekulic, M., Radovanovic, M., & Madic, M. (2018). Evolutionary multi-objective optimization of energy efficiency in electrical discharge machining. *Journal of Mechanical Science and Technology*, *32*(10), 4775–4785. doi:10.100712206-018-0925-y

Gostimirovic, M., Rodic, D., Kovac, P., Pucovsky, V., & Savkovic, B. (2014). Application of neuro-fuzzy systems and genetic programming for modelling surface roughness in electrical discharge machining. *Annals of the Faculty of Engineering Hunedoara*, *12*, 137.

Goswami, C., Bhat, I. K., Bathula, S., Singh, T., & Patnaik, A. (2019). Physico-mechanical and surface wear assessment of magnesium oxide filled ceramic composites for hip implant applications. *Silicon*, *11*(1), 39–49. doi:10.100712633-018-9880-6

Govers, C. P. M. (2001). QFD not just a tool but a way of Quality management. *International Journal of Production Economics*, *69*(2), 151–159. doi:10.1016/S0925-5273(00)00057-8

Gowda, B. M., Ravindra, H. V., Prakash, G. V., Nishanth, P., & Ugrasen, G. (2015). Optimization of process parameters in drilling of epoxy Si3N4 composite material. *Materials Today: Proceedings*, *2*, 2852–2861.

Green, D., Howard, D., Yang, X., Kelly, M., & Oreffo, R. O. C. (2003). Natural Marine Sponge Fiber Skeleton: A Biomimetic Scaffold for Human Osteoprogenitor Cell Attachment, Growth and Differentiation. *Tissue Engineering*, *9*(6), 1159–1166. doi:10.1089/10763270360728062 PMID:14670103

Gromov, D. G., Dubkov, S. V., Savitskiy, A. I., Shaman, Y. P., Polokhin, A. A., Belogorokhov, I. A., & Trifonov, A. Y. (2019). Optimization of nanostructures based on Au, Ag, Au Ag nanoparticles formed by thermal evaporation in vacuum for SERS applications. *Applied Surface Science, 489*(November), 701–707. doi:10.1016/j.apsusc.2019.05.286

Guharaja, S., Haq, A. N., & Karuppannan, K. M. (2006). Optimization of green sand casting process parameters by using Taguchi's method. *International Journal of Advanced Manufacturing Technology*, *30*(11-12), 1040–1048. doi:10.100700170-005-0146-2

Guilemany, J. M., Fernández, J., Delgado, J., Benedetti, A. V., & Climent, F. (2002). Effects of thickness coating on the electrochemical behaviour of thermal spray Cr3C2–NiCr coatings. *Surface and Coatings Technology*, *153*(2–3), 107–113. doi:10.1016/S0257-8972(01)01679-6

Gullu, H. H., Isik, M., & Gasanly, N. M. (2018). Structural and optical properties of thermally evaporated Cu-Ga-S (CGS) thin films. *Physica B, Condensed Matter*, *547*, 92–96. doi:10.1016/j.physb.2018.08.015

Günay, M., & Yücel, E. (2013). Application of Taguchi method for determining optimum surface roughness in turning of high-alloy white cast iron. *Measurement*, *46*(2), 913–919. doi:10.1016/j.measurement.2012.10.013

Guo, E. Y., Zheng, Q., & Jing, T. (2012). Numerical Simulation of Solidification of Thick-Wall Stainless Steel Pipe in Horizontal Centrifugal Casting Process. *Materials Science Forum*, *706*, 1427–1432. doi:10.4028/www.scientific.net/MSF.706-709.1427

Gupta, A., Singh, H., & Aggarwal, A. (2011). Taguchi-fuzzy multi output optimization (MOO) in high speed CNC turning of AISI P-20 tool steel. *Expert Systems with Applications*, *38*(6), 6822–6828. doi:10.1016/j.eswa.2010.12.057

Gupta, V., & Takhi, S. (2015). Effects of rice husk ash particulates on the mechanical and tribological properties of the aluminum metal composite reinforced with aluminum oxide. *International Journal for Scientific Research and Development*, *3*(3), 2321–0613.

Hamada, K., Ogawa, T., Okumura, H., & Ishihara, K. N. (2019). The effect of substrate roughness on the properties of RF sputtered AZO thin film. *MRS Communications*, *9*(2), 697–701. doi:10.1557/mrc.2019.66

Hamilton, K. A. (2010). *Friction and External Surface Roughness in Single Point Incremental forming: A study of Surface friction, Contact Area and The 'Orange Peel' Effect.* Queen's University.

Hamilton, R. W., See, D., Butler, S., & Lee, P. D. (2003). Multiscale modeling for the prediction of casting defects in investment cast aluminum alloys. *Materials Science and Engineering A*, *343*(1-2), 290–300. doi:10.1016/S0921-5093(02)00376-3

Han, M. S., Woo, Y., Ko, S.-C., Jeong, Y.-J., Jang, S.-K., & Kim, S.-J. (2009). Effects of thickness of Al thermal spray coating for STS 304. *Transactions of Nonferrous Metals Society of China*, *19*(4), 925–929. doi:10.1016/S1003-6326(08)60379-9

Han, Y., Zhang, X.-B., Yu, E., Sun, L., & Gao, Y. (2017). Numerical analysis of temperature field and structure field in horizontal continuous casting process for copper pipes. *International Journal of Heat and Mass Transfer*, *115*, 294–306. doi:10.1016/j.ijheatmasstransfer.2017.08.037

Haq, A. N., Guharaja, S., & Karuppannan, K. M. (2009). Parameter optimization of CO 2 casting process by using Taguchi method. *International Journal on Interactive Design and Manufacturing*, *3*(1), 41–50. doi:10.100712008-008-0054-4

Hasçalık, A., & Çaydaş, U. (2008). Optimization of turning parameters for surface roughness and tool life based on the Taguchi method. *International Journal of Advanced Manufacturing Technology*, *38*(9-10), 896–903. doi:10.100700170-007-1147-0

Hauser, J. R., & Clausing, D. (1988). The house of quality. *Harvard Business Review*, 63–73.

Hazra, A., & Gogtay, N. (2016). Biostatistics series module 6: Correlation and linear regression. *Indian Journal of Dermatology*, *61*(6), 593. doi:10.4103/0019-5154.193662

Heragu, S. S. (2008). *Facilities design.* CRC Press.

Higo, M., Fujita, K., Mitsushio, M., Yoshidome, T., & Kakoi, T. (2007). Epitaxial growth and surface morphology of aluminum films deposited on mica studied by transmission electron microscopy and atomic force microscopy. *Thin Solid Films*, *516*(1), 17–24. doi:10.1016/j.tsf.2007.04.046

Hocheng, H., & Tsao, C. C. (2005). The path towards delamination-free drilling of composite materials. *Journal of Materials Processing Technology*, *167*(2-3), 251–264. doi:10.1016/j.jmatprotec.2005.06.039

Hocheng, H., & Tsao, C. C. (2006). Effects of special drill bits on drilling-induced delamination of composite materials. *International Journal of Machine Tools & Manufacture*, *46*(12-13), 1403–1416. doi:10.1016/j.ijmachtools.2005.10.004

Houshyar, A., & McGinnis, L. F. (1990). A heuristic for assigning facilities to locations to minimize WIP travel distance in a linear facility. *International Journal of Production Research*, *28*(8), 1485–1498. doi:10.1080/00207549008942807

Ho, Y. C., & Moodie, C. L. (1998). Machine layout with a linear single-row flow path in an automated manufacturing system. *Journal of Manufacturing Systems*, *17*(1), 1–22. doi:10.1016/S0278-6125(98)80006-X

Hwang, C. L., & Yoon, K. (1981). *Multiple Attribute Decision Making–Methods and Applications. Springer-Verlag.*

Islam, M. M., Li, C. P., & Ko, T. J. (2017). Dry electrical discharge machining for deburring drilled holes in CFRP composite. *International Journal of Precision Engineering and Manufacturing-Green Technology*, *4*(2), 149–154. doi:10.100740684-017-0018-x

Jain, V. K. (2002). *Advanced machining processes*. New Delhi: Allied Publishers Private Limited.

Jalali, T., Jafari, M., & Mohammadi, A. (2019). Genetic algorithm optimization of antireflection coating consisting of nanostructured thin films to enhance silicon solar cell efficacy. *Materials Science and Engineering: B*, *247*(January), 114354. doi:10.1016/j.mseb.2019.05.016

Jiang, M., Luo, Y. P., & Yang, S. Y. (2007). Stochastic convergence analysis and parameter selection of the standard particle swarm optimization algorithm. *Information Processing Letters*, *102*(1), 8–16. doi:10.1016/j.ipl.2006.10.005

Jilani, A., Abdel-wahab, M. S., & Hammad, A. H. (2017). Advance Deposition Techniques for Thin Film and Coating. In *Modern Technologies for Creating the Thin-film Systems and Coatings* (Vol. 2, pp. 137–149). InTech., doi:10.5772/65702

Jinno, D., Otsuki, S., Sugimori, S., Daicho, H., Iwaya, M., Takeuchi, T., Kamiyama, S., & Akasaki, I. (2017). Characterization and optimization of sputtered AlN buffer layer on r -plane sapphire substrate to improve the crystalline quality of nonpolar a -plane GaN. *Journal of Crystal Growth*, *480*, 90–95. doi:10.1016/j.jcrysgro.2017.10.018

Joshi, S., Govindan, P., Malshe, A., & Rajurkar, K. (2011). Experimental characterization of dry EDM performed in a pulsating magnetic field. *CIRP Annals*, *60*(1), 239–242. doi:10.1016/j.cirp.2011.03.114

Kalita, K., Mukhopadhyay, T., Dey, P., & Haldar, S. (2019). Genetic programming-assisted multi-scale optimization for multi-objective dynamic performance of laminated composites: The advantage of more elementary-level analyses. *Neural Computing & Applications*, 1–25.

Kanagarajan, D., Karthikeyan, R., Palanikumar, K., & Davim, J. P. (2009). Application of goal programming technique for electro discharge machining (EDM) characteristics of cemented carbide (WC/Co). *International Journal of Materials & Product Technology*, *35*(1/2), 216–227. doi:10.1504/IJMPT.2009.025228

Kannan, S. M., Sivasubramanian, R., & Jayabalan, V. (2009). Particle swarm optimization for minimizing assembly variation in selective assembly. *International Journal of Advanced Manufacturing Technology*, *42*(7-8), 793–803. doi:10.100700170-008-1638-7

Kant, G., & Sangwan, K. S. (2015). Predictive modeling for power consumption in machining using artificial intelligence techniques. *Procedia CIRP*, *26*, 403–407. doi:10.1016/j.procir.2014.07.072

Kao, C. C., Tao, J., & Shih, A. J. (2007). Near dry electrical discharge machining. *International Journal of Machine Tools & Manufacture*, *47*(15), 2273–2281. doi:10.1016/j.ijmachtools.2007.06.001

Karande, P., & Chakraborty, S. (2012). Application of multi-objective optimization on the basis of ratio analysis (MOORA) method for materials selection. *Materials & Design*, *37*, 317–324. doi:10.1016/j.matdes.2012.01.013

Karande, P., & Chakraborty, S. (2012). Application of PROMETHEE-GAIA method for non-traditional machining process selection. *Management Science Letters*, *2*(6), 2049–2060. doi:10.5267/j.msl.2012.06.015

Kaschnitz, E. (2012). Numerical simulation of cenrifugal casting of pipes. *IOP Conf. Series: Materials Science and Engineering*, 33.

Kavitha, A., Kannan, R., & Rajashabala, S. (2017). Effect of target power on the physical properties of Ti thin films prepared by DC magnetron sputtering with supported discharge. *Materials Science Poland*, *35*(1), 173–180. doi:10.1515/msp-2017-0022

Keskin, Y., Halkacı, H. S., & Kizil, M. (2006). An experimental study for determination of the effects of machining parameters on surface roughness in electrical discharge machining (EDM). *International Journal of Advanced Manufacturing Technology*, *28*(11-12), 1118–1121. doi:10.100700170-004-2478-8

Khachatryan, H., Lee, S., Kim, K.-B., Kim, H.-K., & Kim, M. (2018). Al thin film: The effect of substrate type on Al film formation and morphology. *Journal of Physics and Chemistry of Solids*, *122*(May), 109–117. doi:10.1016/j.jpcs.2018.06.018

Khalaf, M. K., Al-Taay, H. F., & Ali, D. S. (2017). Effect of radio frequency magnetron sputtering power on structural and optical properties of Ti6Al4V thin films. *Photonic Sensors*, *7*(2), 163–170. doi:10.100713320-017-0390-8

Khan, A., & Maity, K. (2016). Parametric Optimization of Some Non-Conventional Machining Processes Using MOORA Method. *International Journal of Engineering Research in Africa*, *20*, 19–40. doi:10.4028/www.scientific.net/JERA.20.19

Khan, W., Wang, Q., & Jin, X. (2018). Effect of target composition and sputtering deposition parameters on the functional properties of nitrogenized Ag-Permalloy flexible thin films deposited on polymer substrates. *Materials (Basel)*, *11*(3), 439. Advance online publication. doi:10.3390/ma11030439 PMID:29562603

Khan, Z. A., Siddiquee, A. N., & Sheikh, M. H. (2012). Selection of optimal condition for finishing of centreless-cylindrical ground parts using grey relational and principal component analyses. *International Journal of Materials & Product Technology*, *43*(1-4), 2–21. doi:10.1504/IJMPT.2012.047636

Khatal, G. D., Borkar, B. R., & M, G. M. (2016). Analytical Study of SPIF Process. *IJARIIE*, *2*(4), 359-363.

Kim, D., Lee, H., Bae, J., Jeong, H., Choi, B., Nam, T., & Noh, J. (2018). Effect of Substrate Roughness on Adhesion and Structural Properties of Ti–Ni Shape Memory Alloy Thin Film. *Journal of Nanoscience and Nanotechnology*, *18*(9), 6201–6205. doi:10.1166/jnn.2018.15636 PMID:29677767

Kim, J. H., Lee, S., & Im, H. S. (1999). Effect of target density and its morphology on TiO2 thin films grown on Si(100) by PLD. *Applied Surface Science*, *151*(1), 6–16. doi:10.1016/S0169-4332(99)00269-X

Kim, J. Y., Lee, J. H., & Nahm, S. H. (2006). Statistical Analysis of Casting Defects in Microstructure for Understanding the Effect on Fatigue Property of 17-4PH Stainless Steel. *Key Engineering Materials*, *321*, 1503–1506. doi:10.4028/www.scientific.net/KEM.321-323.1503

Kim, J.-H., Seong, T.-Y., Ahn, K.-J., Chung, K.-B., Seok, H.-J., Seo, H.-J., & Kim, H.-K. (2018). The effects of film thickness on the electrical, optical, and structural properties of cylindrical, rotating, magnetron-sputtered ITO films. *Applied Surface Science*, *440*, 1211–1218. doi:10.1016/j.apsusc.2018.01.318

Kıvak, T. (2014). Optimization of surface roughness and flank wear using the Taguchi method in milling of Hadfield steel with PVD and CVD coated inserts. *Measurement*, *50*, 19–28. doi:10.1016/j.measurement.2013.12.017

Köksoy, O., & Muluk, F. Z. (2004). Solution to the Taguchi's problem with correlated responses. *Gazi University Journal of Science*, *17*(1), 59–70.

Kou, G., Lu, Y., Peng, Y., & Shi, Y. (2012). Evaluation of classification algorithms using MCDM and rank correlation. *International Journal of Information Technology & Decision Making*, *11*(01), 197–225. doi:10.1142/S0219622012500095

Kouvelis, P., & Chiang, W. C. (1992). A simulated annealing procedure for single row layout problems in flexible manufacturing systems. *International Journal of Production Research*, *30*(4), 717–732. doi:10.1080/00207543.1992.9728452

Kovac, J., Stock, H.-R., & Zoch, H.-W. (2012). Influence of Substrate Bias Voltage on the Properties of Sputtered Aluminum-Scandium Thin Sheets. *Journal of Surface Engineered Materials and Advanced Technology*, *02*(02), 115–119. doi:10.4236/jsemat.2012.22018

Krishnamoorthy, A., Boopathy, S. R., & Palanikumar, K. (2009). Delamination analysis in drilling of CFRP composites using response surface methodology. *Journal of Composite Materials*, *43*(24), 2885–2902. doi:10.1177/0021998309345309

Krishnamoorthy, A., Boopathy, S. R., Palanikumar, K., & Davim, J. P. (2012). Application of grey fuzzy logic for the optimization of drilling parameters for CFRP composites with multiple performance characteristics. *Measurement*, *45*(5), 1286–1296. doi:10.1016/j.measurement.2012.01.008

Krishnaraj, V., Prabukarthi, A., Ramanathan, A., Elanghovan, N., Kumar, M. S., Zitoune, R., & Davim, J. P. (2012). Optimization of machining parameters at high speed drilling of carbon fiber reinforced plastic (CFRP) laminates. *Composites. Part B, Engineering*, *43*(4), 1791–1799. doi:10.1016/j.compositesb.2012.01.007

Krishna, S. M., Shridhar, T. N., & Krishnamurthy, L. (2015). Research significance, applications and fabrication of hybrid metal matrix composites. *International Journal of Innovative Science. Engineering & Technology*, *2*(5), 227–237.

Kucukrende, I., & Hakan Yetg, S. (2013). Coating Parameters Influences on Mechanical Properties of Coating. *Journal of Applied Sciences (Faisalabad)*, *13*(4), 645–649. doi:10.3923/jas.2013.645.649

Kumar, A., Mohanta, K., Kumar, D., & Prakash, O. (2012). *Properties and industrial applications of rice husk: a review*. Academic Press.

Kumar, J., Verma, R. K., & Mondal, A. K. (2020). Predictive modeling and machining performance optimization during drilling of polymer nanocomposites reinforced by graphene oxide/carbon fiber. *Archive of Mechanical Engineering*, *67*.

Kumarasinghe, P. K. K., Dissanayake, A., Pemasiri, B. M. K., & Dassanayake, B. S. (2017). Thermally evaporated CdTe thin films for solar cell applications: Optimization of physical properties. *Materials Research Bulletin*, *96*, 188–195. doi:10.1016/j.materresbull.2017.04.026

Kumar, M. S., Islam, M. N., Lenin, N., Vignesh Kumar, D., & Ravindran, D. (2011). A simple heuristic for linear sequencing of machines in layout design. *International Journal of Production Research*, *49*(22), 6749–6768. doi:10.1080/00207543.2010.535860

Kumar, R. S., Sureshkumar, K., & Velraj, R. (2015). Optimization of biodiesel production from Manilkara zapota (L.) seed oil using Taguchi method. *Fuel*, *140*, 90–96. doi:10.1016/j.fuel.2014.09.103

Kumar, S., Sangwan, P., Dhankhar, R. M. V., & Bidra, S. (2013). Utilization of rice husk and their ash: A review. *Res. J. Chem. Env. Sci*, *1*(5), 126–129.

Kumar, S., Satsangi, P. S., & Prajapati, D. R. (2011). Optimization of green sand casting process parameters of a foundry by using Taguchi's method. *International Journal of Advanced Manufacturing Technology*, *55*(1-4), 23–34. doi:10.100700170-010-3029-0

Kumar, S., Satsangi, P. S., & Prajapati, D. R. (2013). Optimization of process factors for controlling defects due to melt shop using Taguchi method. *International Journal of Quality & Reliability Management*, *30*(1), 4–22. doi:10.1108/02656711311288397

Kunieda, M., Takaya, T., & Nakano, S. (2004). Improvement of dry EDM characteristics using piezoelectric actuator. *CIRP Annals*, *53*(1), 183–186. doi:10.1016/S0007-8506(07)60674-X

Kunleda, M., Miyoshi, Y., Takaya, T., Nakajima, N., ZhanBo, Y., & Yoshida, M. (2003). High speed 3D milling by dry EDM. *CIRP Annals*, *52*(1), 147–150. doi:10.1016/S0007-8506(07)60552-6

Kunz, L., Lukáš, P., Konečná, R., & Fintová, S. (2012). Casting defects and high temperature fatigue life of IN 713LC superalloy. *International Journal of Fatigue*, *41*, 47–51. doi:10.1016/j.ijfatigue.2011.12.002

Kuo, R. J., & Han, Y. S. (2011). A hybrid of genetic algorithm and particle swarm optimization for solving bi-level linear programming problem – A case study on supply chain model. *Applied Mathematical Modelling, 35*(8), 3905–3917. doi:10.1016/j.apm.2011.02.008

Kuppan, P., Rajadurai, A., & Narayanan, S. (2008). Influence of EDM process parameters in deep hole drilling of Inconel 718. *International Journal of Advanced Manufacturing Technology, 38*(1-2), 74–84. doi:10.100700170-007-1084-y

Kuriakose, S., & Shunmugam, M. S. (2005). Multi-objective optimization of wire-electro discharge machining process by non-dominated sorting genetic algorithm. *Journal of Materials Processing Technology, 170*(1-2), 133–141. doi:10.1016/j.jmatprotec.2005.04.105

Kurniawan, R., Kumaran, S. T., Prabu, V. A., Zhen, Y., Park, K. M., Kwak, Y. I., ... Ko, T. J. (2017). Measurement of burr removal rate and analysis of machining parameters in ultrasonic assisted dry EDM (US-EDM) for deburring drilled holes in CFRP composite. *Measurement, 110*, 98–115. doi:10.1016/j.measurement.2017.06.008

Lancaster, L., Lung, M. H., & Sujan, D. (2013, January). Utilization of agro-industrial waste in metal matrix composites: towards sustainability. In Proceedings of World Academy of Science, Engineering and Technology (No. 73, p. 1136). World Academy of Science, Engineering and Technology (WASET).

Laonapakul, T. (2015). Synthesis of hydroxyapatite from biogenic wastes. *KKU Engineering Journal, 42*(3), 269–275.

Lazar, M.-B., & Xirouchakis, P. (2011). Experimental analysis of drilling fiber reinforced composites. *International Journal of Machine Tools & Manufacture, 51*(12), 937–946. doi:10.1016/j.ijmachtools.2011.08.009

Lee, H. C., & Lee, J. Y. (1997). Effect of negative bias voltage on the microstructures of AlN thin films fabricated by reactive r.f. magnetron sputtering. *Journal of Materials Science Materials in Electronics, 8*(6), 385–390. doi:10.1023/A:1018551726015

Lenin, N., Kumar, M. S., Islam, M. N., & Ravindran, D. (2013). Multi-objective optimization in single-row layout design using a genetic algorithm. *International Journal of Advanced Manufacturing Technology, 67*(5-8), 1777–1790. doi:10.100700170-012-4608-z

Lenin, N., Siva Kumar, M., Ravindran, D., & Islam, M. N. (2014). A tabu search for multi-objective single row facility layout problem. *Journal of Advanced Manufacturing Systems, 13*(01), 17–40. doi:10.1142/S0219686714500024

Li, B., Shen, Y., & Hu, W. (2011). Casting defects induced fatigue damage in aircraft frames of ZL205A aluminum alloy–A failure analysis. *Materials & Design, 32*(5), 2570–2582. doi:10.1016/j.matdes.2011.01.039

Li, D. G., & Watson, A. C. (1998). Optical thin film optimization design using genetic algorithms. In *1997 IEEE International Conference on Intelligent Processing Systems (Cat. No.97TH8335)* (Vol. 1, pp. 132–136). IEEE. 10.1109/ICIPS.1997.672752

Lien, L. C., & Cheng, M. Y. (2012). A hybrid swarm intelligence based particle-bee algorithm for construction site layout optimization. *Expert Systems with Applications, 39*(10), 9642–9650. doi:10.1016/j.eswa.2012.02.134

Lim, D. G., Kang, G. S., Kwon, S. I., Park, M. W., & Kwak, D. J. (2007). Influence of Positive Substrate Bias on the Electrical Properties of ZnO:Al Films Prepared by DC Magnetron Sputtering. *Journal of the Korean Physical Society, 50*(6), 1697. doi:10.3938/jkps.50.1697

Lindeke, R. R., & Andrew Liou, Y.-H. (1989). Methods for optimization in the manufacturing system - the taguchi method: An engineering approach to its implementation. *Journal of Mechanical Working Technology, 20*(C), 205–218. doi:10.1016/0378-3804(89)90031-4

Lin, H. C., Su, C. T., Wang, C. C., Chang, B. H., & Juang, R. C. (2012). Parameter optimization of continuous sputtering process based on Taguchi methods, neural networks, desirability function, and genetic algorithms. *Expert Systems with Applications, 39*(17), 12918–12925. doi:10.1016/j.eswa.2012.05.032

Lin, J., Sproul, W. D., Moore, J. J., Wu, Z. L., & Lee, S. L. (2011). Effect of negative substrate bias voltage on the structure and properties of CrN films deposited by modulated pulsed power (MPP) magnetron sputtering. *Journal of Physics. D, Applied Physics, 44*(42), 425305. Advance online publication. doi:10.1088/0022-3727/44/42/425305

Lin, M. T. (2009). The single-row machine layout problem in apparel manufacturing by hierarchical order-based genetic algorithm. *International Journal of Clothing Science and Technology, 21*(1), 31–43.

Lin, Y. C., Chen, Y. F., Lin, C. T., & Tzeng, H. J. (2008). Electrical discharge machining (EDM) characteristics associated with electrical discharge energy on machining of cemented tungsten carbide. *Materials and Manufacturing Processes, 23*(4), 391–399. doi:10.1080/10426910801938577

Liu, J., Zhang, H., He, K., & Jiang, S. (2018). Multi-objective particle swarm optimization algorithm based on objective space division for the unequal-area facility layout problem. *Expert Systems with Applications, 102*, 179–192. doi:10.1016/j.eswa.2018.02.035

Liu, Z. G., Liu, T. T., McConkey, B. G., & Li, X. (2011). Empirical analysis on environmental disclosure and environmental performance level of listed steel companies. *Energy Procedia, 5*, 2211–2218. doi:10.1016/j.egypro.2011.03.382

Li, W., & Kara, S. (2015). Characterising energy efficiency of electrical discharge machining (EDM) processes. *Procedia Cirp, 29*, 263–268. doi:10.1016/j.procir.2015.01.039

Lochner, R. H., & Matar, J. E. (1990). *Designing for quality: an introduction to the best of Taguchi and western methods of statistical experimental design*. Springer.

Low, J. W., Nafarizal, N., Sahdan, M. Z., Abd Kadir, M., bin Ahmad, M. K., Yeon, M. S. A., Ammar, Z., Ahmad Saad, F. S., & Mohd Zain, A. F. (2014). Effect of Substrate Bias in Copper Sputtering Plasma Measured by Langmuir Probe. *Advanced Materials Research, 925*, 238–242. . doi:10.4028/www.scientific.net/AMR.925.238

Luo, Y., He, K., & Du, R. (2010). A new sheet metal forming system based on incremental punching, part 2: Machine building and experiment results. *International Journal of Advanced Manufacturing Technology, 54*, 481–491. doi:10.100700170-010-2634-2

Madić, M., Petković, D., & Radovanović, M. (2015). Selection of non-conventional machining processes using the OCRA method. *Serbian Journal of Management, 10*(1), 61–73. doi:10.5937jm10-6802

Madic, M., & Radovanović, M. (2015). Ranking of some most commonly used nontraditional machining processes using ROV and CRITIC methods. *UPB Sci. Bull., Series D, 77*(2), 193–204.

Mahajan, R., Krishna, H., Singh, A. K., & Ghadai, R. K. (2018, June). A Review on Copper and its alloys used as electrode in EDM. *IOP Conference Series. Materials Science and Engineering, 377*(1).

Mahmood Ali, S. (2019). Optimization of Centrifugal Casting Parameters of AlSi Alloy by using the Response Surface Methodology. *International Journal of Engineering, 32*, 1516–1526.

Mahmoodzadeh, S., Shahrabi, J., Pariazar, M., & Zaeri, M. S. (2007). Project selection by using fuzzy AHP and TOPSIS technique. *World Academy of Science, Engineering and Technology, 30*, 333–338.

Maji, K., Pratihar, D. K., & Patra, S. (2010). Modelling of electrical discharge machining process using regression analysis, adaptive neuro-fuzzy inference system and genetic algorithm. *International Journal of Data Mining. Modelling and Management, 2*, 75–94.

Mali, A., & Sonawane, S. A. (2014). Review on Effect of Hybrid Reinforcement on Mechanical Behavior of Aluminium Matrix Composite. *International Journal of Engine Research*, *3*(5).

Malviya, R., & Pratihar, D. K. (2011). Tuning of neural networks using particle swarm optimization to model MIG welding process. *Swarm and Evolutionary Computation*, *1*(4), 223–235. doi:10.1016/j.swevo.2011.07.001

Mandal, D., Pal, S. K., & Saha, P. (2007). Modeling of electrical discharge machining process using back propagation neural network and multi-objective optimization using non-dominating sorting genetic algorithm-II. *Journal of Materials Processing Technology*, *186*(1-3), 154–162. doi:10.1016/j.jmatprotec.2006.12.030

Mandal, N., Doloi, B., Mondal, B., & Das, R. (2011). Optimization of flank wear using Zirconia Toughened Alumina (ZTA) cutting tool: Taguchi method and Regression analysis. *Measurement*, *44*(10), 2149–2155. doi:10.1016/j.measurement.2011.07.022

Mandal, P., & Mondal, S. C. (2017). An application of artificial neural network and particle swarm optimisation technique for modelling and optimisation of centreless grinding process. *International Journal of Productivity and Quality Management*, *20*(3), 344–362. doi:10.1504/IJPQM.2017.082637

Mandal, P., & Mondal, S. C. (2019). Development and application of Cu-SWCNT nanocomposite–coated 6061Al electrode for EDM. *International Journal of Advanced Manufacturing Technology*, *103*(5-8), 3067–3076. doi:10.100700170-019-03710-5

Mandal, P., & Mondal, S. C. (2019). Investigation on the Performance of Copper-Coated 6061 Aluminium Alloy Electrode in Electric Discharge Machining. In *Research into Design for a Connected World* (pp. 345–355). Springer. doi:10.1007/978-981-13-5974-3_30

Mandal, P., & Mondal, S. C. (2019). Surface characteristics of mild steel using EDM with Cu-MWCNT composite electrode. *Materials and Manufacturing Processes*, *34*(12), 1326–1332. doi:10.1080/10426914.2019.1605179

Martínez-Alvarado, R., Alba, G., Leyva-Bravo, J., Praga-Alejo, R., Chias-Sánchez, P., & Hernández-Rodríguez, A. (2018). Radial Basis Function Neural Network for modelling an Electrical Discharge Machining drilling process. *2018 IEEE International Autumn Meeting on Power, Electronics and Computing (ROPEC)*, 1–6. 10.1109/ROPEC.2018.8661380

Martinez, G., Garnier, M., & Durand, F. (1987). Stirring phenomena in centrifugal casting of pipes. In *Modelling the Flow and Solidification of Metals* (pp. 225–239). Springer. doi:10.1007/978-94-009-3617-1_14

Ma, S. H., Wen, Z. G., Chen, J. N., & Wen, Z. C. (2014). Mode of circular economy in China's iron and steel industry: A case study in Wu'an city. *Journal of Cleaner Production*, *64*, 505–512. doi:10.1016/j.jclepro.2013.10.008

Mia, M., Morshed, M. S., Kharshiduzzaman, M., Razi, M. H., Mostafa, M. R., Rahman, S. M. S., Ahmad, I., Hafiz, M. T., & Kamal, A. M. (2018). Prediction and optimization of surface roughness in minimum quantity coolant lubrication applied turning of high hardness steel. *Measurement*, *118*(January), 43–51. doi:10.1016/j.measurement.2018.01.012

Mirkoohi, E., Bocchini, P., & Liang, S. Y. (2019). Analytical temperature predictive modeling and non-linear optimization in machining. *International Journal of Advanced Manufacturing Technology*, *102*(5-8), 1557–1566. doi:10.100700170-019-03296-y

Mishra, P. K., Prasad, J. N., Dave, V., Chandra, R., & Choudhary, A. K. (2015). The significant effect of film thickness on the properties of chalcopyrite thin absorbing films deposited by RF magnetron sputtering. *Materials Science in Semiconductor Processing*, *34*, 350–358. doi:10.1016/j.mssp.2015.02.047

Mondal, S. C., & Mandal, P. (2014). Application of artificial neural network for modeling surface roughness in centerless grinding operation. In *5th International & 26th All India Manufacturing Technology, Design and Research Conference*. IIT Guwahati.

Mondal, S. C., & Mandal, P. (2015). An Application of Particle Swarm Optimization Technique for Optimization of Surface Roughness in Centerless Grinding Operation. In *ICoRD'15–Research into Design Across Boundaries* (Vol. 2, pp. 687–697). Springer. doi:10.1007/978-81-322-2229-3_59

Mondal, S. C., Mandal, P., & Ghosh, G. (2017). Application of genetic algorithm for the optimization of process parameters in keyway milling. In *International Conference on Research into Design* (pp. 71-86). Springer. 10.1007/978-981-10-3518-0_7

Mukherjee, I., & Ray, P. K. (2006). A review of optimization techniques in metal cutting processes. *Computers & Industrial Engineering*, *50*(1–2), 15–34. doi:10.1016/j.cie.2005.10.001

Muzammil, M., Singh, P. P., & Talib, F. (2003). Optimization of gear blank casting process by using Taguchi's robust design technique. *Quality Engineering*, *15*(3), 351–359. doi:10.1081/QEN-120018033

Mwema, F. M., Akinlabi, E. T., Oladijo, O. P., & Dutta Majumdar, J. (2019). Evolution of microstructure and wear properties of aluminum thin films with sputtering substrate temperature. In *2019 IEEE 10th International Conference on Mechanical and Intelligent Manufacturing Technologies (ICMIMT)* (pp. 31–36). IEEE. 10.1109/ICMIMT.2019.8712046

Mwema, Oladijo, & Akinlabi. (2018). Effect of Substrate Temperature on Aluminium Thin Films Prepared byRF-Magnetron Sputtering. *Materials Today: Proceedings*, *5*(9), 20464–20473. doi:10.1016/j.matpr.2018.06.423

Mwema, F. M., Akinlabi, E. T., & Oladijo, O. P. (2019). Effect of Substrate Type on the Fractal Characteristics of AFM Images of Sputtered Aluminium Thin Films. *Materials Science*, *26*(1), 49–57. doi:10.5755/j01.ms.26.1.22769

Mwema, F. M., Akinlabi, E. T., & Oladijo, O. P. (2019). *Two-Dimensional Fast Fourier Transform Analysis of Surface Microstructures of Thin Aluminium Films Prepared by Radio-Frequency (RF) Magnetron Sputtering. In Advances in Material Sciences and Engineering. Lecture Notes in Mechanical Engineering*. Springer. doi:10.1007/978-981-13-8297-0_27

Mwema, F. M., Akinlabi, E. T., & Oladijo, O. P. (2019a). Correction of Artifacts and Optimization of Atomic Force Microscopy Imaging. In K. Kumar & J. Paulo Davim (Eds.), *Title: Design, Development, and Optimization of Bio-Mechatronic Engineering Products* (pp. 158–179). IGI Global. doi:10.4018/978-1-5225-8235-9.ch007

Mwema, F. M., Akinlabi, E. T., & Oladijo, O. P. (2019b). Fractal analysis of hillocks: A case of RF sputtered aluminum thin films. *Applied Surface Science*, *489*(May), 614–623. doi:10.1016/j.apsusc.2019.05.340

Mwema, F. M., Akinlabi, E. T., Oladijo, O. P., & Majumdar, J. D. (2019). Effect of varying low substrate temperature on sputtered aluminium films. *Materials Research Express*, *6*(5), 056404. doi:10.1088/2053-1591/ab014a

Mwema, F. M., Oladijo, O. P., Akinlabi, S. A., & Akinlabi, E. T. (2018). Properties of physically deposited thin aluminium film coatings: A review. *Journal of Alloys and Compounds*, *747*, 306–323. doi:10.1016/j.jallcom.2018.03.006

Nag & Banerjee. (2012). Fundamentals of Medical Implant Materials. ASM Handbook: Materials for Medical Devices, 23, 6-17.

Nagarajan, L., Mahalingam, S. K., Gurusamy, S., & Dharmaraj, V. K. (2018). Solution for bi-objective single row facility layout problem using artificial bee colony algorithm. *European Journal of Industrial Engineering*, *12*(2), 252–275. doi:10.1504/EJIE.2018.090619

Nalbant, M., Gökkaya, H., & Sur, G. (2007). Application of Taguchi method in the optimization of cutting parameters for surface roughness in turning. *Materials & Design*, *28*(4), 1379–1385. doi:10.1016/j.matdes.2006.01.008

Narayanan, G., Joshi, M., Dutta, P., & Kalita, K. (2019). PSO-tuned support vector machine metamodels for assessment of turbulent flows in pipe bends. *Engineering Computations*, *37*(3), 981–1001. doi:10.1108/EC-05-2019-0244

Nieschlag, J., Ruhland, P., Daubner, S., Koch, S.-F., & Fleischer, J. (2018). Finite element optimisation for rotational moulding with a core to manufacture intrinsic hybrid frp metal pipes. *Production Engineering*, *12*(2), 239–247. doi:10.100711740-017-0788-6

Nilsson, K.-F., & Vokál, V. (2009). Analysis of ductile cast iron tensile tests to relate ductility variation to casting defects and material microstructure. *Materials Science and Engineering A*, *502*(1-2), 54–63. doi:10.1016/j.msea.2008.09.082

Nkonyana, T., Sun, Y., Twala, B., & Dogo, E. (2019). Performance evaluation of data mining techniques in steel manufacturing industry. *Procedia Manufacturing*, *35*, 623–628. doi:10.1016/j.promfg.2019.06.004

Odhiambo, J. G., Li, W., Zhao, Y., & Li, C. (2019). Porosity and Its Significance in Plasma-Sprayed Coatings. *Coatings*, *9*(7), 460. doi:10.3390/coatings9070460

Oleksik, V., Pascu, A., Deac, C., Fleaca, R., Bologa, O., & Racz, G. (2010). Experimental Study on The Surface Quality Of The Medical Implants Obtained By Single Point Incremental Forming. *International Journal of Material Forming*, *3*(1), 935–938. doi:10.100712289-010-0922-x

Olson, D. L. (2004). Comparison of weights in TOPSIS models. *Mathematical and Computer Modelling*, *40*(7-8), 721–727. doi:10.1016/j.mcm.2004.10.003

Önüt, S., Tuzkaya, U. R., & Doğaç, B. (2008). A particle swarm optimization algorithm for the multiple-level warehouse layout design problem. *Computers & Industrial Engineering*, *54*(4), 783–799. doi:10.1016/j.cie.2007.10.012

Opricovic, S., & Tzeng, G. H. (2004). Compromise solution by MCDM methods: A comparative analysis of VIKOR and TOPSIS. *European Journal of Operational Research*, *156*(2), 445–455. doi:10.1016/S0377-2217(03)00020-1

Oraon, M., & Sharma, V. (2018). Predicting force in single point incremental forming by using artificial neural network. *International Journal of Engineering, 3*(1), 88-95.

Oraon, M., & Sharma, V. (2010). Sheet Metal Micro Forming: Future ResearchPotentials. *Int. J. on Production and Industrial Engineering*, *1*, 31–35.

Oraon, M., & Sharma, V. (2018). Prediction of surface roughness in single point incremental forming of AA3003-O using artificial neural network. *International Journal of Materials Engineering. Innovation*, *9*, 1–19.

Otten, K. (1997). *Method for producing Ceramic Implant materials including Hydroxyl Apatite.* US Patent: 5,667,796, 1-4.

Ouis, A., & Cailler, M. (2013). Effects of substrate bias voltage on adhesion of DC magnetron-sputtered copper films on E24 carbon steel: Investigations by Auger electron spectroscopy. *Journal of Adhesion Science and Technology*, *27*(21), 2367–2386. doi:10.1080/01694243.2013.777320

Padhee, S., Nayak, N., Panda, S. K., Dhal, P. R., & Mahapatra, S. S. (2012). Multi-objective parametric optimization of powder mixed electro-discharge machining using response surface methodology and non-dominated sorting genetic algorithm. *Sadhana*, *37*(2), 223–240. doi:10.100712046-012-0078-0

Park, J. J., & Kim, Y. H. (2003). Fundamental studies on the incremental sheet metal forming technique. *Journal of Materials Processing Technology*, *140*(1-3), 447–453. doi:10.1016/S0924-0136(03)00768-4

Park, S. (1996). *Robust design and analysis for quality engineering*. Boom Koninklijke Uitgevers.

Patel, J. R., Samvatsar, K. S., Prajapati, H. P., & Rangrej, S. S. (2015). Optimization of Process Parameters for Reducing Surface Roughness Produced During Single Point Incremental Forming Process. *International Journal on Recent Technologies in Mechanical and Electrical Engineering*, *2*(9), 19–23.

Patil, G. G. (2014). *Optimization of casting process parameters using Taguchi Method.* Academic Press.

Peña-Parás, L., Maldonado-Cortés, D., Rodríguez-Villalobos, M., Romero-Cantú, A. G., Montemayor, O. E., & Herrera, M. (2018, September). … Hugler, W. (2019). Optimization of milling parameters of 1018 steel and nanoparticle additive concentration in cutting fluids for enhancing multi-response characteristics. *Wear*, *426–427*, 877–886. doi:10.1016/j.wear.2019.01.078

Peng, X., Zhang, X., & Luo, Z. (2020). Pythagorean fuzzy MCDM method based on CoCoSo and CRITIC with score function for 5G industry evaluation. *Artificial Intelligence Review*, *53*(5), 3813–3847. doi:10.100710462-019-09780-x

Peng, Y., Kou, G., Wang, G., & Shi, Y. (2011). FAMCDM: A fusion approach of MCDM methods to rank multiclass classification algorithms. *Omega*, *39*(6), 677–689. doi:10.1016/j.omega.2011.01.009

Peters, M., Kumpfert, J., Ward, C. H., & Leyens, C. (2003). Titanium alloys for aerospace applications. *Advanced Engineering Materials*, *5*(6), 419–427. doi:10.1002/adem.200310095

Phipon, R., Shivakoti, I., Kalita, K., Kibria, G., Sharma, A., & Ghadai, R. K. (2020). Laser beam micro engraving on silicon carbide. *Materials and Manufacturing Processes*, 1–11.

Pillai, V. M., & Gudivada, B. S. (2005, August). A simulated annealing algorithm for linear sequencing of machines for layout design. In *6th international conference on operations and quantitative management* (pp. 9-11). Academic Press.

Piquet, R., Ferret, B., Lachaud, F., & Swider, P. (2000). Experimental analysis of drilling damage in thin carbon/epoxy plate using special drills. *Composites. Part A, Applied Science and Manufacturing*, *31*(10), 1107–1115. doi:10.1016/S1359-835X(00)00069-5

Pivinskii, Y. E., Litovskaya, T. I., Samarina, O. N., Volchek, I. B., & Kaplan, F. S. (1991). Centrifugal casting of ceramics. Main parameters and the regularities of the process. *Refractories*, *32*(11-12), 551–558. doi:10.1007/BF01280846

Pohlak, M., Majak, J., & Kuttner, R. (2007). Manufacturability and limitations in incremental sheet forming. *Proc. Estonian Acad. Sci. Eng*, *13*(2), 129–139.

Pragadish, N., & Kumar, M. P. (2016). Optimization of dry EDM process parameters using grey relational analysis. *Arabian Journal for Science and Engineering*, *41*(11), 4383–4390. doi:10.100713369-016-2130-6

Prasad, K., & Chakraborty, S. (2014). A decision-making model for non-traditional machining processes selection. *Decision Science Letters*, *3*(4), 467–478. doi:10.5267/j.dsl.2014.7.002

Prasad, K., & Chakraborty, S. (2018). A decision guidance framework for non-traditional machining processes selection. *Ain Shams Engineering Journal*, *9*(2), 203–214. doi:10.1016/j.asej.2015.10.013

Priyadarshini, B. G., Aich, S., & Chakraborty, M. (2014). Substrate bias voltage and deposition temperature dependence on properties of rf-magnetron sputtered titanium films on silicon (100). *Bulletin of Materials Science*, *37*(7), 1691–1700. doi:10.100712034-014-0722-x

Puthumana, G., & Joshi, S. S. (2011). Investigations into performance of dry EDM using slotted electrodes. *International Journal of Precision Engineering and Manufacturing*, *12*(6), 957–963. doi:10.100712541-011-0128-2

Quan, Y., & Zhong, W. (2009). Investigation on drilling-grinding of CFRP. *Frontiers of Mechanical Engineering in China*, *4*(1), 60–63. doi:10.100711465-009-0008-y

Quero, J. M., Perdigones, F., & Aracil, C. (2018). Microfabrication technologies used for creating smart devices for industrial applications. In *Smart Sensors and MEMs* (pp. 291–311). Elsevier., doi:10.1016/B978-0-08-102055-5.00011-5

Radu, M. C., & Cristea, I. (2013). Processing Metal Sheets by SPIF and Analysis of Parts Quality. *Materials and Manufacturing Processes*, *28*(3), 287–293. doi:10.1080/10426914.2012.746702

Rahimipour, M. R., & Sobhani, M. (2013). Evaluation of centrifugal casting process parameters for in situ fabricated functionally gradient Fe-TiC composite. *Metallurgical and Materials Transactions. B, Process Metallurgy and Materials Processing Science*, *44*(5), 1120–1123. doi:10.100711663-013-9903-z

Ramarajan, R., Kovendhan, M., Phan, D. T., & Thangaraju, K. (2018). Optimization of Zn 2 SnO 4 thin film by post oxidation of thermally evaporated alternate Sn and Zn metallic multi-layers. *Applied Surface Science*, *449*, 68–76. doi:10.1016/j.apsusc.2018.01.029

Ramasawmy, H., & Blunt, L. (2004). Effect of EDM process parameters on 3D surface topography. *Journal of Materials Processing Technology*, *148*(2), 155–164. doi:10.1016/S0924-0136(03)00652-6

Rao, S. R., & Padmanabhan, G. (2012). Fabrication and mechanical properties of aluminium-boron carbide composites. *International Journal of Materials and Biomaterials Applications*, *2*(3), 15-18.

Rao, R. V., Pawar, P. J., & Shankar, R. (2008). Multi-objective optimization of electrochemical machining process parameters using a particle swarm optimization algorithm, Proceedings of the Institution of Mechanical Engineers. *Proceedings of the Institution of Mechanical Engineers. Part B, Journal of Engineering Manufacture*, *222*(8), 949–958. doi:10.1243/09544054JEM1158

Raviteja, T., Radhika, N., & Raghu, R. (2014). Fabrication and mechanical properties of stir cast Al-Si12Cu/B4C composites. *International Journal of Research in Engineering and Technology*, *3*(07), 343–346. doi:10.15623/ijret.2014.0307058

Rebelo, J. C., Dias, A. M., Kremer, D., & Lebrun, J. L. (1998). Influence of EDM pulse energy on the surface integrity of martensitic steels. *Journal of Materials Processing Technology*, *84*(1-3), 90–96. doi:10.1016/S0924-0136(98)00082-X

Ren, L., Zhang, Y., Wang, Y., & Sun, Z. (2007). Comparative analysis of a novel M-TOPSIS method and TOPSIS. *Applied Mathematics Research Express*, •••, 2007.

Renold Elsen, S., Ramesh, T., & Aravinth, B. (2014). Optimization of process parameters of zirconia reinforced alumina by powder forming process using Response Surface Method. *Advanced Materials Research*, *984-985*, 129–139. doi:10.4028/www.scientific.net/AMR.984-985.129

Rivera-Munoz. (2011). Hydroxyapatite-Based Materials: Synthesis and Characterization. *Biomedical Engineering - Frontiers and Challenges*, 75-98.

Ronoh, N. K., Mwema, F. M., Akinlabi, S. A., Akinlabi, E. T., Karuri, N. W., & Ngetha, H. T. (2019). Effects of cooling conditions and grinding depth on sustainable surface grinding of Ti-6Al-4V: Taguchi approach. *AIMS Materials Science*, *6*(5), 697–712. doi:10.3934/matersci.2019.5.697

Roszkowska, E. (2011). Multi-criteria decision making models by applying the TOPSIS method to crisp and interval data. *Multiple Criteria Decision Making/University of Economics in Katowice*, *6*, 200-230.

Roualdes, O., Duclos, M.-E., Gutknecht, D., Frappart, L., Chevalier, J., & Hartmann, D. J. (2010). In vitro and in vivo evaluation of an alumina-zirconia composite for arthoplasty applications. *Biomaterials*, *31*(8), 2043–2054. doi:10.1016/j.biomaterials.2009.11.107 PMID:20053439

Rout, M. K., Badgayan, N. D., Pattnaik, S. K., & Rout, M. K. (2013). An Empirical Study on Energy Management Standard (ISO 50000): Effectiveness Performance Evaluation for Steel Plants and Comparison with ISO 9000, ISO 14000, ISO 18000 & ISO. *International Journal of Engineering Research and Technology*, *2*, 2278–0181.

Roy, T. (2013). Analysis of casting defects in foundry by computerised simulations (CAE)-A new approach along with some industrial case studies. *Transactions of 61stIndian Foundry Congress*, 1–9.

Roy, M. K., Ray, A., & Pradhan, B. B. (2014). Non-traditional machining process selection using integrated fuzzy AHP and QFD techniques: A customer perspective. *Production & Manufacturing Research*, *2*(1), 530–549. doi:10.1080/21693277.2014.938276

Saaty, T. L. (1988). *The Analytic Hierarchy Process: Planning, Priority Setting, Resource Allocation.* RWS Publication.

Saaty, T. L. (1994). How to make decision: The analytic hierarchy process. *Interfaces*, *24*(6), 19–43. doi:10.1287/inte.24.6.19

Safari, A., & Shayeghi, H. (2011). Iteration particle swarm optimization procedure for economic load dispatch with generator constraints. *Expert Systems with Applications*, *38*(5), 6043–6048. doi:10.1016/j.eswa.2010.11.015

Saha, S. K., & Choudhury, S. K. (2009). Experimental investigation and empirical modeling of the dry electric discharge machining process. *International Journal of Machine Tools & Manufacture*, *49*(3-4), 297–308. doi:10.1016/j.ijmachtools.2008.10.012

Şahin, R., & Türkbey, O. (2009). A simulated annealing algorithm to find approximate Pareto optimal solutions for the multi-objective facility layout problem. *International Journal of Advanced Manufacturing Technology*, *41*(9-10), 1003–1018. doi:10.100700170-008-1530-5

Salman, Ö., & Kayacan, M. C. (2008). Evolutionary programming method for modeling the EDM parameters for roughness. *Journal of Materials Processing Technology*, *200*, 347–355.

Samanta, S., & Chakraborty, S. (2011). Parametric optimization of some non-traditional machining processes using artificial bee colony algorithm. *Engineering Applications of Artificial Intelligence*, *24*(6), 946–957. doi:10.1016/j.engappai.2011.03.009

Samarghandi, H., & Eshghi, K. (2010). An efficient tabu algorithm for the single row facility layout problem. *European Journal of Operational Research*, *205*(1), 98–105. doi:10.1016/j.ejor.2009.11.034

Samarghandi, H., Taabayan, P., & Jahantigh, F. F. (2010). A particle swarm optimization for the single row facility layout problem. *Computers & Industrial Engineering*, *58*(4), 529–534. doi:10.1016/j.cie.2009.11.015

Sánchez, J. A., Izquierdo, B., Ortega, N., Pombo, I., Plaza, S., & Cabanes, I. (2009). Computer simulation of performance of electrical discharge machining operations. *International Journal of Computer Integrated Manufacturing*, *22*(8), 799–811. doi:10.1080/09511920902741125

Saravanan, M., Ramalingam, D., Manikandan, G., & Kaarthikeyen, R. R. (2012). Multi objective optimization of drilling parameters using genetic algorithm. *Procedia Engineering*, *38*, 197–207. doi:10.1016/j.proeng.2012.06.027

Saravanan, S. D., & Senthilkumar, M. (2014). Mechanical Behavior of Aluminum (AlSi10Mg)-RHA Composite. *IACSIT International Journal of Engineering and Technology*, *5*(6), 4834–4840.

Sardinas, R. Q., Reis, P., & Davim, J. P. (2006). Multi-objective optimization of cutting parameters for drilling laminate composite materials by using genetic algorithms. *Composites Science and Technology*, *66*(15), 3083–3088. doi:10.1016/j.compscitech.2006.05.003

Sarıkaya, M., & Güllü, A. (2014). Taguchi design and response surface methodology based analysis of machining parameters in CNC turning under MQL. *Journal of Cleaner Production*, *65*, 604–616. doi:10.1016/j.jclepro.2013.08.040

Sarıkaya, M., & Güllü, A. (2015). Multi-response optimization of MQL parameters using Taguchi-based GRA in turning of difficult-to-cut alloy Haynes 25. *Journal of Cleaner Production*, *91*, 347–357. doi:10.1016/j.jclepro.2014.12.020

Sarkar, B. R., Doloi, B., & Bhattacharyya, B. (2006). Parametric analysis on electrochemical discharge machining of silicon nitride ceramics. *International Journal of Advanced Manufacturing Technology*, *28*(9-10), 873–881. doi:10.100700170-004-2448-1

Sasikumar, K. S. (2015). Optimization of drilling parameter on delamination based on taguchi method in drilling of natural fiber reinforced (Agave) composite. *International Journal of Recent Technology and Engineering*, *4*, 53–55.

Satheesh Kumar, R. M., Asokan, P., & Kumanan, S. (2010). An artificial immune system-based algorithm to solve linear and loop layout problems in flexible manufacturing systems. *International Journal of Product Development*, *10*(1-3), 165–179. doi:10.1504/IJPD.2010.029991

Semaltianos, N. G. (2001). Thermally evaporated aluminium thin films. *Applied Surface Science*, *183*(3–4), 223–229. doi:10.1016/S0169-4332(01)00565-7

Seo, D. K., Kim, Y. H., Eo, Y. D., Park, W. Y., & Park, H. C. (2017). Generation of radiometric, phenological normalized image based on random forest regression for change detection. *Remote Sensing*, *9*(11), 1163. doi:10.3390/rs9111163

Shah, H., & Chaudhary, S. (2016). Optimization of Process Parameters for Incremental Sheet Metal Forming Process. *International Journal For Technological Research In Engineering*, *3*(7), 1432–1435.

Shailesh, P., Sundarrajan, S., & Komaraiah, M. (2014). Optimization of process parameters of Al-Si alloy by centrifugal casting technique using Taguchi design of experiments. *Procedia Materials Science*, *6*, 812–820.

Shailesh, P., Kumar, B. P., Sundarrajan, S., & Komariahia, M. (2012). Experimental investigation on centrifugal casting of 5500 alloy: A Taguchi approach. *Scientific Research and Essays*, *7*, 3797–3808.

Shankar, G., Jayashree, P. K., Shetty, R., Kini, A., & Sharma, S. S. (2013). Individual and combined effect of reinforcements on stir cast aluminium metal matrix composites-a review. *International Journal of Current Engineering and Technology*, *3*(3), 922–934.

Sharma, N., Khanna, R., & Gupta, R. D. (2015). WEDM process variables investigation for HSLA by response surface methodology and genetic algorithm. *Engineering Science and Technology, an International Journal*, *18*, 171–177.

Sharma, P., Verma, A., Sidhu, R. K., & Pandey, O. P. (2005). Process parameter selection for strontium ferrite sintered magnets using Taguchi L9 orthogonal design. *Journal of Materials Processing Technology*, *168*(1), 147–151. doi:10.1016/j.jmatprotec.2004.12.003

Shetty, R., Pai, R. B., Rao, S. S., & Nayak, R. (2009). Taguchi's technique in machining of metal matrix composites. *Journal of the Brazilian Society of Mechanical Sciences and Engineering*, *31*(1), 12–20. doi:10.1590/S1678-58782009000100003

Shieh, H. P. D., Kryder, M. H., & Lee, J. W. (1988). Effects of bias-sputtering on magnetron-sputtered magneto-optical recording media. *IEEE Transactions on Magnetics*, *24*(6), 2446–2448. doi:10.1109/20.92136

Shivakoti, I., Pradhan, B. B., Diyaley, S., Ghadai, R. K., & Kalita, K. (2017). Fuzzy TOPSIS-based selection of laser beam micro-marking process parameters. *Arabian Journal for Science and Engineering*, *42*(11), 4825–4831. doi:10.100713369-017-2673-1

Shi, Y., & Eberhart, R. C. (1999). *Emperical study of particle swarm optimization*. Congress on Evolutionary Computation.

Simon, A. H. (2018). Sputter Processing. In *Handbook of Thin Film Deposition* (pp. 195–230). Elsevier. doi:10.1016/B978-0-12-812311-9.00007-4

Singh, A. K., Mahajan, R., Tiwari, A., Kumar, D., & Ghadai, R. K. (2018). Effect of dielectric on electrical discharge machining: A review. *Journal of Marine Science and Engineering*, *377*, 012184.

Singh, C., & Wang, L. (2007). Environmental/economic power dispatch using a fuzzified multi-objective particle swarm optimization algorithm. *Electric Power Systems Research*, *77*(12), 1654–1664. doi:10.1016/j.epsr.2006.11.012

Singh, N. K., Singh, Y., Kumar, S., & Sharma, A. (2019). Comparative study of statistical and soft computing-based predictive models for material removal rate and surface roughness during helium-assisted EDM of D3 die steel. *SN Applied Sciences*, *1*(6), 529. doi:10.100742452-019-0545-x

Singh, R. K., Murty, H. R., Gupta, S. K., & Dikshit, A. K. (2007). Development of composite sustainability performance index for steel industry. *Ecological Indicators*, *7*(3), 565–588. doi:10.1016/j.ecolind.2006.06.004

Singh, S. P., & Sharma, R. R. (2006). A review of different approaches to the facility layout problems. *International Journal of Advanced Manufacturing Technology*, *30*(5-6), 425–433. doi:10.100700170-005-0087-9

Sivakumar, K., Balamurugan, C., & Ramabalan, S. (2011). Simultaneous optimal selection of design and manufacturing tolerances with alternative manufacturing process selection. *Computer Aided Design*, *43*(2), 207–218. doi:10.1016/j.cad.2010.10.001

Sivaram, S. (1995). *Chemical Vapor Deposition. Appl. Phys. A* (Vol. 73). Boston, MA: Springer US. doi:10.1007/978-1-4757-4751-5

Solimanpur, M., Vrat, P., & Shankar, R. (2005). An ant algorithm for the single row layout problem in flexible manufacturing systems. *Computers & Operations Research*, *32*(3), 583–598. doi:10.1016/j.cor.2003.08.005

Somashekhar, K. P., Ramachandran, N., & Mathew, J. (2010). Optimization of material removal rate in micro-EDM using artificial neural network and genetic algorithms. *Materials and Manufacturing Processes*, *25*(6), 467–475. doi:10.1080/10426910903365760

Somdock, N., Kianwimol, S., Harnwunggmoung, A., Sakulkalavek, A., & Sakdanuphab, R. (2019). Simultaneous stoichiometric composition and highly (00l) orientation of flexible Bi2Te3thin films via optimising the DC magnetron sputter-deposition process. *Journal of Alloys and Compounds*, *773*, 78–85. doi:10.1016/j.jallcom.2018.09.216

Sonkar, V., Abhishek, K., Datta, S., & Mahapatra, S. S. (2014). Multi-objective optimization in drilling of GFRP composites: a degree of similarity approach. *Procedia Materials Science, 6*, 538-543.

Soren, S., Alexander, P., Peter, S., Sebastian, M., Dieter, W., & Zdenek, R. (2019). Incremental sheet metal forming on the example of car exterior skin part. *Proceedia Manufacturing*, *29*, 105–111. doi:10.1016/j.promfg.2019.02.112

Souissi, N., Souissi, S., Niniven, C. L., Amar, M. B., Bradai, C., & Elhalouani, F. (2014). Optimization of squeeze casting parameters for 2017 a wrought al alloy using Taguchi method. *Metals*, *4*(2), 141–154. doi:10.3390/met4020141

Soundhar, A., Zubar, H., Sultan, M., & Kandasamy, J. (n.d.). Dataset on optimization of EDM machining parameters by using central composite design. *Data in Brief, 23*, 103671.

Sousa, M. G., & da Cunha, A. F. (2019). Optimization of low temperature RF-magnetron sputtering of indium tin oxide films for solar cell applications. *Applied Surface Science*, *484*, 257–264. doi:10.1016/j.apsusc.2019.03.275

Srivastava, V., & Pandey, P. M. (2012). Effect of process parameters on the performance of EDM process with ultrasonic assisted cryogenically cooled electrode. *Journal of Manufacturing Processes*, *14*(3), 393–402. doi:10.1016/j.jmapro.2012.05.001

Stanković, M., Stević, Ž., Das, D. K., Subotić, M., & Pamučar, D. (2020). A new fuzzy MARCOS method for road traffic risk analysis. *Mathematics*, *8*(3), 457. doi:10.3390/math8030457

Steinmuller, S. J., Vaz, C. A. F., Ström, V., Moutafis, C., Tse, D. H. Y., Gürtler, C. M., Kläui, M., Bland, J. A. C., & Cui, Z. (2007). Effect of substrate roughness on the magnetic properties of thin fcc Co films. *Physical Review B: Condensed Matter and Materials Physics*, *76*(5), 1–8. doi:10.1103/PhysRevB.76.054429

Stephens, A. W., Vossen, J. L., & Kern, W. (1976). The Effect of Substrate Bias on the Properties of Reactively Sputtered Silicon Nitride. *Journal of the Electrochemical Society*, *123*(2), 303–305. doi:10.1149/1.2132809

Stević, Ž., & Brković, N. (2020). A Novel Integrated FUCOM-MARCOS Model for Evaluation of Human Resources in a Transport Company. *Logistics*, *4*(1), 4. doi:10.3390/logistics4010004

Stević, Ž., Pamučar, D., Puška, A., & Chatterjee, P. (2020). Sustainable supplier selection in healthcare industries using a new MCDM method: Measurement of alternatives and ranking according to COmpromise solution (MARCOS). *Computers & Industrial Engineering*, *140*, 106231. doi:10.1016/j.cie.2019.106231

Subrahmanyam, A. P. S. V. R., Narsaraju, G., & Rao, B. S. (2015). Effect of rice husk ash and fly ash reinforcements on microstructure and mechanical properties of aluminium alloy (AlSi10Mg) matrix composites. *International Journal of Advanced Science and Technology*, *76*, 1–8. doi:10.14257/ijast.2015.76.01

Subramanyam, T. K., Goutham, P., Pavan Kumar, S., Yadhuraj, S. R., & Geetha, K. S. (2018). Optimization of Sputtered AZO Thin Films for Device Application. *Materials Today: Proceedings*, *5*(4), 10851–10859. doi:10.1016/j.matpr.2017.12.373

Sun, Z., Hu, H., & Chen, X. (2008). Numerical optimization of gating system parameters for a magnesium alloy casting with multiple performance characteristics. *Journal of Materials Processing Technology*, *199*(1-3), 256–264. doi:10.1016/j.jmatprotec.2007.08.036

Suresh, G., & Sahu, S. (1993). Multiobjective facility layout using simulated annealing. *International Journal of Production Economics*, *32*(2), 239–254. doi:10.1016/0925-5273(93)90071-R

Syrcos, G. P. (2003). Die casting process optimization using Taguchi methods. *Journal of Materials Processing Technology*, *135*(1), 68–74. doi:10.1016/S0924-0136(02)01036-1

Taghavi, A., & Murat, A. (2011). A heuristic procedure for the integrated facility layout design and flow assignment problem. *Computers & Industrial Engineering*, *61*(1), 55–63. doi:10.1016/j.cie.2011.02.011

Taguchi, G. (1990). *Introduction to Quality Engineering. Asian Productivity Organization*. APO.

Takakuda, K., Koyama, Y., Matsumoto, H. N., Shirahama, N., Akita, K., Shoji, D., Ogawa, T., Kikuchi, M., & Tanaka, J. (2007). Material Design of Bioabsorbable Inorganic / Organic Composites for Bone Regeneration. *Journal of Nanoscience and Nanotechnology*, *7*(3), 738–741. doi:10.1166/jnn.2007.502 PMID:17450826

Tao, J., Shih, A. J., & Ni, J. (2008). Experimental study of the dry and near-dry electrical discharge milling processes. *Journal of Manufacturing Science and Engineering*, 130.

Teimouri, R., & Baseri, H. (2012). Improvement of dry EDM process characteristics using artificial soft computing methodologies. *Production Engineering*, *6*(4-5), 493–504. doi:10.100711740-012-0398-2

Teimouri, R., & Baseri, H. (2013). Experimental study of rotary magnetic field-assisted dry EDM with ultrasonic vibration of workpiece. *International Journal of Advanced Manufacturing Technology*, *67*(5-8), 1371–1384. doi:10.100700170-012-4573-6

Temuçin, T., Tozan, H., Vayvay, Ö., Harničárová, M., & Valíček, J. (2014). A fuzzy based decision model for non-traditional machining process selection. *International Journal of Advanced Manufacturing Technology*, *70*(9-12), 2275–2282. doi:10.100700170-013-5474-z

Teo, Y. T., & Ponnambalam, S. G. (2008, August). A hybrid ACO/PSO heuristic to solve single row layout problem. In *2008 IEEE International Conference on Automation Science and Engineering* (pp. 597-602). IEEE. 10.1109/COASE.2008.4626491

Thornton, J. A. (1986). The microstructure of sputter-deposited coatings. *Journal of Vacuum Science & Technology. A, Vacuum, Surfaces, and Films*, *4*(6), 3059–3065. doi:10.1116/1.573628

Tillmann, W., Kokalj, D., & Stangier, D. (2018). Optimization of the deposition parameters of Ni-20Cr thin films on thermally sprayed Al2O3 for sensor application. *Surface and Coatings Technology*, *344*(October), 223–232. doi:10.1016/j.surfcoat.2018.03.029

Tiwari, S. K., Singh, R. K., & Srivastava, S. C. (2016). Optimisation of green sand casting process parameters for enhancing quality of mild steel castings. *International Journal of Productivity and Quality Management*, *17*(2), 127–141. doi:10.1504/IJPQM.2016.074446

Tompkins, J. A., White, J. A., Bozer, Y. A., & Tanchoco, J. M. A. (2010). *Facilities planning*. John Wiley & Sons.

Tondare, R. S., Shivaraj, B. W., Narasimhamurthy, H. N., Krishna, M., & Subramanyam, T. K. (2018). Effect of deposition time on structural, electrical and optical properties of Aluminium doped ZnO thin films by RF magnetron sputtering. *Materials Today: Proceedings*, *5*(1), 2710–2715. doi:10.1016/j.matpr.2018.01.052

To, T. B. T., de Sousa, V. B., & Aarão Reis, F. D. A. (2018). Thin film growth models with long surface diffusion lengths. *Physica A*, *511*, 240–250. Advance online publication. doi:10.1016/j.physa.2018.07.024

Tóth, L. (1987). A model of substrate surface roughness effect on the electrical properties of thin films. *Vacuum*, *37*(1–2), 103–106. doi:10.1016/0042-207X(87)90094-7

Tripathi, P. K., Bandyopadhyay, S., & Pal, S. K. (2007). Multi-objective particle swarm optimization with time variant inertia and acceleration coefficients. *Information Sciences*, *177*(22), 5033–5049. doi:10.1016/j.ins.2007.06.018

Tsao, C. C., & Hocheng, H. (2004). Taguchi analysis of delamination associated with various drill bits in drilling of composite material. *International Journal of Machine Tools & Manufacture*, *44*(10), 1085–1090. doi:10.1016/j.ijmachtools.2004.02.019

Tuş, A., & Adalı, E. A. (2019). The new combination with CRITIC and WASPAS methods for the time and attendance software selection problem. *Opsearch*, *56*(2), 528–538. doi:10.100712597-019-00371-6

Tzeng, C.-J., & Chen, R.-Y. (2013). Optimization of electric discharge machining process using the response surface methodology and genetic algorithm approach. *International Journal of Precision Engineering and Manufacturing*, *14*(5), 709–717. doi:10.100712541-013-0095-x

Upadhyay, P. C., & Lyons, J. S. (1999). On the evaluation of critical thrust for delamination-free drilling of composite laminates. *Journal of Reinforced Plastics and Composites*, *18*(14), 1287–1303. doi:10.1177/073168449901801402

Usman, A. M., Raji, A., Hassan, M. A., & Waziri, N. H. (2014). A comparative study on the properties of Al-7% Si-Rice husk ash and Al-7% Si-Bagasse ash composites produced by stir casting. *International Journal of Engineering Science, 3*(8), 1–7.

Usmani, B., Vijay, V., Chhibber, R., & Dixit, A. (2017). Optimization of sputtered zirconium thin films as an infrared reflector for use in spectrally-selective solar absorbers. *Thin Solid Films, 627*, 17–25. doi:10.1016/j.tsf.2017.02.055

Vajd, A., & Samadi, A. (2019). Optimization of Centrifugal Casting Parameters to Produce the Functionally Graded Al–15wt% Mg 2 Si Composites with Higher Tensile Properties. *International Journal of Metalcasting*, 1–12.

Varthini, R., & Gandhinathan, R., Pandivelan, C., & Jeevanantham, A. K. (2014). Modelling and optimization of process parameters of The single point incremental forming of aluminium 5052 alloy sheet using genetic algorithm-back Propagation neural network. *International Journal of Mechanical And Production Engineering, 2*(5), 55–62.

Vengatesh, D., & Chandramohan, V. (2014). Aluminium alloy metal matrix composite: Survey paper. *International Journal of Engineering Research and General Sciences, 2*(6), 792–796.

Venkata Ramana, G., Saravanan, P., Kamat, S. V., & Aparna, Y. (2012). Optimization of sputtering parameters for SmCo thin films using design of experiments. *Applied Surface Science, 261*, 110–117. doi:10.1016/j.apsusc.2012.07.109

Vijian, P., Arunachalam, V. P., & Charles, S. (2007). *Study of surface roughness in squeeze casting LM6 aluminium alloy using Taguchi method.* Academic Press.

Vijian, P., & Arunachalam, V. P. (2006). Optimization of squeeze cast parameters of LM6 aluminium alloy for surface roughness using Taguchi method. *Journal of Materials Processing Technology, 180*(1-3), 161–166. doi:10.1016/j.jmatprotec.2006.05.016

Vijian, P., & Arunachalam, V. P. (2007). Optimization of squeeze casting process parameters using Taguchi analysis. *International Journal of Advanced Manufacturing Technology, 33*(11-12), 1122–1127. doi:10.100700170-006-0550-2

Vinayagamoorthy, R., Manoj, I. V., Kumar, G. N., Chand, I. S., Kumar, G. V., & Kumar, K. S. (2018). A central composite design based fuzzy logic for optimization of drilling parameters on natural fiber reinforced composite. *Journal of Mechanical Science and Technology, 32*(5), 2011–2020. doi:10.100712206-018-0409-0

Viswanathan, R., Ramesh, S., & Subburam, V. (2018). Measurement and optimization of performance characteristics in turning of Mg alloy under dry and MQL conditions. *Measurement, 120*(November), 107–113. doi:10.1016/j.measurement.2018.02.018

Wang, P., Takeno, T., Fontaine, J., Aono, M., Adachi, K., Miki, H., & Takagi, T. (2014). Effects of substrate bias voltage and target sputtering power on the structural and tribological properties of carbon nitride coatings. *Materials Chemistry and Physics, 145*(3), 434–440. doi:10.1016/j.matchemphys.2014.02.032

Wang, Q. G., Apelian, D., & Lados, D. A. (2001). Fatigue behavior of A356-T6 aluminum cast alloys. Part I. Effect of casting defects. *Journal of Light Metals, 1*(1), 73–84. doi:10.1016/S1471-5317(00)00008-0

Wang, T. Y., Lin, H. C., & Wu, K. B. (1998). An improved simulated annealing for facility layout problems in cellular manufacturing systems. *Computers & Industrial Engineering, 34*(2), 309–319. doi:10.1016/S0360-8352(97)00318-5

Wang, Y., Wu, S. P., Xue, X., & Guo, J. J. (2013). Optimization of Tilt-Casting of Aluminum Alloy Automobile Drain Pipes through Numerical Simulation. *Materials Science Forum, 762*, 633–638. doi:10.4028/www.scientific.net/MSF.762.633

Wasserman, G. S. (1993). On how to prioritize design requirements during the QFD planning process. *IIE Transactions, 25*(3), 59–65. doi:10.1080/07408179308964291

Wassilkowska, A. (2017). Microstructure inhomogeneity of centrifugally cast ductile iron pipes and its effect on mechanical properties. *Kovove Mater*, *55*(05), 311–316. doi:10.4149/km_2017_5_311

Wu, C. S., Lin, C. T., & Lee, C. (2010). Optimal marketing strategy: A decision-making with ANP and TOPSIS. *International Journal of Production Economics*, *127*(1), 190–196. doi:10.1016/j.ijpe.2010.05.013

Wu, C.-K., Huang, J.-H., & Yu, G.-P. (2019). Optimization of deposition processing of VN thin films using design of experiment and single-variable (nitrogen flow rate) methods. *Materials Chemistry and Physics*, *224*, 246–256. doi:10.1016/j.matchemphys.2018.12.038

Wu, H., Lv, K., Liang, L., & Hu, H. (2017). Measuring performance of sustainable manufacturing with recyclable wastes: A case from China's iron and steel industry. *Omega*, *66*, 38–47. doi:10.1016/j.omega.2016.01.009

Yadav, V. K., Kumar, P., & Dvivedi, A. (2019). Effect of tool rotation in near-dry EDM process on machining characteristics of HSS. *Materials and Manufacturing Processes*, *34*(7), 779–790. doi:10.1080/10426914.2019.1605171

Yang, W. P., & Tarng, Y. S. (1998). Design optimization of cutting parameters for turning operations based on the Taguchi method. *Journal of Materials Processing Technology*, *84*(1-3), 122–129. doi:10.1016/S0924-0136(98)00079-X

Yan, M.-T., & Fang, C.-C. (2008). Application of genetic algorithm-based fuzzy logic control in wire transport system of wire-EDM machine. *Journal of Materials Processing Technology*, *205*(1-3), 128–137. doi:10.1016/j.jmatprotec.2007.11.091

Yan, X., Zheng, J., Zheng, L., Lin, G., Lin, H., Chen, G., Du, B., & Zhang, F. (2018). Optimization of sputtering NiO films for perovskite solar cell applications. *Materials Research Bulletin*, *103*, 150–157. doi:10.1016/j.materresbull.2018.03.027

Yeh, J.-W., & Jong, S.-H. (1994). The cast structure of a 7075 alloy produced by a water-cooling centrifugal casting method. *Metallurgical and Materials Transactions. A, Physical Metallurgy and Materials Science*, *25*(3), 643–650. doi:10.1007/BF02651606

Ye, W., Shiping, W., Lianjie, N., Xiang, X., Jianbing, Z., & Wenfeng, X. (2014). Optimization of low-pressure die casting process parameters for reduction of shrinkage porosity in ZL205A alloy casting using Taguchi method. *Proceedings of the Institution of Mechanical Engineers. Part B, Journal of Engineering Manufacture*, *228*(11), 1508–1514. doi:10.1177/0954405414521065

Yildiz, A. R. (2013). Optimization of cutting parameters in multi-pass turning using artificial bee colony-based approach. *Information Sciences*, *220*, 399–407. doi:10.1016/j.ins.2012.07.012

Yoon, S. G., Kim, H. K., Kim, M. J., Lee, H. M., & Yoon, D. H. (2005). Effect of substrate temperature on surface roughness and optical properties of Ta2O5 using ion-beam sputtering. In Thin Solid Films (Vol. 475, pp. 239–242). doi:10.1016/j.tsf.2004.07.043

Yurdakul, M., & Cogun, C. (2003). Development of a multi-attribute selection procedure for non-traditional machining processes. *Proceedings of the Institution of Mechanical Engineers. Part B, Journal of Engineering Manufacture*, *217*(7), 993–1009. doi:10.1243/095440503322011515

Yusup, N., Zain, A. M., & Hashim, S. Z. (2012). Evolutionary techniques in optimizing machining parameters: Review and recent applications (2007–2011). *Expert Systems with Applications*, *39*(10), 9909–9927. doi:10.1016/j.eswa.2012.02.109

Zain, A. M., Haron, H., & Sharif, S. (2010). Application of GA to optimize cutting conditions for minimizing surface roughness in end milling machining process. *Expert Systems with Applications*, *37*(6), 4650–4659. doi:10.1016/j.eswa.2009.12.043

Zavadskas, E. K., Turskis, Z., & Kildienė, S. (2014). State of art surveys of overviews on MCDM/MADM methods. *Technological and Economic Development of Economy*, *20*(1), 165–179. doi:10.3846/20294913.2014.892037

Zeng, Y., Tan, Z., Zhou, L., Jiang, M., Qiu, Y., Fang, F., Huang, H., Zhang, X., & Jiang, J. (2015). Effects of Bias Voltage on Fen Films Prepared by Magnetron Sputtering. *Materials Research, 18*(suppl 1), 115–119. doi:10.1590/1516-1439.334614

Zhang, C., Sun, J., Zhu, X., & Yang, Q. (2008). An improved particle swarm optimization algorithm for flow shop scheduling problem. *Information Processing Letters, 108*(4), 204–209. doi:10.1016/j.ipl.2008.05.010

Zhang, G., Zhang, Z., Guo, J., Ming, W., Li, M., & Huang, Y. (2013). Modeling and optimization of medium-speed WEDM process parameters for machining SKD11. *Materials and Manufacturing Processes, 28*(10), 1124–1132. doi:10.1080/10426914.2013.773024

Zhang, J. Z., & Chen, J. C. (2009). Surface roughness optimization in a drilling operation using the Taguchi design method. *Materials and Manufacturing Processes, 24*(4), 459–467. doi:10.1080/10426910802714399

Zhang, X., & Vecchio, K. S. (2013). Conversion of natural marine skeletons as scaffolds for bone tissue engineering. *Frontiers of Materials Science, 7*(2), 103–117. doi:10.100711706-013-0204-x

Zhang, Y., Gong, D., & Ding, Z. (2012). A bare-bones multi-objective particle swarm optimization algorithm for environmental/economic dispatch. *Information Sciences, 192*, 213–227. doi:10.1016/j.ins.2011.06.004

Zhao, S., Avendano, E., Gelin, K., Lu, J., & Wackelgard, E. (2006). Optimization of an industrial DC magnetron sputtering process for graded composition solar thermal absorbing layer. *Solar Energy Materials and Solar Cells, 90*(3), 308–328. doi:10.1016/j.solmat.2005.03.018

Zhao, S.-Z., Iruthayarajan, M. W., Baskar, S., & Suganthan, P. N. (2011). Multi-objective robust PID controller tuning using two lbests multi-objective particle swarm optimization. *Information Sciences, 181*(16), 3323–3335. doi:10.1016/j.ins.2011.04.003

Zhu, B. L., Wang, J., Zhu, S. J., Wu, J., Zeng, D. W., & Xie, C. S. (2012). Optimization of sputtering parameters for deposition of Al-doped ZnO films by rf magnetron sputtering in Ar + H 2 ambient at room temperature. *Thin Solid Films, 520*(23), 6963–6969. doi:10.1016/j.tsf.2012.07.049

Zhu, D. L., Wang, Q., Han, S., Cao, P. J., Liu, W. J., Jia, F., Zeng, Y. X., Ma, X. C., & Lu, Y. M. (2014). Optimization of process parameters for the electrical properties in Ga-doped ZnO thin films prepared by r.f. magnetron sputtering. *Applied Surface Science, 298*, 208–213. doi:10.1016/j.apsusc.2014.01.163

About the Contributors

Kanak Kalita received his B.E in mechanical engineering from RGTU, Bhopal, India; M.E and Ph.D. in aerospace engineering and applied mechanics from Indian Institute of Engineering, Science & Technology, Shibpur, India. He has over 6 years of teaching, research and industrial experience. Currently, he is with Vel Tech Rangarajan Dr. Sagunthala R&D Institute of Science and Technology, India as assistant professor in mechanical engineering department. He is on the editorial board of 2 international journals and has reviewed 170+ manuscripts for 30+ journals and conferences. He has been awarded thrice by Publons for his reviewing efforts. He has published 20 SCI and 42 SCOPUS research articles and edited 1 book volume for IOP publishing. His areas of interests include metamodeling, process optimization, finite element method and composites.

Ranjan Kumar Ghadai received his B. Tech in Mechanical Engineering from Biju Patnaik University of Technology, Odisha, India, M.E and PhD from Indian Institute of Engineering, Science & Technology, Shibpur, India. He has over 6 years of teaching and research experience. Currently, he is working as an assistant professor in the mechanical engineering department of Sikkim Manipal Institute of Technology, Sikkim. He has published more than 35 SCI/Scopus indexed research articles. His areas of interests include thin-film coatings and its characterization, heat treatment, optimization of coatings and machining process parameters. He is on the editorial board of several peer-reviewed journals. He also serves as reviewer of many peer-reviewed journals. He has given several expert talks in many conference and workshop as a resource person.

Xiao-Zhi Gao received his B.Sc. and M.Sc. from Harbin Institute of Technology, China; PhD degree from the Helsinki University of Technology (currently Aalto University), Finland in 1999. He has 20+ years of teaching and research experience and is currently a professor at University of Eastern Finland, Finland. He serves as chief editor/associate editor/editorial board member in several leading soft-computing journals like Swarm and Evolutionary Computation, Information Sciences, Applied Soft Computing. He has published 350+ SCI/SCOPUS research articles and 2 books. He has also edited 4 books for leading publishers like Springer and IGI Global. His current research interests are nature-inspired computing methods with applications in optimization, prediction, data mining, signal processing, and control.

* * *

Esther Titilayo Akinlabi is a Full Professor of Mechanical Engineering. She assumed office as the Director of the Pan African University for Life and Earth Sciences Ibadan (PAULESI), Nigeria in July

2020. Prior to joining PAULESI, she had a decade of meritorious service at the Department of Mechanical Engineering Science, Faculty of Engineering and the Built Environment, University of Johannesburg (UJ), South Africa. During her period of service at UJ, she had the privilege to serve as the Head of Department of the Department of Mechanical Engineering Science and as the Vice Dean for Teaching and Learning for the Faculty. Her research interest is in the field of modern and advanced manufacturing processes – Friction stir welding and additive manufacturing. Her research in the field of laser based additive manufacturing include laser material processing and surface engineering. She also conducts research in the field of renewable energy, and biogas production from waste. She is a rated National Research Foundation (NRF) researcher in South Africa and has demonstrated excellence in all fields of endeavors. Prof Akinlabi has supervised to completion 22 PhD candidates and 28 Masters Students. Her leadership, mentorship and research experience is enviable as she guides her team of postgraduate students through the research journey. She is a recipient of several research grants and has received many awards of recognition to her credit. She is a registered member of the Engineering Council of South Africa (ECSA), Council for the Regulation of Engineering Profession in Nigeria (COREN), South African Institution of Mechanical Engineers (SAIMechE), America Society of Mechanical Engineers (ASME), and the Nigerian Society of Engineers (NSE). Prof Akinlabi has filed two patents, edited four books, published seven books and authored/co-authored over 500 peer reviewed publications.

Partha Das recieved his B.E. in Mechanical Engineering from Annamalai University, Tamil Nadu, India, M.Tech in Design & Manufacturing from NIT Silchar, Assam,India and pursuing Ph.D. in Production Engineering from Jadavpur University, Kolkata, India. Currently, he is working as an Assistant professor in Mechanical Engineering Department of Sikkim Manipal Institute of Technology,Sikkim. He has published numerous papers in national and international journal and is also a regular reviewer of various international journals of repute. His area of of intrests include Production control and management, Multi-criteria decision making, Soft computing techniques, Quality control, Manufacturing.

Arindam Debroy is a senior research fellow in the Department of Industrial & Systems Engineering at Indian Institute of Technology, India. His current research interests include non-traditional machining, process optimization, supply chain management, logistics and operation research.

Ashnut Dutt is a graduate student in the mechanical department at Sikkim Manipal Institute of Technology, Sikkim, India.

Rishi Dwivedi is currently working as an Assistant Professor in the Department of Finance of XISS, Ranchi. Before joining XISS, Ranchi, he has worked as an Assistant Professor in Management Department at Central University of Jharkhand. He has obtained his PhD degree from Jadavpur University, Kolkata. Dr. Rishi has been awarded with esteemed "UGC-BSR Research Fellowship in Science for Meritorious Students". He has also almost one year of corporate experience as an Assistant Manager at ICICI Bank Limited, Kolkata, West Bengal. He has published several research papers in international journals of high repute as well as presented numerous papers in international conferences. He is regular reviewer of many international journals of high repute like, International Journal of Production Research, Journal of Cleaner Production, OPSEARCH etc. His research interests include development of integrated activity based costing and quality function deployment models for various industries in order to attain competitive edge. Of late he is also interested in exploring the ways in which MCDM model can be applied to achieve sustainable advantage for assorted industries.

Selvakumar Gurusamy, born in 1978, received his PhD in Engineering from Jadavpur University, Kolkata, India in 2013. He has 18 years of professional experience in teaching and research. Currently, he is working as Associate Professor in the Mechanical Engineering Department at the Sri Sivasubramaniya Nadar (SSN) College of Engineering, Chennai, India. He has published 36 international journal papers, two national journal papers, 21 conference papers, edited three books and a reference book in the area of advanced manufacturing processes. He has successfully completed One research project for a tune of 32 Lakh, sponsored by SERB, New Delhi.

Salil Haldar is a Professor in the Department of Aerospace Engineering and Applied Mechanics, Indian Institution of Engineering Science and Technology, Shibpur, Howrah, India. He obtained his PhD in Composite structures from the Bengal Engineering & Science University, Shibpur, Howrah, India, in 2002. He has obtained his Master's degree in engineering with a specialization in Production from Jadavpur University, Kolkata, India, in 1991. His field of research is free vibration of plates and shells, composite materials, and optimization, etc.

Prashant Kumar Jha is currently working as Assistant Professor at XISS, Ranchi. He had worked at BSE Ltd for 7 years and for 2 years as Equity Research analyst at a prominent wealth management fund before that. He has a background in capital market specialising in equities and portfolio management.

Jayakrishna Kandasamy is an Associate professor in the School of Mechanical Engineering at the Vellore Institute of Technology University, India. Dr. Jayakrishna's research is focused on the design and management of manufacturing systems and supply chains to enhance efficiency, productivity, and sustainability performance. He has mentored undergraduate and graduate students (1 M.Tech Thesis and 24 B.Tech Thesis) which have so far led to 45 journal publications in leading SCI/ SCOPUS Indexed journals, 22 book chapters and 85 refereed conference proceedings, 7 books in CRC/Springer Series.

Dhruva Kumar is currently working as an Asst. Professor in Dept. of Mechanical Engineering. He obtained his B.Tech in Mechanical Engineering from RGPV, Bhopal and M.Tech in Applied Mechanics from Jadavpur University, Kolkata. He has also submitted his PhD thesis from Sikkim Manipal University. He has published more than 10 research papers. He is member of Board of Studies of Mechanical engineering dept and admission committee of the institute.

Shamvavi Kumari is currently pursuing her post graduate diploma in management in the Finance department of XISS, Ranchi. Before that she completed her graduation in Bachelor's of computer application from BIT Mesra, Ranchi. Her research interest includes Multi criteria decision making in business.

Siva Kumar Mahalingam is working as a Professor in the Mechanical Engineering Department at Vel Tech Rangarajan Dr. Sagunthala R&D Institute of Science and Technology. He obtained his Ph.D. from Anna University, Chennai in the year 2009. He is having 26 years of teaching and 16 years of research experience. His areas of research interest include tolerance allocation, selective assembly, and manufacturing optimization. He published 43 papers in International peer-reviewed journals and presented around 32 papers in international and national conferences. He delivered expert lectures in different forums organized by various engineering institutions in the title of 'Applications of Evolutionary Algorithms for Sustainable Manufacturing using MATLAB programming'.

Premchand Kumar Mahto has graduated in Mechanical Engineering from Sikkim Manipal Institute of Technology, Sikkim in 2011 and completed M.Tech in Heat Power Engineering from BIT Sindri in 2013. He is currently working as Assistant Professor in SMIT, Sikkim, India. Mr. Mahto has 7+ years of rich experience with a blend of teaching and research. He has a number of research publications in his name and also a reviewer for various journals.

Prosun Mandal is a full time PhD Research Scholar in the Department of Mechanical Engineering, Indian Institute of Engineering Science and Technology, Shibpur, West Bengal, India. He received his BE (Mechanical) from Jadavpur University in 2010 and M. Tech (Mechanical) from Indian Institute of Engineering Science and Technology, Shibpur in 2014. His areas of interest and research include unconventional or advanced manufacturing processes, carbon nanotube, composite material, precision machining, optimisation, modelling and simulation of production processes. He has authored more than 10 research papers in reputed international journals.

Fredrick Mwema is a lecturer and research at the departments of mechanical engineering at the Dedan Kimathi University of Technology, Kenya, and the University of Johannesburg, South Africa respectively. He is currently Chair of the Department (COD) of mechanical engineering at the Dedan Kimathi University of Technology, Nyeri Kenya. He is also the director for the centre for nanomaterials and nanoscience research (CSNR) at the same university. In terms of teaching, he has over 7 years of undergraduate teaching at the university level in mechanical engineering and 1 year of postgraduate teaching. He mentors several postgraduate students in the field of manufacturing and materials. His research interests are in the field of manufacturing and materials, specifically in surface engineering and coating technologies, severe plastic deformation, advanced materials, and characterizations. He has extensively worked on thin-film preparation through RF magnetron sputtering and published his works in several peer-reviewed journals, book chapters, and conferences. He has over 60 publications. He also has interests in additive manufacturing technologies including laser-based manufacturing and fused deposition modeling. He is currently supervising 4 masters and 2 Ph.D. students on additive manufacturing, surface engineering, and advanced recycling of polymer-based composites for green building technologies. He has written three books, out of which one has been published, and two under production. Dr. FM Mwema has written several research proposals for funding some of which have been successful in Kenya.

Lenin Nagarajan is the Professor of the Mechanical Engineering Department, Vel Tech Rangarajan Dr. Sagunthala R&D Institute of Science and Technology. He is having 14 years of teaching experience and 11 years of research experience. He obtained his Ph.D. from Anna University, Chennai in the year 2015. Manufacturing optimization, facility layout design, selective assembly and machinability studies on composite materials are the areas of his interest and specialization. He has 25 journal publications in International repute and 30 papers in international and national conferences. He organized 03 International conferences and 06 national conferences. He edited 02 books that were published by Springer and Lap Lambert.

Amer Nasr A. Elghaffar is currently associated with the Faculty of Engineering, Minia University, Egypt. He has published 30+ peer-reviewed journals. He is active in the area of protection and control system.

Oluseyi Oladijo is with the Department of Chemical, Material and Metallurgical Engineering at Botswana International University of Science and Technology, (BIUST), Botswana. Prior to that, he completed his postdoctoral fellowship in the School of Chemical and Metallurgical Engineering at the University of Witwatersrand, South Africa. He received his Ph.D. degree in Material Science and Engineering field from the University of the Witwatersrand, South Africa in 2013. Until now, he has published several peer-reviewed publications and research projects. He is a member of several professional Societies including the South African Institute of Tribology, SAIT, Powder Metallurgy Association of South Africa and European Thermal Spray Association. He has guided several honors and postgraduate students in his field.

Manish Oraon did his B Tech from National Institute of Foundry and Forge Technology, NIFFT Ranchi. He then completed his M.Tech and PhD from Production department of Birla Institute of Technology, Mesra Ranchi. He is having more than a decade experience in teaching and Research. Also he has published good number of journal papers and conference proceedings.

Rajesh P. V. completed B.E in Mechanical Engineering at M.A.M College of Engineering and Technology, Trichy with First Class with Distinction in 2014. Completed M.E in Manufacturing Engineering at M.I.E.T Engineering College, Trichy with an aggregate of 8.72 out of 10 in 2016. Working as an Assistant Professor in the Department of Mechanical Engineering at Saranathan College of Engineering, Trichy. Written articles in 10 journals so far. Presented papers in 15 national and international conferences.

Subham Pal received his M.Tech with specialization Engineering Mechanics from Indian Institute of Engineering Science and Technology, Shibpur, Howrah, India, in 2019. His research area includes Composites, Vibration analysis of plates and shells, Optimization, etc. He is doing PhD in the Department of Aerospace Engineering & Applied Mechanics, Indian Institute of Engineering Science & Technology.

Kanika Prasad is an Assistant Professor at National Institute of Technology, Jamshedpur, India. She graduated from Jadavpur University in Production Engineering after completing her undergraduate study in Mechanical Engineering from SMIT, Sikkim in India, earning distinction in both degrees. She earned prestigious DST INSPIRE fellowship to pursue her doctoral research from Jadavpur University. With many publications in journal and conferences of international repute so early in her academic career, much is expected of her research career. She specializes in application of quality function deployment technique and multi-criteria decision-making to develop expert systems for selection and design problems in manufacturing. Of late sustainability in manufacturing domain interests her and she is working in that area. She has guided several UG and PG dissertations. She is also a regular reviewer of several journals of international repute.

Manish Roy recieved his B.E. degree from BIT Mesra in Mechanical Engineering, post-gradute degree also from BIT Mesra in Mechanical Engineering department with specialization in Heat power. He finished his his Ph.D from Sikkim Manipal University in the year 2017 in the field of Non-Traditional Machining process incorporating decision making approaches. He has more than 16 years of teaching experience and has published a few papers in journals of repute and international conferences.

Saurabh Sharma is working as an Assistant Professor in the Mechanical Engineering Department at Sikkim Manipal Institute of Technology, Majhitar, Sikkim, India. He has completed his Diploma in Mechatronics from Advanced Technical Training Centre, Bardang, Sikkim, India, B. Tech in Mechanical Engineering and M. Tech in Production from Sikkim Manipal Institute of Technology, Majhitar, Sikkim, India. At present, he is pursuing PhD from Sikkim Manipal University, Sikkim, India. He has more than 10 years of teaching experience. His active area of research is metal cutting and process parameter optimization. He has published more than 10 research articles in peer-reviewed journal and conference proceedings.

Vinay Sharma is professor and in-charge of production engineering, BIT Mesra off-campus Deoghar. He is also working as in-charge training and placement, BIT Mesra off-campus Deoghar. He guided three PhD and published more than 30 papers in different journals.

Dinesh Shinde is working as a Assistant professor in Mechanical Engineering Department of Mulesh Patel School of Technology Management and Engineering, SVKM's NMIMS Shirpur campus, Dist-Dhule, MS-India. He received his B.E. in Mechanical Engineering & M.E.in Machine Design from North Maharashtra University, Jalgaon-India and pursuing PhD from Gujarat Technological University, Ahmedabad-India. He has 8+ years of academic and research experience. His research interest includes tribology of composite material, metamodel-based optimization, finite element analysis. He published more than 20 research papers in reputed international journals. He is reviewer of some peer review journals.

Ankit Kumar Singh is working as a Research Assistant in Department of Mechanical Engineering at Indian Institute of Science, Bangalore. He obtained his B.Tech in Mechanical Engineering from Manipal Institute of Technology, Sikkim. His active area of research is process parameter optimization, combustion, rocket propulsion and laser diagnosis.

Index

T

W

www.ingramcontent.com/pod-product-compliance
Lightning Source LLC
LaVergne TN
LVHW080525300926
849162LV00007B/1536